Titles in This Series

16 **Sergei Gelfand and Simon Gindikin, editors,** I. M. Gelfand seminar, Parts 1 and 2, 1993
15 **A. T. Fomenko, editor,** Minimal surfaces, 1993
14 **Yu. S. Il′yashenko, editor,** Nonlinear Stokes phenomena, 1992
13 **V. P. Maslov and S. N. Samborskiĭ, editors,** Idempotent analysis, 1992
12 **R. Z. Khasminskiĭ, editor,** Topics in nonparametric estimation, 1992
11 **B. Ya. Levin, editor,** Entire and subharmonic functions, 1992
10 **A. V. Babin and M. I. Vishik, editors,** Properties of global attractors of partial differential equations, 1992
9 **A. M. Vershik, editor,** Representation theory and dynamical systems, 1992
8 **E. B. Vinberg, editor,** Lie groups, their discrete subgroups, and invariant theory, 1992
7 **M. Sh. Birman, editor,** Estimates and asymptotics for discrete spectra of integral and differential equations, 1991
6 **A. T. Fomenko, editor,** Topological classification of integrable systems, 1991
5 **R. A. Minlos, editor,** Many-particle Hamiltonians: spectra and scattering, 1991
4 **A. A. Suslin, editor,** Algebraic K-theory, 1991
3 **Ya. G. Sinaĭ, editor,** Dynamical systems and statistical mechanics, 1991
2 **A. A. Kirillov, editor,** Topics in representation theory, 1991
1 **V. I. Arnold, editor,** Theory of singularities and its applications, 1990

Part 2

Photograph courtesy of C. MacPherson

Advances in
SOVIET
MATHEMATICS

Volume 16, Part 2

I. M. Gelfand Seminar

Sergei Gelfand
Simon Gindikin
Editors

American Mathematical Society
Providence, Rhode Island

1991 *Mathematics Subject Classification.* Primary 00B15.

Library of Congress Catalog Card Number: 91-640741

ISBN 0-8218-4118-1 (Part 1)
ISBN 0-8218-4119-X (Part 2)
ISBN 0-8218-4117-3 (Set)

ISSN 1051-8037

The paper used in this book is acid-free and falls within the guidelines
established to ensure permanence and durability. ♾

This publication was typeset using $\mathcal{A}\mathcal{M}\mathcal{S}$-TEX,
the American Mathematical Society's TEX macro system.

10 9 8 7 6 5 4 3 2 1 98 97 96 95 94 93

Contents

Photographs courtesy of C. MacPherson

Foreword

September 2, 1993 is Israel Moiseevich Gelfand's 80th birthday. This date practically coincides with the 50th anniversary of the Gelfand Seminar in Functional Analysis at Moscow University. The present volume consists of papers written by "young" participants of this seminar. One of the reasons for introducing an age limit was to keep the volume's size within reasonable bounds, another was Gelfand's constant orientation to the younger participants of his seminar. This collection is intended to be a surprise to the man whose birthday we are celebrating, and I hope that he will learn of the book's existence only after its publication. However, we have tried to imitate Gelfand's own preferences as much as possible. All of the invited authors were participants and welcome speakers at the seminar; if one imagines its Golden Jubilee Session, it may be safely conjectured that these mathematicians would have been invited to participate. I hope that Gelfand will approve our choice and will enjoy seeing articles written by these remarkable mathematicians. The invited authors were free to choose their topics and, if they so desired, their coauthors.

The problem of defining the notion of "young" scientist, in particular, of "young mathematician", is one of the most difficult unsolved problems, and is also of the utmost importance for the applications. In the given case its solution was formalized in the following way. For the upper bound of the age of an invited author to this collection, we chose the age of Serezha, Gelfand's older son, who incidentally played a key role in the appearance of this book. As paradoxal as this may sound, there are serious grounds for this choice. When Serezha first appeared at the seminar in 1961, significant changes in it took place. The orientation to the younger participants that had always been important, but concerned only Ph.D.-track students (occasionally younger graduate students) until then, was now drastically amended to include freshmen and sophomores, and later even some high school students. Among the students of this "first draft" one should note Dima Kazhdan and, somewhat later, Ossya Bernstein. Both became important participants in the seminar for many years. Progressively the other authors of this volume also appeared there. By 1962 the seminar left the relatively small auditorium 13-11 to move to the roomier 14-08. By then no less than half the participants were undergraduates. At that time Gelfand liked to repeat that the seminar was open to all students of the lower courses and to the most talented professors.

I think that it is only natural to include in this foreword some information about the seminar itself. I understand that it was well known in the West. Usually mathematical voyagers from Western Europe or the US felt compelled to visit it (to do that, they had to overcome the vigilance of the Moscow University guards, by no means an easy task even for Muscovites). It is difficult to explain to the Western reader what the seminar meant to "Soviet mathematical life". Surprisingly, that life, in many respects, was not at all so bad, despite the almost unwavering antisemitism and the constantly increasing control by the mathematical rabble with communist party background over the key positions in mathematics. Mathematics was an Oasis of sorts, very attractive to young people with strong interest in science and without career aspirations in the communist hierarchy. If one were lucky, you were able to live an intense intellectual life and write articles free of any references to the classics of Marxism-Leninism. Fortunately for mathematics, Stalin did not find time for the subject (unlike economics, biology, and linguistics).

The Gelfand seminar was always an important event in the very vivid mathematical life in Moscow, and, doubtless, one of its leading centers. A considerable number of the best Moscow mathematicians participated in it at one time or another. Mathematicians from other cities used all possible pretexts to visit it. I recall how a group of Leningrad students agreed to take turns to come to Moscow on Mondays (the day of the seminar, to which other events were linked), and then would retell their friends what they heard there. There were several excellent and very popular seminars in Moscow, but nevertheless the Gelfand seminar was always an event.

I would like to point out that, on the other hand, the seminar was very important in Gelfand's own personal mathematical life. Many of us witnessed how strongly his activities were focused on the seminar. When, in the early fifties, at the peak of antisemitism, Gelfand was chased out of Moscow University, he applied all his efforts to save the seminar. The absence of Gelfand at the seminar, even because of illness, was always something out of the ordinary.

One cannot avoid mentioning that the general attitute to the seminar was far from unanimous. Criticism mainly concerned its style, which was rather unusual for a scientific seminar. It was a kind of a theater with a unique stage director, simultaneously playing the leading role in the performance and organizing the supporting cast, most of whom had the highest qualifications. I use this metaphor with the utmost seriousness, without any intension to mean that the seminar was some sort of a spectacle. Gelfand had chosen the hardest and most dangerous genre: to demonstrate in public how he understood mathematics. It was an open lesson in the grasping of mathematics by one of the most amazing mathematicians of our time. This role could only be played under the most favorable conditions: the genre dictates the rules of the game, which are not always very convenient for the listeners. This means, for example, that the leader follows only his own intuition in the final choice of the topics of the talks, interrupts them with comments and questions (a priviledge not granted to other participants), organizes their "understanding", mainly by the younger participants (parts of the reports are repeated, participants are summoned to the blackboard). All this is done with extraordinary generosity, a true passion for mathematics.

Let me recall some of the stage director's stratagems. An important feature were improvisations of various kinds. The course of the seminar could change dramatically at any moment. Another important *mise en scène* involved the "trial listener" game, in which one of the participants (this could be a student as well as a professor) was instructed to keep informing the seminar of his understanding of the talk, and whenever the information was negative, that part of the report would be repeated. A well-qualified trial listener could usually feel when the head of the seminar wanted an occasion for such a repetition. Also, Gelfand himself had the faculty of being "unable to understand" in situations when everyone around was sure that everything is clear. What extraordinary vistas were opened to the listeners, and sometimes even to the mathematician giving the talk, by this ability not to understand. Gelfand liked that old story of the professor complaining about his students: "Fantastically stupid students—five times I repeat proof, already I understand it myself, and still they don't get it."

It has remained beyond my understanding how Gelfand could manage all that physically for so many hours. Formally the seminar was supposed to begin at 6 P.M., but usually started with about an hour's delay. I am convinced that the free conversations before the actual beginning of the seminar were part of the scenario. The seminar would continue without any break until 10 or 10:30 (I have heard that before my time it was even later). The end of the seminar was in constant conflict with the rules and regulations of Moscow State University. Usually at 10 P.M. the cleaning woman would make her appearance, trying to close the proceedings to do her job. After the seminar, people wishing to talk to Gelfand would hang around. The elevator would be turned off, and one would have to find the right staircase, so as not to find oneself stuck in front of a locked door on the ground floor, which meant walking back up to the 14th (where else but in Russia is the locking of doors so popular!). The next riddle was to find the only open exit from the building. Then the last problem (of different levels of difficulty for different participants)—how to get home on public transportation, at that time in the process of closing up. Seeing Gelfand home, the last mathematical conversations would conclude the seminar's ritual. Moscow at night was still safe and life seemed so unbelievably beautiful!

The Gelfand seminar exerted a great influence on many mathematicians, not only on his direct pupils. The seminar for all of us was a unique opportunity to witness Gelfand's informal understanding of mathematics, to learn the usually hidden mysteries of the craft. We are all glad that Israel Moiseevich approaches his 80th anniversary in fine health and spirits, has preserved his workaholic ways, and, as always, is bristling with new plans and ideas, on the seventh decade of his service to Mathematics.

Simon Gindikin

ADVANCES IN SOVIET MATHEMATICS
Volume 16, Part 2, 1993

Homotopy Lie Algebras

VLADIMIR HINICH AND VADIM SCHECHTMAN

Dedicated to Prof. I. M. Gelfand

§1. Introduction

In this work we study, much in the spirit of [HS], the notion of a homotopy Lie algebra, which means "a complex which is a Lie algebra up to higher homotopies". The simplest example of a homotopy Lie algebra is a DG Lie algebra. Roughly speaking, a homotopy Lie algebra has a skew symmetric bracket, but the Jacobi identity is satisfied up to homotopies; these homotopies must themselves satisfy some identities up to some higher homotopies, etc.

There are two ways of formulating precise definitions. One of them, due to Drinfeld [D] and Stasheff, uses the standard chain complex construction. The other one, which we use here, is the language of algebras over operads, cf. [HS]. Homotopy Lie (and, dually, commutative) algebras in the sense of Drinfeld appear as algebras over certain important "standard operads". They are studied in Section 4. The key result (Proposition 4.5.2) says that the homology groups of the standard chain complex of such an algebra are certain Tor 's in the appropriate category of modules.

The main result of this article (Theorem 5.2.1) says that the complex associated with a cosimplicial DG Lie algebra admits a structure of a homotopy Lie algebra. In particular, the Čech complex of a sheaf of Lie algebras is a homotopy Lie algebra.

Based on the previous results, we present a sketch of the homology theory for such sheaves. In the next publication we will show that such homology groups arise in deformation theory as formal completions of local rings of moduli spaces.

Some proofs in this paper are omitted or sketched. Full details will appear in a later publication.

We are greately indebted to V. Drinfeld whose (four years old) letter to one of us was one of our main sources of inspiration.

1991 *Mathematics Subject Classification.* Primary 55U35; Secondary 18D35, 18G30.
The second author was supported in part by an NSF grant.

We are greatly indebted to L. Avramov who explained to us how to define the homology of unbounded DG algebras and to J. Stasheff who read the manuscript and made some helpful comments.

Part of this work was done while the second author was visiting the Weizmann Institute of Science and the Harvard University. He would like to thank these institutions for their hospitality.

The notion of operad and that of operad algebra (in the category of spaces) are due to May [Ma].

§2. Operads and operad algebras

In this section we give some basic definitions and some important examples. All definitions are given for a basic tensor abelian category $\mathscr{A}$, an abelian monoidal category with an involutive commutativity constraint (see [DM, Section 1] or [M2, VII.7]). The following examples of $\mathscr{A}$ appear throughout the paper:

— $\mathrm{mod}(k)$, the category of modules over a commutative ring k;
— $C(k)$, the category of complexes over k;
— $C(X)$, the category of complexes of sheaves of $\mathscr{O}_X$-modules on a locally noetherian scheme X;
— $\Delta C(X)$, the category of cosimplicial objects in $C(X)$.

2.1. First definitions.

2.1.1. DEFINITION. An operad $\mathscr{O}$ over an abelian tensor category $(\mathscr{A}, \otimes, 1)$ consists of:

— a collection of objects $\mathscr{O}(n)$ in $\mathscr{A}$ endowed with a (right) action of symmetric groups Σ_n, numbered by $n \in \mathbb{N} \cup \{0\}$;
— a collection of composition maps

$$\gamma: \mathscr{O}(n) \otimes \mathscr{O}(m_1) \otimes \cdots \otimes \mathscr{O}(m_n) \longrightarrow \mathscr{O}(m),$$

given for any n-tuple $(m_1, \dots, m_n)$ with $m = \sum m_i$;
— a *unit* $e: 1 \to \mathscr{O}(1)$.

The composition maps are supposed to be associative: the following diagram is commutative.

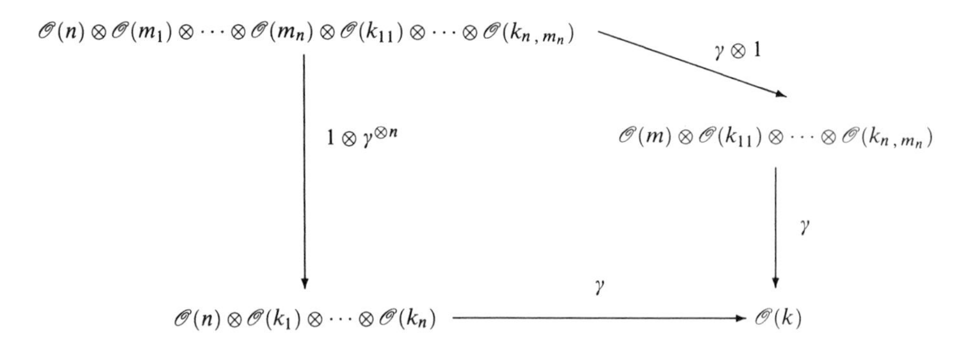

where $k_i = \sum_j k_{ij}$. The map γ should also be invariant with respect to the map $\Sigma_n \times \Sigma_{m_1} \times \cdots \times \Sigma_{m_n} \longrightarrow \Sigma_m$ which sends the collection $\sigma, \tau_1, \ldots, \tau_n$ to the element $\tilde{\sigma}(\tau_1 \times \cdots \times \tau_n) \in \Sigma_m$ obtained as the composition of the direct product of τ_i with the element $\tilde{\sigma}$ which permutes the blocks of length $m_1, \ldots, m_n$ in the way that σ permutes $1, \ldots, n$.

The unit axioms claim that $\gamma(e \otimes \mathrm{id}) = \mathrm{id}$ and $\gamma(\mathrm{id} \otimes e^{\otimes n}) = \mathrm{id}$.

2.1.2. Definition. An algebra over an operad $\mathscr{O}$ is an object A of $\mathscr{A}$ endowed with multiplication operations $\nu_n \colon \mathscr{O}(n) \otimes A^{\otimes n} \longrightarrow A$, satisfying the following conditions:

— associativity: the diagram

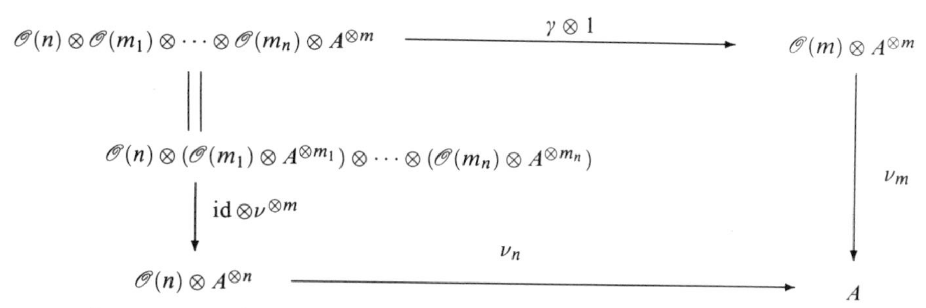

is commutative;

— ν_n is Σ_n-invariant, i.e., the diagram

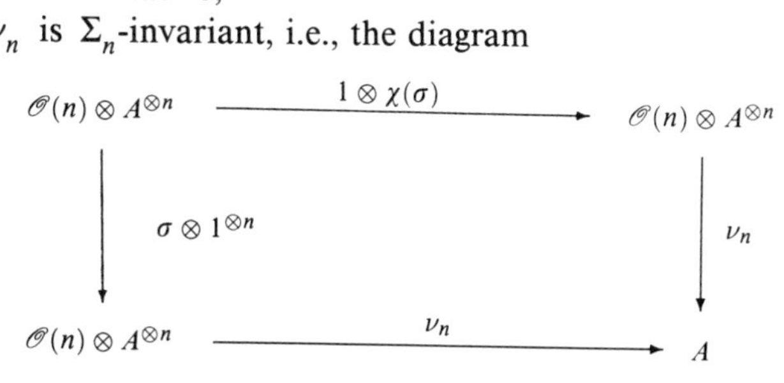

is commutative where $\chi(\sigma) \colon A^{\otimes n} \to A^{\otimes n}$ is the composition of commutativity constraints in $\mathscr{A}$ in a standard way;

— the unit condition:

$$A \xrightarrow{e \otimes \mathrm{id}} \mathscr{O}(1) \otimes A \xrightarrow{\nu_1} A$$

is the identity map.

2.2. Examples. In the examples 2.2.1–2.2.5 below k is a commutative ring and $\mathscr{A} = C(k)$ is the category of complexes of k-modules.

2.2.1. *Trivial commutative operad*. Put $\mathscr{C}(n) = k$ for all $n \geqslant 0$ with the trivial Σ_n-action and obvious multiplication operations. $\mathscr{C}$ is called the *trivial commutative operad* and $\mathscr{C}$-algebras are just commutative DG algebras with unit.

2.2.2. *Trivial associative operad*. Put $\mathscr{O}(n) = k\Sigma_n$ for $n \geqslant 0$. The definition of multiplication maps is left to the reader. The $\mathscr{O}$-algebras are just DG algebras over k with unit.

2.2.3. *Trivial Lie operad*. There exists also an operad $\mathscr{L}$ such that $\mathscr{L}$-algebras are exactly DG Lie k-algebras. Here is the construction of $\mathscr{L}$.

Let L_n be the free Lie k-algebra generated by the elements $x_1, \dots, x_n$. Denote by $\mathscr{L}(n)$ the free k-submodule generated by homogeneous monomials in $x_1, \dots, x_n$ having degree 1 on each variable. The symmetric group Σ_n acts on $\mathscr{L}_n$ on the right by permutation of variables. The multiplication

$$\gamma\colon \mathscr{L}(n) \otimes \mathscr{L}(m_1) \otimes \cdots \otimes \mathscr{L}(m_n) \longrightarrow \mathscr{L}(m)$$

is defined as follows: $\gamma(f \otimes g_1 \otimes \cdots \otimes g_n)$ is a result of substitution into $f \in \mathscr{L}(n)$ of $g_1(x_1, \dots, x_{m_1})$ instead of x_1, $g_2(x_{m_1+1}, \dots, x_{m_1+m_2})$ instead of x_2, etc.

EXERCISE. Check that $\mathscr{L}$-algebras are precisely differential graded Lie k-algebras.

2.2.4. *Operad of endomorphisms*. Let $A \in C(k)$. Define the operad $\mathscr{E}_A$ of endomorphisms of A by the formula

$$\mathscr{E}_A(n) = \underline{\mathrm{Hom}}(A^{\otimes n}, A), \qquad n > 0.$$

The composition in $\mathscr{E}_A$ and the action of Σ_n on $\mathscr{E}_A(n)$ are defined in a straightforward way. The complex A is obviously an $\mathscr{E}_A$-algebra and the structure of an $\mathscr{O}$-algebra on A is given by an operad morphism $\mathscr{O} \to \mathscr{E}_A$.

2.2.5. *Twisting*. Let $\mathscr{O}$ be an operad. We will define now a *twisted* operad $t_i(\mathscr{O})$ such that for any $\mathscr{O}$-algebra A the complex $A[i]$ admits a natural structure of $t_i(\mathscr{O})$-algebra.

For this we put $t_i(\mathscr{O})(n) = \mathscr{O}(n)[i - ni]$. The composition in $t_i(\mathscr{O})$ is defined by the composition in $\mathscr{O}$ if one remembers well that $X[i] = k[i] \otimes X$ and applies the usual sign rule to define natural isomorphisms. There is a subtlety in the definition of the Σ_n-action on $t_i(\mathscr{O})$: if u^i is the generator of $k[i]$, $o \in \mathscr{O}(n)$, and $\sigma \in \Sigma_n$, then one has

$$(u^{i-ni} \otimes o)\sigma = (-1)^{i|\sigma|} u^{i-ni} \otimes o\sigma.$$

EXERCISE. Check that $t_i(\mathscr{E}_A) = \mathscr{E}_{A[i]}$.

2.2.6. *Free algebras*. Let $\mathscr{O}$ be an operad over $\mathscr{A}$. The forgetful functor from the category of $\mathscr{O}$-algebras to $\mathscr{A}$ admits a left adjoint: the free algebra functor. For $V \in \mathscr{A}$ the free algebra $F(V)$ is defined as the graded algebra in $\mathscr{A}$ with homogeneous components

$$F^n(V) = (\mathscr{O}(n) \otimes V^{\otimes n})/\Sigma_n.$$

Here the right action of $\sigma \in \Sigma_n$ on $\mathscr{O}(n) \otimes V^{\otimes n}$ is given by the tensor product $\sigma \otimes \chi(\sigma^{-1})$, where $\chi(\sigma)\colon V^{\otimes n} \to V^{\otimes n}$ is the composition of commutativity constraints in $\mathscr{A}$ in a standard way. The multiplication in $F(V)$ is given by obvious formulas.

2.3. Weak versions.

2.3.1. DEFINITION. An operad $\mathcal{O}$ endowed with an operad quasi-isomorphism $\mathcal{O} \to \mathcal{L}$ (resp., $\mathcal{O} \to \mathcal{C}$) is called *a Lie* (*resp., commutative*) *operad.*

Morphisms of Lie (resp., commutative) operads are the morphisms of operads commuting with the above quasi-isomorphisms.

2.3.2. Algebras over commutative operads represent commutative algebras in the homotopy category of complexes over k (see [HS] where they are called May algebras). We call these algebras *commutative May algebras* and the algebras over a Lie operad are called *Lie May algebras.*

Properties of commutative May algebras are discussed in [HS]. There, in particular, the Eilenberg-Zilber commutative operad is built, which makes the normalization of a cosimplicial commutative algebra a May algebra.

We will construct in Section 4 "versal" commutative and Lie operads which admit a quasi-isomorphism into any commutative (resp., Lie) operad.

Finally, a Lie analog of the Eilenberg-Zilber operad will be constructed in Section 5.

§3. Modules

In this section we define the category of modules over an operad algebra $(A, \mathcal{O})$. This category is naturally equivalent to that of modules over the *enveloping algebra* $\mathcal{U}(A, \mathcal{O})$ of $(A, \mathcal{O})$, see 3.3. We give an explicit construction of the enveloping algebra and define the homology of an operad algebra.

3.1. Modules over an operad algebra. Fix an abelian tensor category $\mathcal{A}$. Let $\mathcal{O}$ be an operad over $\mathcal{A}$ and A be an $\mathcal{O}$-algebra.

3.1.1. DEFINITION. An $(A, \mathcal{O})$-module consists of the following data:

— an object $M \in \mathrm{Ob}\mathcal{A}$;

— a collection of multiplication maps

$$\mu_n: \mathcal{O}(n+1) \otimes A^{\otimes n} \otimes M \longrightarrow M.$$

The following axioms must satisfy

(Mod 1) (unit) $\mu_0(e \otimes \mathrm{id}_M) = \mathrm{id}_M$.

(Mod 2) (associativity) The following diagram is commutative:

$$\begin{array}{ccc}
\mathcal{O}(n) \otimes \mathcal{O}(m_1) \otimes \cdots \otimes \mathcal{O}(m_n) \otimes A^{\otimes m-1} \otimes M & \xrightarrow{\gamma \otimes 1 \otimes 1} & \mathcal{O}(m) \otimes A^{\otimes m-1} \otimes M \\
\| & & \\
\mathcal{O}(n) \otimes (\mathcal{O}(m_1) \otimes A^{\otimes m_1}) \otimes \cdots \otimes (\mathcal{O}(m_{n-1}) \otimes A^{\otimes m_{n-1}}) \otimes \mathcal{O}(m_n) \otimes A^{\otimes m_n - 1} \otimes M & & \Big\downarrow \mu_{m-1} \\
\Big\downarrow \mathrm{id} \otimes \nu^{\otimes m-1} \otimes \mu_{m_n - 1} & & \\
\mathcal{O}(n) \otimes A^{\otimes n-1} \otimes M & \xrightarrow{\mu_{n-1}} & M
\end{array}$$

(Mod 3) (symmetricity) For any $\sigma \in \Sigma_n$ the following diagram is commutative:

$$\begin{array}{ccc} \mathcal{O}(n+1)\otimes A^{\otimes n}\otimes M & \xrightarrow{1\otimes\chi(\sigma)\otimes 1} & \mathcal{O}(n+1)\otimes A^{\otimes n}\otimes M \\ \downarrow{\scriptstyle \sigma'\otimes 1^{\otimes n}\otimes 1} & & \downarrow{\scriptstyle \mu_n} \\ \mathcal{O}(n+1)\otimes A^{\otimes n}\otimes M & \xrightarrow{\mu_n} & M \end{array}$$

where σ' is the image of $\sigma \in \Sigma_n$ under the obvious injection $\Sigma_n \subseteq \Sigma_{n+1}$ and it acts on the right on $\mathcal{O}(n+1)$; the isomorphism $\chi(\sigma): A^{\otimes n} \to A^{\otimes n}$ is as in 2.2.6 the composition of commutativity constraints.

3.1.2. The category $\mathrm{mod}(A, \mathcal{O})$ has as objects the $(A, \mathcal{O})$-modules defined above. The morphisms are morphisms in $\mathcal{A}$ preserving all the structures.

3.1.3. Example. Let $\mathcal{O} = \mathcal{C}$ be the trivial commutative operad over a commutative ring k (see (2.2.1)). Then $\mathcal{O}$-algebras are just commutative k-algebras and $(A, \mathcal{L})$-modules are just A-modules.

3.1.4. Example. Let $\mathcal{O}$ be the trivial associative operad (see (2.2.2)) over a commutative ring k. Then $\mathcal{O}$-algebras are just the associative k-algebras and the $(A, \mathcal{O})$-modules are A-bimodules.

3.1.5. Example. If $\mathcal{O} = \mathcal{L}$ is the trivial Lie operad (see (2.2.3)) over k then $\mathcal{O}$-algebras are just Lie algebras and $(A, \mathcal{O})$-modules are just modules over the Lie algebra A.

3.1.6. Example. The categories $\mathrm{mod}(A, \mathcal{O})$ and $\mathrm{mod}(A[i], t_i(\mathcal{O}))$ are naturally equivalent.

3.1.7. *The trivial module.* An operad $\mathcal{O}$ is called *augmented* if $\mathcal{O}(0) = 0$ (this is not the case for $\mathcal{O} = \mathcal{C}$) and $\mathcal{O}(1)$ is augmented by a map $\pi: \mathcal{O}(1) \to 1$, so that $\pi e = \mathrm{id}_1$.

If $\mathcal{O}$ is augmented and A is any $\mathcal{O}$-algebra, the maps $\mu_0 = \pi$ and $\mu_i = 0$ for $i > 0$ define the structure of an $(A, \mathcal{O})$-module on 1. This is the *trivial* $(A, \mathcal{O})$*-module.*

3.2. Tensor algebra. As above, $\mathcal{A}$ is a base abelian tensor category.

3.2.1. Definition. Let $\mathcal{O}$ be an operad and A be an object in $\mathcal{A}$. *The tensor algebra* $T(A, \mathcal{O})$ of A is the graded algebra in $\mathcal{A}$ defined by

$$T_n(A, \mathcal{O}) = (\mathcal{O}(n+1)\otimes A^{\otimes n})/\Sigma_n \quad (n \geqslant 0).$$

The right action of Σ_n on $\mathcal{O}(n+1)\otimes A^{\otimes n}$ is given by the tensor product $\sigma' \otimes \chi(\sigma^{-1})$ (compare with the axiom (Mod 3)).

The multiplication in T is induced by the composition

$$\begin{aligned} &(\mathcal{O}(n+1)\otimes A^{\otimes n})\otimes(\mathcal{O}(m+1)\otimes A^{\otimes m}) \\ &\quad \xrightarrow{e^{\otimes n}} \mathcal{O}(n+1)\otimes\mathcal{O}(1)^{\otimes n}\otimes\mathcal{O}(m+1)\otimes A^{\otimes(m+n)} \qquad (1) \\ &\quad \xrightarrow{\gamma} \mathcal{O}(m+n+1)\otimes A^{\otimes m+n}. \end{aligned}$$

The correctness and the associativity are verified immediately. The unit element in $T_0(A, \mathscr{O}) = \mathscr{O}(1)$ is e.

3.2.2. EXERCISE. If $\mathscr{O} = \mathscr{L}$ then $T(A, \mathscr{O})$ is the usual tensor algebra of A.

3.2.3. EXAMPLE. If $\mathscr{O} = \mathscr{C}$ the tensor algebra $T(A, \mathscr{O})$ is just the symmetric algebra of A.

3.2.4. Let now A be an $\mathscr{O}$-algebra and M be an $(A, \mathscr{O})$-module. Then M admits a canonical structure of a $T(A, \mathscr{O})$-module—this is an immediate consequence of the definitions. This identifies the category $\mathrm{mod}(A, \mathscr{O})$ with a full subcategory of $\mathrm{mod}(T(A, \mathscr{O}))$.

3.3. Enveloping algebra. An algebra morphism $T(A, \mathscr{O}) \to U$ in $\mathscr{A}$ is *acceptable* if any U-module considered as a $T(A, \mathscr{O})$-module, belongs (essentially) to $\mathrm{mod}(A, \mathscr{O})$. All acceptable morphisms form a category whose initial object is called *the enveloping algebra* $\mathscr{U}(A, \mathscr{O})$.

The enveloping algebra always exists and is easy to construct. The following straightforward result describes $\mathscr{U}(A, \mathscr{O})$ explicitly.

3.3.1. PROPOSITION. *The enveloping algebra $\mathscr{U}(A, \mathscr{O})$ is the quotient of the tensor algebra $T(A, \mathscr{O})$ by the ideal generated by the relations*

$$\begin{aligned} \gamma(o \otimes o_1 \otimes \cdots \otimes o_n) \otimes a_1 \otimes \cdots \otimes a_{m-1} \\ = (o \otimes \nu_{m_1}(o_1 \otimes a_1 \cdots \otimes o_{m_1}) \\ \otimes \cdots \otimes \nu_{m_{n-1}}(o_{n-1} \otimes \cdots \otimes a_{m-m_n}))(o_n \otimes \cdots \otimes a_{m-1}), \end{aligned} \tag{2}$$

where $o \in \mathscr{O}(n)$, $o_i \in \mathscr{O}_{m_i}$, $m = \sum m_i$, and γ, ν mean the multiplication in $\mathscr{O}$ and the $\mathscr{O}$-algebra structure on A respectively.

The described relation can be considered as a direct generalization of the relation defining the usual universal enveloping algebra.

One immediately sees that the categories of modules over $(A, \mathscr{O})$ and over $\mathscr{U}(A, \mathscr{O})$ are equivalent.

3.4. Examples. The base tensor category in 3.4.1–3.4.4 is $\mathrm{mod}(k)$ or $C(k)$ for a commutative ring k.

3.4.1. EXERCISE. If $\mathscr{O}$ is the trivial commutative operad, then $\mathscr{U}(A, \mathscr{O})$ is isomorphic to the algebra A.

3.4.2. EXERCISE. If $\mathscr{O}$ is the trivial associative operad, then $\mathscr{U}(A, \mathscr{O})$ is isomorphic to the algebra $A^{\mathrm{op}} \otimes A$.

3.4.3. EXERCISE. If $\mathscr{O}$ is the trivial Lie operad, then $\mathscr{U}(A, \mathscr{O})$ is isomorphic to the usual enveloping algebra of A.

3.4.4. Note that one has a natural isomorphism

$$\mathscr{U}(A, \mathscr{O}) \longrightarrow \mathscr{U}(A[i], t_i(\mathscr{O})),$$

compatible with the equivalence of categories 3.1.6.

3.4.5. *Enveloping algebra of a free algebra.* Let $\mathcal{O}$ be an operad over $\mathcal{A}$, $V \in \mathcal{A}$, and let $F(V)$ be the free algebra of V, see 2.2.6. Then there is a natural isomorphism

$$\mathcal{U}(FV, \mathcal{O}) = T(V, \mathcal{O}),$$

defined by the composition

$$\begin{aligned} &\mathcal{O}(n+1) \otimes \mathcal{O}(m_1) \otimes \cdots \otimes \mathcal{O}(m_n) \\ &\xrightarrow{e} \mathcal{O}(n+1) \otimes \mathcal{O}(m_1) \otimes \cdots \otimes \mathcal{O}(m_n) \otimes \mathcal{O}(1) \\ &\xrightarrow{\gamma} \mathcal{O}(\sum m_i + 1). \end{aligned}$$

3.5. PBW Theorem. In the rest of this section we consider the case $\mathcal{A} = C(k)$ for a fixed commutative ring k.

For enveloping algebras of operad algebras a direct analog of the Poincaré-Birkhoff-Witt theorem takes place.

First of all, let us define an analog of the symmetric algebra functor.

3.5.1. *$\mathcal{O}$-symmetric algebra.* Let $\mathcal{O}$ be an operad. Denote by $D'_{k+1}(n+1)$ the subcomplex of $\mathcal{O}(n+1)$ generated by the images of the multiplication maps

$$\begin{aligned} &\mathcal{O}(k+1) \otimes \mathcal{O}(n_1) \otimes \cdots \otimes \mathcal{O}(n_k) \\ &\quad \xrightarrow{e} \mathcal{O}(k+1) \otimes \mathcal{O}(n_1) \otimes \cdots \otimes \mathcal{O}(n_k) \otimes \mathcal{O}(1) \xrightarrow{\gamma} \mathcal{O}(n+1) \end{aligned} \tag{3}$$

where $k < n = \sum n_i$. Set $D_k(n) = \sum_{j \leqslant k} D'_j(n)$. One has $D_k \subseteq D_{k+1}$; put $D(n+1) = D_n(n+1)$. This is a Σ_n-invariant subcomplex, so that the following definition makes sence.

DEFINITION. For $X \in C(k)$, the $\mathcal{O}$-symmetric algebra $S(X, \mathcal{O})$ is defined as

$$\sum (\mathcal{O}(n+1)/D(n+1) \otimes X^{\otimes n})/\Sigma_n.$$

The multiplication in $S(X, \mathcal{O})$ is defined as in the tensor algebra and one has an obvious map $T(X, \mathcal{O}) \to S(X, \mathcal{O})$.

We must warn the reader that an $\mathcal{O}$-symmetric algebra is in no sense commutative in general.

3.5.2. *A filtration of the enveloping algebra.* Let

$$F^n T(A, \mathcal{O}) = \bigoplus_{i \leqslant n} T^i(A, \mathcal{O}).$$

Define the filtration $\{F^n U\}$ on $U = \mathcal{U}(A, \mathcal{O})$ as the image of the corresponding filtration on $T(A, \mathcal{O})$. One immediately sees that $D(n+1) \otimes A^n$ lies in the kernel of the map $\mathcal{O}(n+1) \otimes A^n \to \mathrm{gr}^n U$ and so that a natural map

$$\varepsilon \colon S(A, \mathcal{O}) \longrightarrow \mathrm{gr}^{\cdot} \mathcal{U}(A, \mathcal{O})$$

is defined.

Let us calculate this morphism in the case $A = F(V)$, where $V \in C(k)$. According to subsection 3.4.5, $U = \mathscr{U}(FV, \mathscr{O}) = T(X, \mathscr{O})$ and $F^n U = \sum_m D_{n+1}(m+1) \otimes V^{\otimes m}$.

The symmetric algebra $S(FV, \mathscr{O})$ has form

$$\sum_{n, m_i} \mathscr{O}(n+1)/D(n+1) \otimes O(m_1) \otimes \cdots \otimes O(m_n) \otimes V^{\otimes m}$$

with $m = \sum m_i$ and this is easily identified with $\mathrm{gr}(U)$ when all $\mathscr{O}(i)$ consist of flat k-modules.

3.5.3. *The Poincaré-Birkhoff-Witt Theorem.* We are ready now to prove the following

THEOREM. *Suppose that $k \supseteq \mathbb{Q}$. Let an operad $\mathscr{O}$ consist of complexes of flat k-modules and the embeddings $D(n) \longrightarrow \mathscr{O}(n)$ split as maps of graded k-modules (not respecting the differentials). Then for any $\mathscr{O}$-algebra A the morphism*

$$\varepsilon\colon S(A, \mathscr{O}) \longrightarrow \mathrm{gr}\,\mathscr{U}(A, \mathscr{O})$$

is a DG algebra isomorphism.

PROOF. First of all, since all involved constructions do not use the differential and the map ε commutes with the diferentials *a priori*, we can forget the differentials in all our objects-operads, algebras, enveloping and symmetric algebras.

Choose a Σ_n-invariant splitting $\mathscr{O}(n+1)/D(n+1) \to \mathscr{O}(n+1)$; this is possible since $k \supseteq \mathbb{Q}$. This defines a map $\varepsilon'\colon S(A, \mathscr{O}) \to \mathscr{U}(A, \mathscr{O})$ and our aim is to prove that ε' is an isomorphism of graded k-modules.

We have already checked the assertion in the case of free $\mathscr{O}$-algebras. In the general case we will make use of the following simplicial resolution of A. Recall that the free algebra functor $F\colon C(k) \to \{\mathscr{O}\text{-algebras}\}$ is left adjoint to the forgetful functor $\sharp$. The composition $F\sharp$ gives rise to a simplicial resolution of A (this is the "unnormalized" bar resolution, see [M1, IX.6]) of which we write only the last terms:

$$\begin{array}{c} \xleftarrow{s_0} \\ (F\sharp)^2(A) \xrightarrow{d_1} F\sharp(A) \xrightarrow{d_0} A. \\ \xrightarrow[d_0]{} \end{array}$$

We apply to this resolution the functorial map $\varepsilon'\colon S(\_, \mathscr{O}) \longrightarrow \mathscr{U}(\_, \mathscr{O})$. The assertion will be proved if we show that the functors $S(\_, \mathscr{O})$ and $\mathscr{U}(\_, \mathscr{O})$ transform the simplicial resolution above to an exact sequence of graded k-modules. The symmetric algebra functor gives an exact sequence of graded k-modules since it is factored through the forgetful functor and the simplicial resolution of A splits as a sequence of graded k-modules.

The exactness of the sequence of DG algebras

$$\mathcal{U}((F\natural)^2(A),\mathcal{O}) \underset{d_0}{\overset{d_1}{\underset{\longrightarrow}{\overset{\overset{s_0}{\longleftarrow}}{\longrightarrow}}}} \mathcal{U}(F\natural(A),\mathcal{O}) \xrightarrow{d_0} \mathcal{U}(A,\mathcal{O})$$

immediately follows from Proposition 3.3.1. To prove that this sequence is also exact as the sequence of graded k-modules, we must verify that the k-submodule of $\mathcal{U}(F\natural(A),\mathcal{O})$ generated by the elements $d_0(x)-d_1(x)$ is in fact a two-sided ideal. The existence of the degeneracy s_0 immediately implies this.

3.5.4. EXAMPLE. In the case $\mathcal{O}=\mathcal{L}$ the quotient $\mathcal{O}(n+1)/D(n+1)$ is one-dimensional with the trivial action of Σ_n—it is represented in $\mathcal{O}(n+1)$ by the element $x_1,\dots,x_n,y\mapsto \mathrm{ad}(x_1)\cdots\mathrm{ad}(x_n)y$. Thus $S(\_,\mathcal{L})$ is the usual symmetric algebra and our PBW is the usual one.

3.6. Homology. Since our enveloping algebras are often unbounded complexes, we cannot use the standard definition of homology for DG algebras as it is given in, say, [Sm, Section 1].

In 3.6.1–3.6.7 below we sketch some definitions and results from [AFH].

Let R be a (not necessarily bounded) DG algebra. In the sequel differential graded R-modules will be called simply R-modules. The functor $R\mapsto R^\sharp$, $M\mapsto M^\sharp$ to the category of graded algebras (modules) forgets the differential.

3.6.1. DEFINITION. 1. An R-module M is *free* if $M^\sharp$ has an $R^\sharp$-basis consisting of cycles.

2. An R-module M is *semi-free* if M admits a filtration

$$0=M(-1)\subseteq\cdots\subseteq M(i)\subseteq M(i+1)\subseteq\cdots$$

such that $M=\bigcup M(i)$ and each quotient $M(i+1)/M(i)$ is free.

3.6.2. PROPOSITION. *A semi-free R-module P has the following lifting property: each diagram*

$$\begin{array}{ccc} & & M \\ & & \downarrow{\scriptstyle\alpha} \\ P & \longrightarrow & N \end{array}$$

with α a surjective quasi-isomorphism can be completed to a commutative diagram:

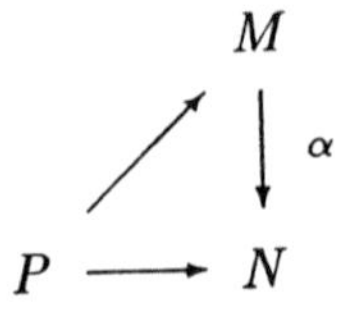

3.6.3. Proposition. *If an R-module F is semi-free then the functor $-\otimes_R F$ transforms quasi-isomorphisms into quasi-isomorphisms.*

3.6.4. Definition. A surjective quasi-isomorphism $P \to X$ with semi-free P is called *a semi-free resolution* for an R-module X.

3.6.5. Proposition. *Any R-module X admits a semi-free resolution $P \to X$.*

3.6.6. If M is a right R-module, set $\mathrm{Tor}^R(M, N) = H(P \otimes_R N)$ where $P \to M$ is a semi-free resolution. It is a graded $\mathbb{Z}$-module.

3.6.7. Theorem. *The functor Tor^R is correctly defined. Moreover, the following comparison theorem takes place:*

Let $f\colon R \to R'$ be a quasi-isomorphism of DG algebras, $g\colon M \to M'$ and $h\colon N \to N'$ be quasi-isomorphisms over f. Then the induced map $\mathrm{Tor}^R(M, N) \to \mathrm{Tor}^{R'}(M', N')$ is an isomorphism.

In the case when $R^i = 0$ for $i > 0$ the usual techniques of proper resolutions [Sm] can be applied.

We are now able to define homology of operad algebras.

Let $\mathscr{O}$ be an augmented operad in $C(k)$ and A be an $\mathscr{O}$-algebra.

3.6.8. Definition. The homology of an $\mathscr{O}$-algebra A with coefficients in $M \in \mathrm{mod}(A, \mathscr{O})$ is defined as

$$H^{\mathscr{O}}_{\bullet}(A, M) = \mathrm{Tor}^{\mathscr{U}}_{\bullet}(1, M)$$

where $\mathscr{U} = \mathscr{U}(A, \mathscr{O})$.

3.6.9. *Comparison theorem.* The following result is extremely important in the applications.

Theorem. *Suppose that $k \supseteq \mathbb{Q}$. Let $f\colon \mathscr{O} \to \mathscr{O}'$ be a quasi-isomorphism of operads in $C^-(k)$. Suppose that $\mathscr{O}$ and $\mathscr{O}'$ satisfy the conditions of Theorem 3.5.3 and are connected, i.e., that the augmentation maps $\mathscr{O}(1) \to 1$, $\mathscr{O}'(1) \to 1$ are isomorphisms.*

Let, moreover, $g\colon A \to A'$ be a quasi-isomorphism of algebras over f in $C^-(k)$ and $h\colon M \to M'$ be a quasi-isomorphism of modules over g. Suppose also that A and A' are complexes of flat k-modules. Then the natural map

$$H^{\mathscr{O}}_{\bullet}(A, M) \to H^{\mathscr{O}'}_{\bullet}(A', M')$$

is an isomorphism.

Proof. This follows from Theorem 3.6.7 and Proposition 3.6.12 at the end of the section, which claims that Theorem 3.6.7 is applicable.

3.6.10. *Bar resolutions.* Recall that in 3.5.3 we constructed a simplicial resolution of an $\mathscr{O}$-algebra A whose object of n-simplices has the form $\beta_n = (F\natural)^{n+1}(A)$ (a nonnormalized bar-resolution).

If the operad $\mathcal{O}$ is connected, there exists a "normalized" analog of this resolution.

For $X \in C(k)$ define $\overline{F}X$ by the formula

$$\overline{F}X = \sum_{n\geqslant 2}(\mathcal{O}(n) \otimes X^{\otimes n})/\Sigma_n$$

so that one has a short exact sequence $0 \to X \to \sharp FX \to \overline{F}X \to 0$. Then put $B_n = F\overline{F}^n \sharp A$. The face maps $d_i \colon B_n \to B_{n-1}$ are defined as in the unnormalized case and also as in the unnormalized case the augmented complex $\sharp B.$ is contractible in the sense that there exist a system of maps $s_{-1} \colon \sharp A \to \sharp B_0$, $s_n \colon \sharp B_n \to \sharp B_{n+1}$ such that $d_0 s = 1$, $d_i s = s d_{i-1}$ for $i > 0$.

LEMMA. *The augmented simplicial complexes* $\{\mathcal{U}(\beta_n, \mathcal{O})\}$, $\{\mathcal{U}(B_n, \mathcal{O})\}$ *are acyclic over* $C(k)$.

PROOF. According to the PBW, it suffices to prove the acyclicity of the complexes $\{S(\beta_n, \mathcal{O})\}$ and $\{S(B_n, \mathcal{O})\}$ obtained using the symmetric algebra functor. But the symmetric algebra functor is factored through the forgetful functor $\sharp$: $\{\mathcal{O}\text{-algebras}\} \to C(k)$ which transforms the simplicial resolution of A into the contractible simplicial complex $\{\sharp(F\sharp)^{n+1}(A)\}$ in the first case, and into a contractible normalized complex in the second case. It is clear that any functor transforms these complexes into contractible ones.

3.6.11. We apply now Lemma 3.6.10 to a very special case of $\mathcal{O}$-algebras.

Let $\mathcal{O}$ be a connected operad and $\mathbb{V}$ be a free k-module. We endow $\mathbb{V}$ with the structure of "commutative" $\mathcal{O}$-algebra: the map $\nu_1 \colon \mathcal{O}(1) \otimes \mathbb{V} \to \mathbb{V}$ is induced by the augmentation map $\pi \colon \mathcal{O}(1) \to 1$ and the maps ν_i are zero for $i > 1$. One checks immediately that

$$\mathcal{U}(\mathbb{V}, \mathcal{O}) = S(\mathbb{V}, \mathcal{O}) = \sum(\mathcal{O}(n+1)/D(n+1) \otimes \mathbb{V}^{\otimes n})/\Sigma_n.$$

Then the normalized resolution for $\mathcal{U}(\mathbb{V}, \mathcal{O})$ has members $\mathcal{U}(F\overline{F}^n \sharp A, \mathcal{O})$. The differentials in this resolution preserve the natural grading and for each integer n the degree n component of the resolution is finite.

This implies the following important

LEMMA. *Let* $f \colon \mathcal{O} \to \mathcal{O}'$ *be a quasi-isomorphism of connected operads and let* D *and* D' *be the subcomplexes of* $\mathcal{O}$ *and* $\mathcal{O}'$ *as in* 3.5.1. *Then for any* $n \in \mathbb{N}$ *the natural map* $\mathcal{O}(n)/D(n) \to \mathcal{O}'(n)/D'(n)$ *is a quasi-isomorphism.*

PROOF. The complex $\mathcal{U}(F\overline{F}^n \sharp(\mathbb{V}), \mathcal{O})$ is a direct sum of complexes of form $\mathcal{O}(T) \otimes \mathbb{V}^{\otimes k}/\Sigma$, where T is a tree of height $n+1$, $\mathcal{O}(T)$ is the tensor product of $\mathcal{O}(o(v))$ with v running through the set of vertices of T and $o(v)$ being the number of outgoing edges of $v \in T$, and Σ is a subgroup of Σ_k defined by T.

It is clear that the quasi-isomorphism $f \colon \mathcal{O} \to \mathcal{O}'$ induces quasi-isomorphisms of the corresponding members of the resolutions of $\mathcal{U}(\mathbb{V}, \mathcal{O})$ and

$\mathscr{U}(\mathbb{V}, \mathscr{O}')$ and this proves that the natural map

$$(\mathscr{O}(n+1)/D(n+1) \otimes \mathbb{V}^{\otimes n})/\Sigma_n \longrightarrow (\mathscr{O}'(n+1)/D'(n+1) \otimes \mathbb{V}^{\otimes n})/\Sigma_n$$

is a quasi-isomorphism for any free k-module $\mathbb{V}$. In other words, the map

$$\mathscr{O}(n+1)/D(n+1) \otimes_{k\Sigma_n} \mathbb{V}^{\otimes n} \longrightarrow \mathscr{O}'(n+1)/D'(n+1) \otimes_{k\Sigma_n} \mathbb{V}^{\otimes n}$$

is a quasi-isomorphism for any free k-module $\mathbb{V}$.

Since any irreducible representation of Σ_n appears in $\mathbb{V}^{\otimes n}$ for $\mathbb{V}$ sufficiently big, we obtain that the map

$$\mathscr{O}(n+1)/D(n+1) \otimes_{\mathbb{Q}\Sigma_n} X \longrightarrow \mathscr{O}'(n+1)/D'(n+1) \otimes_{\mathbb{Q}\Sigma_n} X$$

is a quasi-isomorphism for any representation X of Σ_n over $\mathbb{Q}$. Applying this to $X = \mathbb{Q}\Sigma_n$ we obtain the required assertion.

3.6.12. Proposition. *In the situation of Theorem* 3.6.9 *the natural map* $\mathscr{U}(A, \mathscr{O}) \to \mathscr{U}(A', \mathscr{O}')$ *is a quasi-isomorphism.*

Proof. Immediately follows from PBW and Lemma 3.6.11.

§4. Standard constructions

Let k be a commutative $\mathbb{Q}$-algebra. In this section we are working in the category $\mathscr{A} = C(k)$ of complexes over k.

4.1. Standard Lie operad. Our next aim is to construct a very important Lie operad $\mathscr{S}$ which will be called the *standard Lie operad*. This is in a sense a free resolution of the trivial Lie operad $\mathscr{L}$: any Lie operad $\mathscr{O}$ admits a morphism $\mathscr{S} \to \mathscr{O}$ over $\mathscr{L}$. The algebras over $\mathscr{S}$ are exactly weak Lie algebras in the sense of Drinfeld [D] and Stasheff [St].

The existence of free resolutions in this case follows from the possibility to adjoin a variable that kills a cycle (compare with [T]) as explained below.

4.1.1. *Adjoining a variable to an operad.* Let $\mathscr{O}$ be an operad and let $z \in \mathscr{O}(n)$ be a cycle. Let $G \subseteq \Sigma_n$ be the stabilizer of the line $k \cdot z$ and let a character $\chi: G \to k^*$ satisfy the condition $zg = \chi(g)z$.

Proposition. *There exists an operad* $\mathscr{O}' = \mathscr{O}\langle e;\ de = z,\ eg = \chi(g)e\rangle$ *satisfying the following (versal) property: any operad map* $\mathscr{O} \to \mathscr{P}$ *taking* z *to a boundary in* $\mathscr{P}(n)$, *can be extended to a map* $\mathscr{O}' \to \mathscr{P}$.

4.1.2. Definition. The standard Lie operad $\mathscr{S}$ is generated as an operad by elements δ_n of degree $2-n$, for $n \geqslant 2$, subject to the properties:

$$\begin{aligned} d\delta_n &= -\sum_{i=2}^{n-1}(-1)^{n-i} i \cdot \gamma(\delta_i \otimes \delta_{n-i+1} \otimes e^{\otimes i-1})\,\mathrm{alt}_n; \\ \sigma(\delta_n) &= (-1)^{|\sigma|}\delta_n. \end{aligned} \tag{4}$$

Here $\mathrm{alt}_n = 1/n! \sum_{\sigma\in\Sigma_n}(-1)^{|\sigma|}\sigma$.

REMARK. In order to check that $\mathcal{S}$ is correctly defined, one has to verify that the right-hand side of (4) is a cycle. We leave this to the reader.

Define the map $\alpha: \mathcal{S} \longrightarrow \mathcal{L}$ by the formula $\alpha(\delta_2) = [x_1, x_2]$; $\alpha(\delta_i) = 0$ for $i > 2$.

4.1.3. *$\mathcal{S}(n)$ as a "tree complex".* In order to get a better understanding of what the operad $\mathcal{S}$ looks like, we will give now its more precise description.

To do this we introduce some combinatorial notations.

DEFINITION. A tree is a (directed) graph with exactly one initial (=having no ingoing edges) vertex, such that any noninitial vertex has exactly one ingoing edge. *For each vertex the set of all its outgoing edges is totally ordered.*

Of course, we consider trees only up to isomorphism.

Terminal vertices are those that have no outgoing edges. Internal vertices are those that are not terminal (the initial vertex is included if it is not terminal).

For any tree T the set of its terminal (resp., internal) vertices is denoted by $\mathrm{ter}(T)$ (resp., $\mathrm{int}(T)$), and the numbers of the corresponding vertices by $t(T)$ and $i(T)$. For any vertex $v \in \mathrm{int}(T)$ the set of its outgoing edges is denoted by $\mathrm{out}(v)$ and their number is $o(v)$. *In what follows we consider trees that have no vertices v with $o(v) = 1$.* Equivalently, one could say that we identify trees obtained one from another by adding a vertex inside an edge.

Note that the sets $\mathcal{T}(n) = \{\text{trees } T\colon t(T) = n\}$ form a (nonsymmetric) operad in the category of sets. This operad is freely generated by corollae, that is, trees with exactly one internal (=initial) vertex.

Let S be a finite set of cardinality n. Consider the right action of Σ_n on S. The set of total orders on S is thus a torsor over Σ_n. A total order on S being fixed, we get an identification of Σ_n with the set of total orders of S. We shall use now this identification.

The choice of total orders on the sets of outgoing edges of the vertices of a tree T induce a lexicographic order on $\mathrm{out}(T)$.

These orders being fixed, one identifies the symmetric groups $\Sigma_{o(v)}$, $v \in \mathrm{int}(T)$, and also $\Sigma_{t(T)}$, with the sets of total orders on $\mathrm{out}(v)$ and also on $\mathrm{ter}(T)$ correspondingly. This defines a map (not a homomorphism)

$$\prod_{v \in \mathrm{int}(T)} \Sigma_{o(v)} \longrightarrow \Sigma_n,$$

whose image will be denoted by Σ_T.

The set $\mathcal{T}(n)$ can be identified with the set of monomials on δ_i contained in $\mathcal{S}(n)$: the correspondence takes the element δ_i to the unique corolla with i terminal vertices. The monomial corresponding to a tree T is contained in $\mathcal{S}(t(T))^d$, where $d = \sum_{v \in \mathrm{int}(T)} (2 - o(v))$. The following result is a direct consequence of the discussion above.

Proposition. *The k-module $\mathcal{S}(n)^d$ is generated by expressions $T\cdot\sigma$ with $T\in\mathcal{T}(n)$, $d=\sum_{v\in\mathrm{int}(T)}(2-o(v))$, and $\sigma\in\Sigma_n$. The relations are generated by the following relation*

$$T\cdot\sigma=T^{\sigma}.$$

Here $\sigma\in\Sigma_T$ and T^{σ} is obtained from T by the appropriate change of the orderings of outgoing edges.

4.1.4. **Proposition.** (i) *$(\mathcal{S},\alpha)$ is a Lie operad;*

(ii) *For any Lie operad $(\mathcal{O},\beta)$ there exists a morphism $\mathcal{S}\to\mathcal{O}$ of Lie operads.*

Proof. The claim (i) means that the map $\alpha_n\colon\mathcal{S}(n)\to\mathcal{L}(n)$ is a quasi-isomorphism for any n. This was proven by Ginzburg (private communication). Another proof may be given using methods of [SV]. The details will appear later.

(ii) By induction using 4.1.1. First, choose a cycle $z\in\mathcal{O}(2)^0$ representing the cohomology class $\beta^{-1}([x_1,x_2])$ and put $d_2=z\cdot\mathrm{alt}_2$. Let $d_i=\alpha(\delta_i)\in\mathcal{O}(n)^{2-n}$ be already chosen for $i<n$. To use 4.1.1 one must check that

$$\sum_{i=2}^{n-1}(-1)^{n-i}i\cdot\gamma(d_i\otimes d_{n-i+1}\otimes e^{\otimes i-1})\,\mathrm{alt}_n$$

is a boundary in $\mathcal{O}(n)$. In any case this is a cycle (see Remark in 4.1.2), so it suffices to check that the corresponding cohomology class vanishes. For $n>3$ this is trivial since then $H^{3-n}(\mathcal{O}(n))=0$.

For $n=3$ the cycle in question is $2\gamma(d_2\otimes d_2\otimes e)\,\mathrm{alt}_3$ which represents the zero cohomology class by the Jacobi formula.

4.1.5. *Shift by one.* If $\mathfrak{g}$ is a $\mathcal{S}$-algebra, $\mathfrak{g}[1]$ is a $t_1(\mathcal{S})$-algebra. We will use below the following explicit description of the operad $\widetilde{\mathcal{S}}=t_1(\mathcal{S})$. Denote by u^i the generator of $k[i]$, so that $\deg u^i=-i$. Then the elements $\widetilde{\delta}_i=u^{1-i}\otimes\delta$ freely generate $\widetilde{\mathcal{S}}$ and have degree one.

The defining relations parallel to (4) for $\mathcal{S}$ take the form

$$\begin{aligned} d\widetilde{\delta}_n&=-\sum_{i=2}^{n-1}i\cdot\gamma(\widetilde{\delta}_i\otimes\widetilde{\delta}_{n-i+1}\otimes e^{\otimes i-1})\,\mathrm{sym}_n,\\ \sigma(\widetilde{\delta}_n)&=\widetilde{\delta}_n, \end{aligned}\tag{5}$$

where $\mathrm{sym}_n=(1/n!)\sum_{\sigma\in\Sigma_n}\sigma$.

4.2. Standard complex.

4.2.1. *Symmetric and exterior powers.* Recall that for any $X\in C(k)$ the symmetric group Σ_n acts on the tensor power $T^nX=X^{\otimes n}$ (respecting the standard sign rule) and the nth symmetric (resp., exterior) power S^nX (resp., $\bigwedge^nX$) of X is defined to be the coinvariants (resp., the anti-coinvariants) of $X^{\otimes n}$.

Since $k \supseteq \mathbb{Q}$, the symmetric and the exterior powers are naturally embedded into the tensor power $T^n X$. Denote by $\varepsilon_p^S \colon S^p X \to T^p X$ and $\pi_p^S \colon T^p X \to S^p X$ the maps given by the formulas

$$\varepsilon_p^S(x_1 \cdots x_p) = \sum_{\sigma \in \Sigma_n} \sigma(x_1 \otimes \cdots \otimes x_n),$$

$$\pi_p^S(x_1 \otimes \cdots \otimes x_p) = \frac{1}{p!} x_1 \cdots x_p.$$

In the same way, $\varepsilon_p^\Lambda \colon \bigwedge^p \mathfrak{g} \to T^p \mathfrak{g}$ and $\pi_p^\Lambda \colon T^p \mathfrak{g} \to \bigwedge^p \mathfrak{g}$ are defined by the formulas

$$\varepsilon_p^\Lambda(x_1 \wedge \cdots \wedge x_p) = \sum_{\sigma \in \Sigma_n} (-1)^{|\sigma|} \sigma(x_1 \otimes \cdots \otimes x_n),$$

$$\pi_p^\Lambda(x_1 \otimes \cdots \otimes x_p) = \frac{1}{p!} x_1 \wedge \cdots \wedge x_p.$$

We will usually omit the superscript since it is always clear which one is meant. One has $\pi_p \varepsilon_p = 1$, $\varepsilon_p^\Lambda \pi_p^\Lambda = \mathrm{alt}_p$, and $\varepsilon_p^S \pi_p^S = \mathrm{sym}_p$.

Note that the identity map $X \to X[1]$ of degree -1 induces an isomorphism between $\bigwedge^n X$ and $S^n(X[1])$ of degree $-n$.

4.2.2. Let $\mathfrak{g}$ be a $\mathscr{S}$-algebra, $V = \mathfrak{g}[1]$. Define a collection of maps

$$d_{qp} \colon S^p V \longrightarrow S^q V, \tag{6}$$

of degree 1, for $p \geqslant q$, as follows.

First of all, put $d_{pp} = d_p$, the differential in $S^p V$.

Define for $q < p$

$$d_{qp} = q \cdot \pi_q (\widetilde{\delta}_{p-q+1} \otimes 1^{\otimes q-1}) \varepsilon_p. \tag{7}$$

Using the isomorphisms between $\bigwedge^n X$ and $S^n(X[1])$ of degree $-n$, one can consider d_{qp} as maps $\bigwedge^p \mathfrak{g} \to \bigwedge^q \mathfrak{g}$ of degree $q - p + 1$.

4.2.3. Definition. The standard complex $C(\mathfrak{g})$ of an algebra $\mathfrak{g}$ over $\mathscr{S}$ is the total complex of the twisted complex $(\bigwedge^{\cdot} \mathfrak{g}, d_{qp})$. In other words, $C(\mathfrak{g}) = S^{\cdot} V$ as a graded k-module, with the differential given by

$$d|_{S^p V} = \sum_q d_{qp}.$$

Exercise. Verify that the condition $d^2 = 0$ follows from (7).

4.2.4. The element $1 \in \bigwedge^0 \mathfrak{g} = S^0 V \subseteq C(\mathfrak{g})$ defines a direct sum decomposition $C(\mathfrak{g}) = k \cdot 1 \oplus \overline{C(\mathfrak{g})}$.

4.3. Drinfeld's weak Lie algebras. According to Drinfeld [D] and Stasheff [St] a homotopy Lie algebra (Drinfeld calls it "a Sugawara Lie algebra", and Stasheff "a strong homotopy Lie algebra") is a graded vector space $\mathfrak{g}$ endowed with a differential on $C(\mathfrak{g})$ which converts the latter into a DG coalgebra. We will show now that Drinfeld-Stasheff homotopy Lie algebras are just $\mathcal{S}$-algebras.

Let $\mathfrak{g}$ be a $\mathcal{S}$-algebra. Forget for a while the differentials in $\mathfrak{g}$ and in $C(\mathfrak{g})$. The latter is just the symmetric algebra of $V = \mathfrak{g}[1]$ and it admits a unique graded (commutative cocommutative) Hopf algebra structure such that V consists of primitive elements. Let $\Delta\colon C(\mathfrak{g}) \to C(\mathfrak{g}) \otimes C(\mathfrak{g})$ be this comultiplication and $\Delta^{(n-1)}\colon C(\mathfrak{g}) \to C(\mathfrak{g})^{\otimes n}$ be the multiple comultiplication. Denote by $\Delta_{k_1,\dots,k_n}$, $k_1 + \cdots + k_n = k$, the homogeneous component of $\Delta^{(n-1)}$ acting from $S^k V$ to $S^{k_1} V \otimes \cdots \otimes S^{k_n} V$.

4.3.1. Lemma. *One has* $\Delta_{k_1,\dots,k_n} = (\pi_{k_1} \otimes \cdots \otimes \pi_{k_n}) \cdot \varepsilon_k$.

Proof. This is a direct consequence of [Q, Appendix B, (3.4)].

4.3.2. Proposition. *The standard complex $C(\mathfrak{g})$ of a $\mathcal{S}$-algebra $\mathfrak{g}$ is a differential graded coalgebra with respect to Δ.*

Proof. One must verify the Leibniz condition, which claims that the comultiplication $\Delta\colon C(\mathfrak{g}) \longrightarrow C(\mathfrak{g})\otimes C(\mathfrak{g})$ commutes with the differentials. The homogeneous components of the Leibniz condition have the form

$$\Delta_{a,b} d_{qp} = (1 \otimes d_{b,p-a})\Delta_{a,p-a} + (d_{a,p-b} \otimes 1)\Delta_{p-b,b},$$

where $q = a + b$. This condition can be immediately verified using the expression (7) for d_{qp} through $d_{1,p-q+1}$ and Lemma 4.3.1, which gives a formula for $\Delta_{a,b}$.

Let, on the other hand, $\mathfrak{g}$ be a graded vector space and put $C(\mathfrak{g})$ to be the (graded) symmetric algebra of $V = \mathfrak{g}[1]$. Let Δ be the comultiplication in $C(\mathfrak{g})$.

4.3.3. Proposition. *For any differential d in $C(\mathfrak{g})$ satisfying the Leibniz condition with respect to Δ, there exists a unique structure of $\mathcal{S}$-algebras on $\mathfrak{g}$ for which $(C(\mathfrak{g}), d)$ is the standard complex of $\mathfrak{g}$.*

Proof. Let d be a differential in $C(\mathfrak{g})$ satisfying the Leibniz condition. Denote by d_{qp} the homogeneous component of d taking $S^p V$ to $S^q V$.

Since the multiple comultiplication $\Delta^{(q-1)}\colon C(\mathfrak{g}) \to C(\mathfrak{g})^{\otimes q}$ commutes with the differential, the following diagrams are commutative:

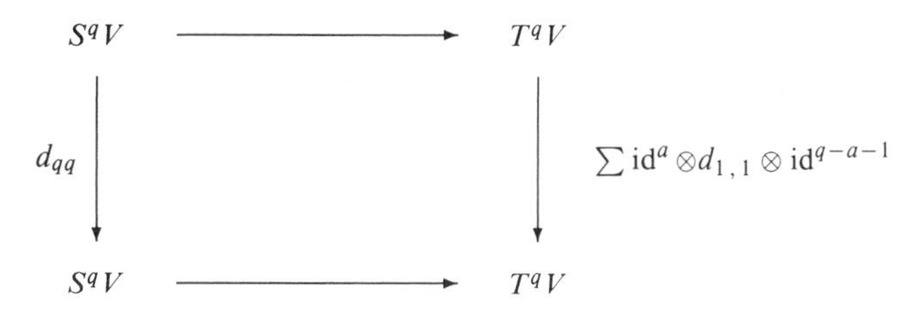

and

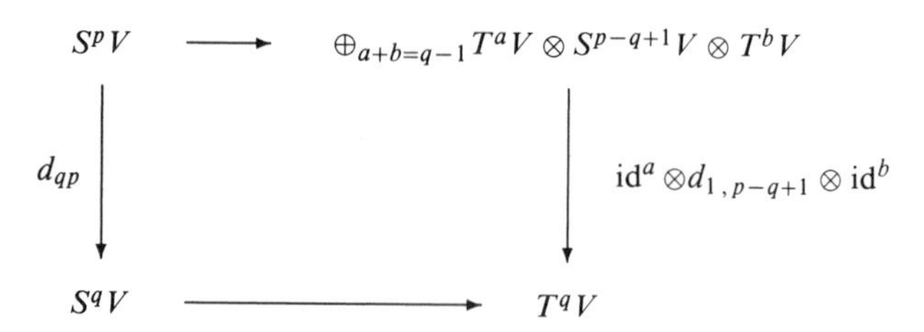

for $p \neq q$. Here the long horizontal arrows are $\Delta_{1,\dots,1}$, i.e., ε_q by Lemma 4.3.1, and the short horizontal arrow is the sum of $\Delta_{1^a,p-q+1,1^b}$.

The first diagram allows one to express d_{qq} in terms of d_{11} and the second one implies immediately the formula (7). If we put now $\widetilde{\delta}_i = d_{1i}\pi_i\colon V^{\otimes i} \to V$, the formula (5) is obtained from $d^2 = 0$ using (7).

4.4. Enveloping algebra of an $\mathcal{S}$-algebra. First of all, recall that we are working in the tensor category $\mathcal{A} = C(k)$, so that the enveloping algebra in question will be a differential graded (DG) k-algebra.

In order to describe the enveloping algebra $\mathcal{U}(\mathfrak{g}, \mathcal{S})$ of an $\mathcal{S}$-algebra $\mathfrak{g}$, we need the following

4.4.1. LEMMA. *Let $T \in \mathcal{T}(n+1)$ and let $\Sigma_n \subseteq \Sigma_{n+1}$ be identified with the subgroup of permutations satisfying the condition $\sigma(n+1) = n+1$. Then one has $\Sigma_{n+1} = \Sigma_T \Sigma_n$.*

PROOF. Identifying Σ_T with the collection of total orders on the sets $o(v)$, $v \in \mathrm{int}(T)$, one obtains the following reformulation of the lemma: for any vertex $w \in \mathrm{ter}(T)$ there exists a collection of total orders on $o(v)$, $v \in \mathrm{int}(T)$, such that w is maximal in the induced lexicographic order on $\mathrm{ter}(T)$. This reformulation is obvious.

According to 3.4.4, the enveloping algebra $\mathcal{U}(\mathfrak{g}, \mathcal{S})$ is naturally isomorphic to $\mathcal{U}(V, \widetilde{\mathcal{S}})$ and it is an image of the tensor algebra

$$T(V, \widetilde{\mathcal{S}}) = \bigoplus_{n \geqslant 0} (\widetilde{\mathcal{S}}(n+1) \otimes V^{\otimes n})/\Sigma_n. \tag{8}$$

Define a map

$$\sum_{n \geqslant 1} S^n V[-1] \longrightarrow \mathcal{U}(V, \widetilde{\mathcal{S}}), \tag{9}$$

as taking the element $u \otimes x_1 \cdots x_n$ to the image of $-(n+1)! \cdot \widetilde{\delta}_{n+1} \otimes x_1 \otimes \cdots \otimes x_n$ under the natural projection $T(V, \widetilde{\mathcal{S}}) \to \mathcal{U}(V, \widetilde{\mathcal{S}})$.

Note that as a graded k-module the left-hand side of (9) is isomorphic to $\overline{C(\mathfrak{g})}[-1]$, see (4.2). The following result claims that the enveloping algebra is freely generated by $\overline{C(\mathfrak{g})}[-1]$.

4.4.2. THEOREM. *The map (9) induces an isomorphism*

$$\alpha\colon T(\overline{C(\mathfrak{g})}[-1]) \xrightarrow{\sim} \mathcal{U}(V, \widetilde{\mathcal{S}})$$

of graded k-modules.

PROOF. We will construct a map $\beta: T(V, \widetilde{\mathscr{S}}) \to T(\overline{C(\mathfrak{g})}[-1])$ that will induce a map $\mathscr{U}(V, \widetilde{\mathscr{S}}) \to T(\overline{C(\mathfrak{g})}[-1])$ inverse to α.

According to Proposition 4.1.3 and Lemma 4.4.1, the "decomposable" elements $T \cdot g_1 \otimes \cdots \otimes g_n$ with $g_i \in V$ and $T \in \mathscr{T}(n+1)$ generate $T(V, \widetilde{\mathscr{S}})$ as a k-module. The value $\beta(T \cdot g_1 \otimes \cdots \otimes g_n)$ is the result of recursive "partial evaluation" using the relations (2): if $T = \gamma(\widetilde{\delta}_k \otimes T_1 \otimes \cdots \otimes T_k)$ with $T_i \in \mathscr{T}(n_i)$, $\sum n_i = n$, then

$$\beta(T \cdot g_1 \otimes \cdots \otimes g_n) = \left(-\frac{1}{k!} u \otimes \nu_1 \otimes \cdots \otimes \nu_{k-1}\right) \cdot \beta(T_k \cdot g_{n-n_k+2} \otimes \cdots \otimes g_n),$$

where

$$\nu_i = \nu(T_i \otimes g_{n_1+\cdots+n_{i-1}+1} \otimes \cdots \otimes g_{n_1+\cdots+n_i}).$$

The Σ_n-invariance of the operation ν implies that the result of the calculation does not depend on the choice of the presentation of a decomposable element of $T(V, \widetilde{\mathscr{S}})$ as $T \cdot g_1 \otimes \cdots \otimes g_n$.

The relations (2) belong to the kernel of β and thus a map $\mathscr{U}(V, \widetilde{\mathscr{S}}) \to T(\overline{C(\mathfrak{g})}[-1])$ is induced by β and is inverse to α.

4.4.3. *Cobar construction.* Let C be a connected DG coalgebra. Recall that the cobar construction $\mathbf{CB}(C)$ of C is defined to be a DG algebra isomorphic to $T^{\cdot}(\overline{C}[-1])$ as a graded algebra, with the differential D given by its value on $\overline{C}[-1]$ as follows:

$$D|_{\overline{C}[-1]} = \widetilde{d} + \widetilde{\Delta},$$

where $\widetilde{d}$ is the differential in $\overline{C}[-1]$ and $\widetilde{\Delta}: \overline{C}[-1] \longrightarrow \overline{C}[-1] \otimes \overline{C}[-1]$ is the map of degree one induced by the comultiplication Δ:

$$\widetilde{\Delta}(x) = \Delta(x) - x \otimes 1 - 1 \otimes x.$$

The following result claims that the isomorphism α from Theorem 4.4.2 identifies $\mathscr{U}(\mathfrak{g}, \mathscr{S}) = \mathscr{U}(V, \widetilde{\mathscr{S}})$ with the cobar construction of $C(\mathfrak{g})$.

4.4.4. THEOREM. *The composition*

$$\mathbf{CB}(C(\mathfrak{g})) = T(\overline{C(\mathfrak{g})}[-1]) \xrightarrow{\alpha} \mathscr{U}(V, \widetilde{\mathscr{S}})$$

is an isomorphism of DG algebras.

PROOF. The only thing to check is that the composition in question restricted to $C(\mathfrak{g})[-1]$ commutes with the differentials. This can be done by a straightforward calculation involving shuffles.

4.5. $(\mathfrak{g}, \mathscr{S})$**-modules.** Let, as above, $\mathfrak{g}$ be an $\mathscr{S}$-algebra, $\mathscr{U} = \mathscr{U}(\mathfrak{g}, \mathscr{S})$ be the enveloping algebra, $C(\mathfrak{g})$ be the standard complex of $\mathfrak{g}$. According to Theorem 4.4.4, $(\mathfrak{g}, \mathscr{S})$-modules are just modules over $\mathbf{CB}(C)$ and so such a module M is defined by a multiplication $\mu\colon C(\mathfrak{g})\otimes M\to M$ of degree one such that $\mu(1)=0$ and the following condition holds:

$$d(\mu)=\mu(1\otimes\mu)(\overline{\Delta}\otimes 1),$$

where $\overline{\Delta}(x)=\Delta(x)-x\otimes 1-1\otimes x$.

4.5.1. Definition. The standard complex of a $\mathscr{U}$-module M, $C(\mathfrak{g},M)$, is the tensor product $C(\mathfrak{g})\otimes M$ as a graded k-module, with the differential defined by the formula:

$$d_{C(\mathfrak{g},M)}=d_{C(\mathfrak{g})\otimes M}+(1\otimes\mu)(\overline{\Delta}\otimes 1).$$

Example. $C(\mathfrak{g},1)=C(\mathfrak{g})$.

4.5.2. Proposition. *The homology of an $\mathscr{S}$-algebra $\mathfrak{g}$ with coefficients in M can be calculated using the standard complex $C(\mathfrak{g},M)$.*

Proof. Note that $C(\mathfrak{g},M)=C(\mathfrak{g},\mathscr{U})\otimes_{\mathscr{U}}M$, so it suffices to prove that $C(\mathfrak{g},\mathscr{U})$ is an admissible $\mathscr{U}$-projective resolution of the trivial module 1. For this we define a homotopy

$$h\colon C(\mathfrak{g},\mathscr{U})\to C(\mathfrak{g},\mathscr{U})[-1],$$

such that $d(h)$ is the map

$$C(\mathfrak{g},\mathscr{U})\to 1\to C(\mathfrak{g},\mathscr{U}),$$

composed of the counit in $C(\mathfrak{g})$, the augmentation in $\mathscr{U}$, and the unit in $C(\mathfrak{g})$.

The homotopy h is given by the formula

$$h(c\otimes x_1\otimes\cdots\otimes x_n)=\eta(c)\cdot(u\otimes x_1)\otimes\cdots\otimes x_n,$$

where $c\in C(\mathfrak{g})$, $x_i\in C(\mathfrak{g})[-1]$, $u\in k[1]$ is the generator of degree -1, and $\eta\colon C(\mathfrak{g})\to 1$ is the counit.

4.6. Standard commutative operad.

4.6.1. Denote by $\mathrm{Op}_{\mathscr{L}}$ (resp., $\mathrm{Op}_{\mathscr{C}}$) the category of Lie (resp., commutative) operads over k. One has an obvious functor

$$\mathscr{O}\mapsto\mathscr{O}\otimes\mathscr{L}$$

from $\mathrm{Op}_{\mathscr{C}}$ to $\mathrm{Op}_{\mathscr{L}}$, where the tensor product is componentwise and the action of the symmetric group is diagonal.

Proposition. *The functor above admits a left adjoint*

$$\mathrm{co}\colon\mathrm{Op}_{\mathscr{L}}\longrightarrow\mathrm{Op}_{\mathscr{C}}.$$

Proof. For a Lie operad $\mathscr{O}$, the operad $\mathrm{co}(\mathscr{O})$ is generated as a symmetric operad by the subcomplexes $\mathscr{O}(n)\otimes\mathscr{L}(n)^*$ modulo the relation imposed by the commutativity of the following diagram:

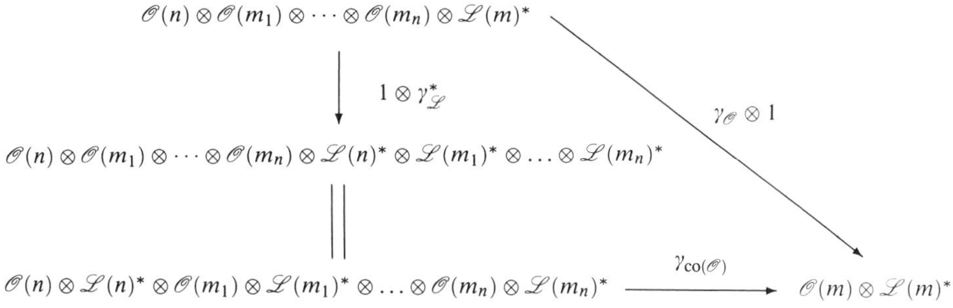

$$\begin{array}{ccc}
\mathcal{O}(n)\otimes\mathcal{O}(m_1)\otimes\cdots\otimes\mathcal{O}(m_n)\otimes\mathcal{L}(m)^* & & \\
\downarrow 1\otimes\gamma^*_{\mathcal{L}} & \searrow \gamma_{\mathcal{O}}\otimes 1 & \\
\mathcal{O}(n)\otimes\mathcal{O}(m_1)\otimes\cdots\otimes\mathcal{O}(m_n)\otimes\mathcal{L}(n)^*\otimes\mathcal{L}(m_1)^*\otimes\ldots\otimes\mathcal{L}(m_n)^* & & \\
\| & & \\
\mathcal{O}(n)\otimes\mathcal{L}(n)^*\otimes\mathcal{O}(m_1)\otimes\mathcal{L}(m_1)^*\otimes\ldots\otimes\mathcal{O}(m_n)\otimes\mathcal{L}(m_n)^* & \xrightarrow{\gamma_{\mathrm{co}(\mathcal{O})}} & \mathcal{O}(m)\otimes\mathcal{L}(m)^*
\end{array}$$

where

$$\gamma^*_{\mathcal{L}}\colon \mathcal{L}(m)^* \longrightarrow \mathcal{L}(n)^*\otimes\mathcal{L}(m_1)^*\otimes\cdots\otimes\mathcal{L}(m_n)^*$$

is dual to the multiplication in $\mathcal{L}$.

4.6.2. Definition. The standard commutative operad $\mathcal{S}_C$ is defined as $\mathrm{co}(\mathcal{S})$.

Therefore, for any commutative operad $\mathcal{O}$ one has

$$\mathrm{Hom}(\mathcal{S}_C, \mathcal{O}) = \mathrm{Hom}(\mathcal{S}, \mathcal{O}\otimes\mathcal{L}).$$

This proves the versal property of $\mathcal{S}_C$.

4.6.3. One can prove that the complexes $\mathcal{S}_C(n)$ coincide with those studied in [BG] under the name "the tree part of Kontsevich complexes".

§5. Cosimplicial Lie algebras and the Eilenberg-Zilber operad

In this section we prove that the total complex of a cosimplicial Lie algebra admits a structure of a Lie May algebra. For this we construct a Lie operad $\mathcal{Y}$, called the Eilenberg-Zilber Lie operad, which acts on any total complex of a cosimplicial Lie algebra. In a similar way, the total complex of any cosimplicial module over a cosimplicial Lie algebra $\mathfrak{g}$ has a structure of module over the total complex of $\mathfrak{g}$.

These results imply the following

Claim (cf. 6.2). *Let $f\colon X\to Y$ be a map of locally noetherian schemes over k and let $\mathfrak{g}$ be a sheaf of Lie $\mathcal{O}_X$-algebras. Then the complex of sheaves $Rf_*(\mathfrak{g})$, the higher direct image if $\mathfrak{g}$, admits a structure of a Lie May algebra over Y.*

In what follows we fix a commutative base ring k. As usual in the subject, Δ denotes the category of finite ordered sets $[n]$, $n\in\mathbb{N}$, and monotonic maps (and not the comultiplication!). For a category $\mathcal{C}$ we denote by $\Delta^0\mathcal{C}$ the category of simplicial objects in $\mathcal{C}$ and by $\Delta\mathcal{C}$ the category of cosimplicial objects.

Thus, $\Delta^0\mathcal{E}ns$ is the category of simplicial sets and we denote by $\Delta^p\in\Delta^0\mathcal{E}ns$ the standard p-simplex.

The base tensor category in this section is $\Delta C(k)$, the category of cosimplicial complexes over k.

5.1. The total complex functor.

5.1.1. *The functor M.* The functor $M: \Delta(k) \longrightarrow C^{\geqslant 0}(k)$ is defined by the formulas

$$(MX)^p = X^p, \qquad d = \sum(-1)^i d^i : X^p \to X^{p+1},$$

where d^i are the cofaces in X.

In the same way, the functor $M^0: \Delta^0(k) \longrightarrow C^{\leqslant 0}(k)$ is defined by the formulas

$$(M^0X)^p = X_{-p}, \qquad d = \sum(-1)^i d_i : X_{-p} \to X_{-p-1}.$$

5.1.2. Define the object $Y \in C^{\leqslant 0}\Delta(k) = \Delta C^{\leqslant 0}(k)$ by the formula

$$Y: [n] \mapsto Y^n = M^0(k\Delta^n) =:: C_{\cdot}(\Delta^n).$$

LEMMA. *The object Y represents the functor M: for $A \in \Delta(k)$ one has $MA = \mathrm{Hom}_\Delta(Y, M)$.*

PROOF. Straightforward.

5.1.3. DEFINITION. The functor $\mathrm{Tot}: \Delta C(k) \longrightarrow C(k)$ sends an object X to the ordinary complex associated with the double complex obtained from X by applying M in the cosimplicial direction.

5.2. The main result.

5.2.1. THEOREM. *Let $\mathfrak{g} \in \Delta C(k)$ be a cosimplicial Lie algebra. Then the complex* $\mathrm{Tot}(\mathfrak{g})$ *admits a natural structure of a Lie May algebra. Moreover, if $M \in \Delta C(k)$ is a $\mathfrak{g}$-module, then* $\mathrm{Tot}(M)$ *admits a natural structure of a* $\mathrm{Tot}(\mathfrak{g})$*-module.*

5.2.2. REMARK. One can prove the following generalization of Theorem 5.2.1 analogous to Theorem (1.6) of [HS].

CLAIM. *Let $\mathscr{C}$ be a small connected category and let $(\mathfrak{g}, \mathscr{O})$ be a Lie May algebra over $\mathscr{C}$. Then the homotopy limit* $\mathrm{holim}_{\mathscr{C}}\, \mathfrak{g}$ *admits a natural structure of a Lie May algebra.*

The definition of a Lie May algebra over a category is a complete analog of [HS, Definition (1.3), 2], where commutative May algebras over a category are defined. The proof in [HS] can also be adapted to our context.

In this paper we present a more transparent proof of a special case which seems to be only important. We leave the proof of the general statement above as an exercise to an interested reader (Hint: take the proof of Theorem (1.6) in [HS] and adapt it).

5.2.3. *Eilenberg-Zilber operad.* Define an operad $\mathscr{Y}$ in $C(k)$ which will act on the total complex of any cosimplicial Lie algebra.

DEFINITION. The (Lie) Eilenberg-Zilber operad $\mathscr{Y}$ is defined by the formula

$$\mathscr{Y}(n) = \mathrm{Hom}_\Delta(Y, Y^{\otimes n}) \otimes \mathscr{L}(n)$$

where $\mathscr{L}$ is the trivial Lie operad.

The right Σ_n-action on $\mathcal{Y}(n)$ is induced by the action on $\mathcal{L}(n)$ and on $Y^{\otimes n}$; The unit in $\mathcal{Y}(1)$ is given by $\mathrm{id}\otimes e \in \mathrm{Hom}(Y, Y)\otimes\mathcal{L}(1)$. Finally, the multiplication is given by the composition

$$\begin{aligned}
&\mathrm{Hom}_\Delta(Y, Y^{\otimes n})\otimes\mathcal{L}(n)\otimes\mathrm{Hom}_\Delta(Y, Y^{\otimes m_1})\otimes\mathcal{L}(m_1)\otimes\dots\\
&\quad\otimes\mathrm{Hom}_\Delta(Y, Y^{\otimes m_n})\otimes\mathcal{L}(m_n)\longrightarrow\mathrm{Hom}_\Delta(Y, Y^{\otimes n})\otimes\mathrm{Hom}_\Delta(Y^{\otimes n}, Y^{\otimes\sum m_i})\\
&\quad\otimes\mathcal{L}(n)\otimes\mathcal{L}(m_1)\otimes\cdots\otimes\mathcal{L}(m_n)\longrightarrow\mathrm{Hom}_\Delta(Y, Y^{\otimes\sum m_i})\otimes\mathcal{L}\left(\sum m_i\right).
\end{aligned}\tag{10}$$

5.2.4. Note. The commutative Eilenberg-Zilber operad $\mathcal{Z}$ can be defined as $\mathcal{Z}(n)=\mathrm{Hom}_\Delta(Y, Y^{\otimes n})$, compare with [HS, Section 2], where the normalization functor N was used instead of the functor M.

One obviously has $\mathcal{Y}=\mathcal{Z}\otimes\mathcal{L}$ in the sense of 4.6.1.

5.2.5. *"External" action of $\mathcal{Y}$.* Let $X, X_1, \dots, X_n\in\Delta(k)$ be cosimplicial k-modules and let a map

$$f\colon\mathcal{L}(n)\otimes X_1\otimes\cdots\otimes X_n\longrightarrow X$$

be given. Then the map

$$M(f)\colon\mathcal{Y}(n)\otimes M(X_1)\otimes\cdots\otimes M(X_n)\longrightarrow M(X)$$

is defined as the composition

$$\begin{aligned}
&\mathrm{Hom}_\Delta(Y, Y^{\otimes n})\otimes\mathcal{L}(n)\otimes\mathrm{Hom}_\Delta(Y, X_1)\otimes\cdots\otimes\mathrm{Hom}_\Delta(Y, X_n)\\
&\quad\longrightarrow\mathrm{Hom}_\Delta(Y, Y^{\otimes n})\otimes\mathrm{Hom}_\Delta(Y^{\otimes n}, \mathcal{L}(n)\otimes X_1\otimes\cdots\otimes X_n)\\
&\quad\longrightarrow M(\mathcal{L}(n)\otimes X_1\otimes\cdots\otimes X_n)\longrightarrow M(X).
\end{aligned}\tag{11}$$

5.2.6. Let $\mathfrak{g}\in\Delta C(k)$ be a cosimplicial DG Lie algebra. Consider $\mathfrak{g}$ as a complex

$$\cdots\longrightarrow\mathfrak{g}^i\longrightarrow\mathfrak{g}^{i+1}\longrightarrow\mathfrak{g}^{i+2}\longrightarrow\cdots$$

of cosimplicial k-modules. The Lie algebra structure on $\mathfrak{g}$ defines a collection of maps

$$\mathcal{L}(n)\otimes\mathfrak{g}^{i_1}\otimes\cdots\otimes\mathfrak{g}^{i_n}\longrightarrow\mathfrak{g}^{\sum i_k},$$

to which we can apply the "external action" formulas, defining in this way a $\mathcal{Y}$-algebra structure on $\mathrm{Tot}\,\mathfrak{g}$.

5.2.7. Exactly as in 5.2.6, a cosimplicial module M over a cosimplicial DG Lie algebra $\mathfrak{g}$ defines a structure of a $(\mathrm{Tot}\,\mathfrak{g}, \mathcal{Y})$-module on $\mathrm{Tot}(M)$. For this one should apply the "external action" formulas to the maps

$$\mathcal{L}(n)\otimes\mathfrak{g}^{i_1}\otimes\cdots\otimes\mathfrak{g}^{i_{n-1}}\otimes M^{i_n}\longrightarrow M^{\sum i_k},$$

defining on M the structure of a cosimplicial $\mathfrak{g}$-module.

5.2.8. *Explicit formulas for $\mathcal{Z}$.* The complex $\mathcal{Z}(n)=\mathrm{Hom}_\Delta(Y, Y^{\otimes n})$ has a very natural presentation as the ordinary complex of the double complex having components

$$\mathcal{Z}(n)^{j,-i} = \mathrm{Hom}_{\Delta}(Y_j, (Y^{\otimes n})_i) = M_j(Y^{\otimes n})_i$$
$$= (C.(\Delta^j))_i^{\otimes n} = \sum_{i=i_1+\cdots+i_n} C_{i_1}(\Delta^j)\otimes\cdots\otimes C_{i_n}(\Delta^j). \tag{12}$$

The complex $\mathcal{Z}(n)$ can be otherwise described as the complex of all n-ary chain operations

$$M(X_1)\otimes\cdots\otimes M(X_n)\longrightarrow M(X_1\otimes\cdots\otimes X_n).$$

It contains elements $\mu_i^{(n)}$ of degree $(-ni, i)$, which correspond to the tensor product in $\Delta(k)$. Let $\mathcal{M} = \prod_i k\mu_i \subseteq \mathcal{Z}(n)$.

PROPOSITION. *The multiplication*

$$\gamma\colon \mathcal{Z}(n)\otimes\mathcal{Z}(1)\otimes\cdots\otimes\mathcal{Z}(1)\longrightarrow\mathcal{Z}(n)$$

induces an isomorphism

$$\mathcal{M}\otimes\mathcal{Z}(1)^{\otimes n}\longrightarrow\mathcal{Z}(n),$$

of graded k-modules.

PROOF. Obvious.

5.2.9. *Proof of* 5.2.1. Define $\pi\colon\mathcal{Z}\to\mathcal{C}$ as the map induced by the map of cosimplicial k-modules $Y^{\otimes n}\to k$ (k is the trivial cosimplicial k-module) induced by projections $\Delta^i\to\{\cdot\}$, followed by the natural projection $C.(\{\cdot\})\to k$.

We will prove that π is a quasi-isomorphism. This implies that

$$\pi\otimes 1\colon\mathcal{Y}=\mathcal{Z}\otimes\mathcal{L}\longrightarrow\mathcal{L}$$

is a quasi-isomorphism, that is, that $\mathcal{Y}$ is a Lie operad.

To prove π is a quasi-isomorphism, let us consider $\mathcal{Z}$ as the ordinary complex corresponding to the double complex $\{\mathcal{Z}^{j,-i}\}$. Any cycle of degree d is represented by a sequence of elements $z_j\in\mathcal{Z}^{j,d-j}$ satisfying some obvious condition. For a fixed j, the complex $\mathcal{Z}^{j,*}=C(\Delta^j)^{\otimes n}$ considered with respect to the vertical differential, is quasi-isomorphic to k. An easy diagram chase finishes the proof.

The action of $\mathcal{Y}$ on $\mathrm{Tot}(\mathfrak{g})$ and of the above pair on $\mathrm{Tot}(M)$ was defined in 5.2.6, 5.2.7.

5.2.10. REMARK. The truncation functor $\tau^{\leqslant 0}$ on $C(k)$ is defined as taking a complex $X=\{X^i, d^i\colon X^i\to X^{i+1}\}$ to

$$\cdots\to X^{-1}\to\ker d^0\to 0.$$

If $\mathcal{O}=\{\mathcal{O}(n)\}$ is an operad in $C(k)$ then $\tau^{\leqslant 0}\mathcal{O}=\{\tau^{\leqslant 0}\mathcal{O}(n)\}$ is also an operad; in particular, since $\pi\colon\mathcal{Z}\to\mathcal{C}$ is a quasi-isomorphism, the induced morphism $\tau^{\leqslant 0}\mathcal{Z}\to\mathcal{C}$ is a homotopy equivalence.

In the same way, the morphism $\tau^{\leqslant 0}\mathcal{Y}\to\mathcal{L}$ is a homotopy equivalence.

§6. Application: homology of sheaves of Lie algebras

In this section our base tensor category is the category of complexes of $\mathcal{O}_X$-modules for a locally noetherian scheme X. This will be denoted by $C(X)$, and the corresponding derived category will be denoted by $D(X)$. The category $C(Qcoh(X))$ consists of complexes of quasi-coherent $\mathcal{O}_X$-modules and $D_{qc}(X)$ is the derived category of complexes with quasi-coherent cohomology. It is well known that for X locally noetherian $D_{qc}(X)$ is the localization of $C(Qcoh(X))$ with respect to the set of quasi-isomorphisms.

As above, we fix a commutative ring $k \supseteq \mathbb{Q}$. Denote by $\mathcal{S}ch$ the category of locally noetherian schemes over k.

Let $f: X \to Y$ be a morphism of schemes in $\mathcal{S}ch$ and let $\mathfrak{g} \in C(Qcoh(X))$ be a sheaf of DG Lie algebras. The *homology direct image* of $\mathfrak{g}$ is an element $Cf_*(\mathfrak{g}) \in D_{qc}(Y)$ obtained, roughly speaking, by the formula

$$Cf_*(\mathfrak{g}) = C(Rf_*(\mathfrak{g})),$$

C in the right-hand side of the formula denoting the standard complex functor studied in Section 4.

To make sense of this formula, we must define the May Lie algebra structure on (a concrete representative of) $Rf_*(\mathfrak{g})$ and prove that the resulting standard complex does not depend up to unique isomorphism on the choice of the representative for $Rf_*(\mathfrak{g})$.

6.1. Operads over a scheme. An operad over $X \in \mathcal{S}ch$ is just an operad in the category $C(X)$.

Let $f: X \to Y$ be in $\mathcal{S}ch$ and $\mathcal{O}$ be an operad over Y. Then $f^*(\mathcal{O})$ admits a natural structure of operad over X. If, in particular, $Y = \operatorname{Spec} k$ and $\mathcal{O} = \mathcal{C}$ or $\mathcal{L}$, we obtain their inverse images $\mathcal{C}_X$, $\mathcal{L}_X$. These are the trivial commutative and the trivial Lie operads over X.

Another example is given by the truncated Eilenberg-Zilber operad. According to 5.2.10 $\overline{\mathcal{Y}} = \tau^{\leqslant 0}\mathcal{Y}$ is an operad over Spec(k) homotopically equivalent to $\mathcal{L}$ and thus the inverse image $\overline{\mathcal{Y}}_X$ is an operad in $C(X)$ homotopically equivalent to $\mathcal{L}_X$. This will be called the (truncated) Lie Eilenberg-Zilber operad over X.

6.1.1. Definition. An operad $\mathcal{O}$ over $X \in \mathcal{S}ch$ endowed with a (componentwise) homotopy equivalence $\pi: \mathcal{O} \to \mathcal{L}_X$ is called *a Lie operad* over X.

In particular, $\overline{\mathcal{Y}}_X$ is a Lie operad over X.

6.1.2. Remark. According to the definition above, a Lie operad over k may not be necessarily a Lie operad over Spec(k); however, the two notions coincide for operads in $C^{\leqslant 0}(k)$—see Remark 5.2.10.

We will need in the sequel the following sheaf analogs of Proposition 4.1.4 and Theorem 5.2.1.

6.1.3. PROPOSITION. *Let $X \in \mathcal{S}ch$. For any Lie operad $\mathcal{O}$ on X there exists a map $\mathcal{S}_X \to \mathcal{O}$ of Lie operads.*

6.1.4. PROPOSITION. *Let $\mathfrak{g} \in \Delta C(X)$ be a cosimplicial Lie algebra over X. Then* $\mathrm{Tot}(\mathfrak{g})$ *admits a natural structure of a $\overline{\mathcal{Y}}_X$-algebra. If $M \in \Delta C(X)$ is a cosimplicial module over $\mathfrak{g}$, then* $\mathrm{Tot}(M)$ *admits a natural structure of a* $(\mathrm{Tot}(\mathfrak{g}), \overline{\mathcal{Y}}_X)$*-module.*

6.2. Construction. Let $\mathcal{U} = \{U_i\}_{i \in I}$ be a set of open subsets of a topological space X. Let, as usual, $[n] = \{0, \dots, n\}$. For $\phi: [n] \to I$ we define $U_\phi = \bigcup_r U_{\phi(r)}$ and put $\imath_\phi : U_\phi \to X$ to be the natural open embedding.

Let $X \in \mathcal{S}ch$. For a sheaf A of $\mathcal{O}_X$-modules define sheaves $\check{C}^{\mathcal{U}}_n(A)$ by

$$\check{C}^{\mathcal{U}}_n(A) = \prod_{\phi\,:\,[n]\to I} \imath_{\phi *}\imath^*_\phi(A),$$

the product being taken over all monotonous ϕ. The structure of a cosimplicial object on $\check{C}^{\mathcal{U}}_\cdot(A)$ is given by maps $\alpha_* : \check{C}^{\mathcal{U}}_m(A) \to \check{C}^{\mathcal{U}}_n(A)$ induced for any monotonous map $\alpha: [m] \to [n]$ by adjunction maps $A \to \imath_*\imath^*(A)$.

The Proposition 6.2.1 below is fairly standard (see [G, II.5.2.1], or [SGA4, Exp. 5, Proposition 1.11]).

6.2.1. PROPOSITION. *Let $\mathcal{U}$ be an open covering of X. Then $A \to \check{C}^{\mathcal{U}}_\cdot(A)$ is a cosimplicial resolution of A.*

Let $f: X \to Y$ be a map in $\mathcal{S}ch$.

6.2.2. DEFINITION. An open covering $X = \bigcup U_i$ is said to be *affine with respect to* $f: X \to Y$ if all U_i are affine and the compositions $U_i \to X \to Y$ are affine morphisms.

6.2.3. Fix an open covering $\mathcal{U}$ of X. For any $M \in C(X)$ $\check{C}^{\mathcal{U}}_\cdot(M)$ is a cosimplicial object in $C(X)$. The direct image $f_*\check{C}^{\mathcal{U}}_\cdot(M)$ is an object in $\Delta C(Y)$ and thus $\mathrm{Tot}(f_*\check{C}^{\mathcal{U}}_\cdot(M)) \in C(Y)$ is defined.

If $\mathcal{U}$ is affine with respect to f and $M \in C(Qcoh(X))$, then $\mathrm{Tot}(f_*\check{C}^{\mathcal{U}}_\cdot(M))$ represents $Rf_*(M) \in D(Y)$.

Let $\mathfrak{g} \in C(Qcoh(X))$ be a sheaf of DG Lie algebras on X. Then $\check{C}^{\mathcal{U}}_\cdot(\mathfrak{g})$ is a cosimplicial DG Lie algebra on X and $\mathrm{Tot}(f_*\check{C}^{\mathcal{U}}_\cdot(\mathfrak{g})) \in C(Y)$ admits a natural structure of a $\overline{\mathcal{Y}}_Y$-algebra by Proposition 6.1.4.

Choose a quasi-isomorphism $\alpha: \mathcal{S}_Y \to \overline{\mathcal{Y}}_Y$, which exists by Proposition 6.1.3. Thus, a $\mathcal{S}_Y$-algebra structure is defined on $\mathrm{Tot}(f_*\check{C}^{\mathcal{U}}_\cdot(\mathfrak{g}))$ and this allows us to construct the corresponding standard complex which will be denoted by $Cf_*(\mathfrak{g})_{\mathcal{U},\alpha}$.

6.3. Uniqueness. We will prove now that $Cf_*(\mathfrak{g})_{\mathcal{U},\alpha}$ as an object of the derived category $D(Y)$ does not depend on the choice of $\mathcal{U}$ and α.

Let $\mathcal{U} = \{U_i\}_{i \in I}$, $\mathcal{V} = \{V_j\}_{j \in J}$ be two coverings of X affine with respect to f and let $\theta: J \to I$ be a map satisfying the condition $V_j \subseteq U_{\theta(j)}$ (so that $\mathscr{V}$ is refinement of $\mathscr{U}$).

These data define a map $\check{C}^{\mathscr{U}}(A) \to \check{C}^{\mathscr{V}}(A)$ for any sheaf A of $\mathscr{O}_X$-modules which does not depend up to homotopy on the choice of θ (cf. [G, II.5.7.1]).

The map defined induces a quasi-isomorphism (since $\mathscr{U}$ and $\mathscr{V}$ are affine with respect to f)

$$\mathrm{Tot}(f_* \check{C}^{\mathscr{U}}_{\cdot}(\mathfrak{g})) \longrightarrow \mathrm{Tot}(f_* \check{C}^{\mathscr{V}}_{\cdot}(\mathfrak{g})),$$

which does not depend up to a homotopy on the choice of $\theta: J \to I$.

Finally, applying the standard complex functor we get a map

$$Cf_*(\mathfrak{g})_{\mathscr{U},\alpha} \longrightarrow Cf_*(\mathfrak{g})_{\mathscr{V},\alpha},$$

which is a quasi-isomorphism by Proposition 3.6.2 and Proposition 4.5.2 and does not depend up to homotopy on θ. Since any pair $\mathscr{U}$, $\mathscr{V}$ of open coverings of X that are affine with respect to f, admit a common affine refinement, for a fixed morphism $\alpha: \mathscr{S}_Y \to \overline{\mathscr{Y}}_Y$ the object $Cf_*(\mathfrak{g})_{\mathscr{V},\alpha}$ from $D(Y)$ is defined uniquely up to a unique isomorphism.

To prove that $Cf_*(\mathfrak{g})_{\mathscr{U},\alpha}$ does not depend on the choice of $\alpha: \mathscr{S}_Y \to \overline{\mathscr{Y}}_Y$, it suffices to note that, by Proposition 3.6.2, α induces a (canonical) isomorphism in $D(Y)$ from $Cf_*(\mathfrak{g})_{\mathscr{V},\alpha}$ to the complex calculating the homology $H^{\overline{\mathscr{Y}}_Y}(T, 1)$, where $T = \mathrm{Tot}(f_* \check{C}^{\mathscr{U}}_{\cdot}(\mathfrak{g}))$.

Thus we have proven the following

6.3.1. THEOREM. *The construction attaching to a sheaf* $\mathfrak{g} \in C(Qcoh(X))$ *of DG Lie algebras an object* $Cf_*(\mathfrak{g}) \in D_{qc}(Y)$ *does not depend, up to unique isomorphism, on the choice of a cover* $\mathscr{U}$ *and a quasi-isomorphism* α.

REFERENCES

[AFH] L. Avramov, H.-B. Foxby, and S. Halperin, *Differential graded homological algebra*, in preparation.

[BG] A. Beilinson and V. Ginsburg, *Infinitesimal structure of moduli spaces of G-bundles*, Internat Math. Res. Notices **4** (1992), 63–74.

[BV] J. Boardman and R. Vogt, *Homotopy invariant algebraic structures on topological spaces*, Lecture Notes in Math., vol. 347, Springer-Verlag, Berlin and New York, 1973.

[DM] P. Deligne and J. Milne, *Tannakian categories*, Lecture Notes in Math., vol. 900, Springer-Verlag, Berlin and New York, 1977, pp. 101–228.

[D] V. Drinfeld, a letter to V. Schechtman, Sept. 1988.

[G] R. Godement, *Topologie algébrique et théorie des faisceaux*, Hermann, Paris, 1958.

[HS] V. Hinich and V. Schechtman, *On homotopy limit of homotopy algebras*, Lecture Notes in Math., vol. 1289, Springer-Verlag, Berlin and New York, 1987, pp. 240–264.

[M1] S. Mac Lane, *Homology*, Springer, 1963.

[M2] ———, *Categories for the working mathematician*, Graduate Texts in Math., vol. 5, Springer-Verlag, Berlin and New York, 1971.

[Ma] J. P. May, *The geometry of iterated loop spaces*, Lecture Notes in Math., vol. 271, Springer-Verlag, Berlin and New York, 1972.

[Q] D. Quillen, *Rational homotopy theory*, Ann. of Math. **90** (1969), no. 2, 205–295.

[SGA4] M. Artin, A. Grothendieck, and J. L. Verdier, *Théorie des topos et cohomologie étale des schémas*, Lecture Notes in Math., vol. 269, Springer-Verlag, Berlin and New York, 1972.

[SV] V. Schechtman and A. Varchenko, *Arrangements of hyperplanes and Lie algebra homology*, Invent. Math. **106** (1991), 139–194.

[Sm] L. Smith, *Homological algebra and the Eilenberg-Moore spectral sequence*, Trans. Amer. Math. Soc. **129** (1967), 58–93.

[St] J. Stasheff, *Differential graded Lie algebras, quasi-Hopf algebras and higher homotopy algebras*, Quantum Groups (P. P. Kulish, ed.), Lecture Notes in Math., vol. 1510, Springer-Verlag, Berlin and New York, 1992, pp. 120–137.

[T] J. Tate, *Homology of noetherian rings and local rings*, Illinois J. Math. **1** (1957), 14–27.

Department of Mathematics and Computer Science, University of Haifa, Mount Carmel, Haifa, 31905 Israel

Department of Mathematics, SUNY at Stony Brook, Stony Brook, New York 11794

ADVANCES IN SOVIET MATHEMATICS
Volume 16, Part 2, 1993

Chow Quotients of Grassmannians. I

M. M. KAPRANOV

To Israil Moiseevich Gelfand

CONTENTS

1991 *Mathematics Subject Classification.* Primary 14M15, 14L30; Secondary 52B20.

Introduction

The study of the action of the maximal torus $H \subset GL(n)$ on the Grassmann variety $G(k, n)$ is connected with numerous questions of geometry and analysis. Among these let us mention the general theory of hypergeometric functions [18], K-theory [7], and combinatorial constructions of characteristic classes [17, 20, 40]. It was noted by I. M. Gelfand and R. D. MacPherson [20, 40] that this problem is equivalent to the classical problem about projective equivalence classes of configurations of n points in the projective space P^{k-1}.

In the present paper we propose a geometric approach to this problem, which is based on the study of the behavior of orbit closures. Namely, the closures of generic orbits are compact varieties "of the same type". Now there is a beautiful construction in algebraic geometry of Chow varieties [50]. It produces compact varieties whose points parametrize algebraic cycles (=positive integral combinations of irreducible subvarieties) in a given variety with given dimension and degree. In particular, any one-parameter family of subvarieties "of the same type" has a limit in the Chow variety. We define the *Chow quotient* $G(k, n)//H$ to be the space of such limits of closures of generic orbits. Any point of $G(k, n)//H$ represents an $(n-1)$-dimensional family of $(k-1)$-dimensional projective subspaces in P^{n-1}.

Families of subvarieties in a variety X whose parameter space has the same dimension as X are classically known as *complexes*. We call closures of generic orbits in $G(k, n)$ Lie complexes and their limit positions generalized Lie complexes. In the simplest case of $G(2, 4)$ Lie complexes are the so-called tetrahedral complexes of lines in P^3, which have a long history (see the bibliography in [20]).

The variety $G(k, n)//H$ can be defined in two more ways:

a) As the space of limits of closures of generic orbits of the group $GL(k)$ in the Cartesian power $(P^{k-1})^n$ (Theorem 2.2.4).

b) As the space of limits of special Veronese varieties in the Grassmannian $G(k-1, n-1) = G(k-1, \mathbf{h})$, where $\mathbf{h}$ is the Lie algebra of the torus H (Theorem 3.3.14).

According to interpretation a), the space $G(k, n)//H$ is obtained by adding some "ideal" elements to the space of projective equivalence classes of configurations of n points (or hyperplanes) in P^{k-1} in general position. These ideal elements are more subtle than just nongeneral configurations: a limit position of closures of generic orbits can be the union of several orbit closures, each representing a certain configuration.

In fact, it turns out that these elements behave in many respects as if they actually were configurations in general position. In particular, there are restriction and projection maps (Theorem 1.6.6)

$$G(k, n-1)//(\mathbb{C}^*)^{n-1} \xleftarrow{\tilde{b}_i} G(k, n)//(\mathbb{C}^*)^n \xrightarrow{\tilde{a}_i} G(k-1, n-1)//(\mathbb{C}^*)^{n-1}.$$

The map $\tilde{a}_i$ corresponds to the restriction of the generic hyperplane configuration $(M_1, \dots, M_n)$ to the hyperplane M_i. The map $\tilde{b}_i$ corresponds to deleting M_i. The maps $\tilde{a}_i$, $\tilde{b}_i$ (regarded on the generic part of a Grassmannian) were at the origin of the use of Grassmannians in the problem of combinatorial calculation of Pontryagin classes [20, 40]. These maps were also used in [7] to define the so-called Grassmannian complexes serving as an approximation to K-theory. Note that $\tilde{a}_i$ cannot be, in general, extended to arbitrary configurations. It is the Chow quotient approach that permits this extension.

Veronese varieties (which make their appearance in the interpretation b) above) are defined classically as images of a projective space in the projective embedding given by all homogeneous polynomials of given degree. In particular, Veronese curves are just rational normal curves and possess a lot of remarkable geometric properties, see [46]. In our situation Veronese varieties in fact lie on a Grassmannian (in the Plücker embedding): a $(k-1)$-dimensional variety lies in $G(k-1, \mathbf{h})$ so that for $k = 2$ we obtain the curve in a projective space.

The Veronese variety associated to a Lie complex Z is obtained as its *visible contour*, i.e., the locus of subspaces from Z that contain a given generic point p, say $p = (1 : \cdots : 1) \in P^{n-1}$. The consideration of visible contours is a classical method of analyzing complexes of subspaces [24, 27], which, to our knowledge, has yet never been applied to Lie complexes.

Along with the visible contour of Z lying in $G(k-1, \mathbf{h})$ we consider the so-called *visible sweep* $\mathrm{Sw}(Z)$. This is a subvariety in the projective space $P(\mathbf{h})$ which is the union of all subspaces from the visible contour. This variety can be found very explicitly. Thus, if our Lie complex has the form $Z = \overline{H \cdot L}$ where $L \in G(k, n)$ is the graph of a linear operator $A = \|a_{ij}\| : \mathbb{C}^k \to \mathbb{C}^{n-k}$, $k \leqslant n-k$ (this assumption does not restrict the generality), then the sweep is the projectivization of the following determinantal cone:

$$\{(t_1, \dots, t_n) \in \mathbf{h} : \operatorname{rank} \|a_{ij}(t_i - t_j)\|_{i=1,\dots,k, j=1,\dots,n-k} \leq k\}.$$

The linear forms $t_i - t_j$ entering the matrix above are roots of $\mathbf{h}$, the Cartan subalgebra of $\mathbf{pgl}(n)$. More precisely, we encounter those roots that enter into the weight decomposition of the parabolic subalgebra defining the Grassmannian.

Veronese varieties in the Grassmannian, which arise naturally in our constructions, seem to be "good" generalizations of Veronese curves in projective spaces. We show in section 3.5 that these varieties admit a Steiner-type construction, which is well known for curves [24].

The homology class in the Grassmannian of such a variety is given by an extremely beautiful formula (Theorem 3.9.8). To state it, recall that the homology of a Grassmannian is freely generated by the Schubert cycle σ_α which corresponds to Young diagrams. It turns out that the multiplicity of the cycle σ_α in the class of the Veronese variety equals the dimension of the space $\Sigma^{\alpha^*}(\mathbb{C}^{n-k})$, the irreducible representation of the group $GL(n-k)$ corresponding to the Young diagram dual to α.

In the particular case $k = 2$ the construction b) realizes points of our variety as limit positions of rational normal curves (Veronese curves, for short) in P^{n-2} through a fixed set of n points in general position. We deduce from this that $G(2, n)//H$ is isomorphic to the Grothendieck-Knudsen moduli space $\overline{M_{0,n}}$ of stable n-pointed curves of genus 0 (Theorem 4.1.8). This is certainly the most natural compactification of the space $M_{0,n}$ of projective equivalence classes of n-tuples of distinct points on P^1. It is smooth and the complement to $M_{0,n}$ is a divisor with normal crossings.

In general, Veronese varieties in Grassmannians arising in construction b) are Grassmannian embeddings of P^{k-1} corresponding to vector bundles of the form $\Omega^1(\log M)$, where $M = (M_1, \dots, M_n)$ is a configuration of hyperplanes in general position. They become Veronese varieties after the Plücker embedding of the Grassmannian. The configuration of hyperplanes on P^{k-1} can be read off the corresponding Veronese variety by intersecting it with natural sub-Grassmannians. This is explained in Chapter 3.

Chow quotients of toric varieties by the action of a subtorus of the defining torus were studied in [30, 31]. It was found that this provides a natural setting for the theory of secondary polytopes introduced in [22, 23] and their generalizations-fiber polytopes [8]. By considering the Plücker embedding of the Grassmannian we can apply the results of [30, 31]. This gives a description of possible degenerations of orbit closures in $G(k, n)$ (i.e., of generalized Lie complexes) in terms of polyhedral decompositions of a certain polytope $\Delta(k, n)$ called the hypersimplex [17, 19, 20]. For the case of $G(2, n)$ these decompositions are in bijection with trees that describe combinatorics of stable curves.

It is now well known [19, 21, 40] that various types of closures of torus orbits in $G(k, n)$ correspond to *matroids,* i.e., types of combinatorial behavior of configurations of n hyperplanes in projective space. To each matroid we can associate the corresponding *stratum* in $G(k, n)$. It is formed by orbits of the given type [18, 19, 21]. Matroids are in one-to-one correspondence with certain polytopes in the hypersimpex [19, 21]. From our point of view, however, a natural object is not an individual matroid but a collection of matroids such that the corresponding polytopes form a polyhedral decomposition of the hypersimplex. We call such collections *matroid decompositions.*

It should be mentioned that our approach differs considerably from that of geometric invariant theory developed by D. Mumford [42]. In particular, Mumford's quotients of $G(2, n)$ by H and of $(P^1)^n$ by $GL(2)$, although isomorphic to each other, do not coincide with the space $\overline{M_{0,n}}$, which provides a finer compactification. Note that Mumford's quotient depends upon a choice of a projective embedding (and this is felt for varieties like $(P^{k-1})^n$ with large Picard group) and upon the choice of linearization, i.e., of the extension of the action to the graded coordinate ring of the embedding (and this is felt for groups like the torus). We prove in Chapter 0 that the Chow quotient always maps to any Mumford quotient by a regular birational map.

Instead of Chow variety one can use Hilbert schemes and obtain a different compactification. Such a construction was considered in 1985 by A. Bialynicki-Birula and A. J. Sommese [10] and later by Y. Hu [26]. The advantage of Hilbert schemes is that they represent an easily described functor so it is easy to construct morphisms into them. In our particular example of the torus action on $G(k, n)$ both constructions lead to the same answer.

In the forthcoming second part of this paper we shall study the degenerations of Veronese varieties which provide a higher-dimensional analog of stable curves of Grothendieck and Knudsen. The main idea of stable pointed curves is that points are never allowed to coincide. When they try to do so, the topology of the curve changes in such a way that the points remain distinct. In higher dimensions instead of a collection of points we have a divisor on a variety. The analog of the condition that points are distinct is that the divisor has normal crossings. In particular, a generic configuration of hyperplanes defines a divisor with normal crossings. When the hyperplanes "try" to intersect nonnormally, the corresponding Veronese variety degenerates in such a way as to preserve the normal crossing.

I am grateful to W. Fulton who suggested that the space $\overline{M_{0,n}}$ might be related to Chow quotients and informed me on his joint work with R. MacPherson [15] on a related subject. I am also grateful to Y. Hu for informing me about his work [26] and about earlier work of A. Bialynicki-Birula and A. J. Sommese [10].

I am happy to be able to dedicate this paper to Israil Moiseevich Gelfand.

Chapter 0. Chow quotients

(0.1). Chow varieties and Chow quotients. Let H be an algebraic group acting on a complex projective variety X. We shall describe in this section an approach to constructing the algebraic "coset space" of X by H, which was introduced in [31]. (A similar approach was introduced earlier in [10], see §0.5 below.)

(0.1.1). *Setup of the approach.* For any point $x \in X$ we consider the orbit closure $\overline{H \cdot x}$ which is a compact subvariety in X. For some sufficiently small Zariski open subset $U \subset X$ of "generic" points all these varieties have the same dimension, say r, and represent the same homology class $\delta \in H_{2r}(X, \mathbb{Z})$. The set U may be supposed to be H-invariant. Moreover, since we are free to delete bad orbits from U, the construction of the quotient U/H presents no difficulty and the problem is to construct a "right" compactification of U/H. A natural approach to this is to study the limit positions of the varieties $\overline{H \cdot x}$, when x tends to the infinity of U (i.e., ceases to be generic). One of the precise ways to speak about such "limits" is provided by Chow varieties of algebraic cycles. Before proceeding further we recall the main definitions (cf. [30, 50]).

(0.1.2). By a (positive) r-dimensional algebraic cycle on X we shall understand a finite formal nonnegative integral combination $Z = \sum c_i Z_i$, where $c_i \in \mathbb{Z}_+$ and Z_i are irreducible r-dimensional closed algebraic subvarieties in X. Denote by $\mathscr{C}_r(X, \delta)$ the set of all r-dimensional algebraic cycles in X which have the homology class δ. It is known that $\mathscr{C}_r(X, \delta)$ is canonically equipped with a structure of a *projective* (in particular, compact) algebraic variety (called the Chow variety). In this form this result is due to D. Barlet [3].

(0.1.3). A more classical approach to Chow varieties is that of Chow forms [50]. This approach first gives the projective embedding of $\mathscr{C}_r(P(V), d)$, the Chow variety of r-dimensional cycles of degree d in the projective space $P(V)$. The Chow form of any cycle Z, $\dim(Z) = r$, $\deg(Z) = d$, is a polynomial $R_Z(l_0, \dots, l_r)$ in the coefficients of $r+1$ indeterminate linear forms $l_i \in V^*$, which is defined, up to a constant factor, by the following properties, see [30, 50]:

(0.1.3.1). $R_{Z+W} = R_Z \cdot R_W$.

(0.1.3.2). If Z is an irreducible subvariety then R_Z is an irreducible polynomial, which vanishes for given $l_0, \dots, l_r$ if and only if the projective subspace $\Pi(l_0, \ldots, l_r) = \{l_0 = \cdots = l_r = 0\}$ of codimension $r+1$ intersects Z.

(0.1.4). It is now classical [50] that the correspondence $Z \mapsto R_Z$ identifies $\mathscr{C}_r(P(V), d)$ with a Zariski closed subset of the projective space of polynomials $F(l_0, \ldots, l_r)$ homogeneous of degree d in each l_i.

(0.1.5). If $X \subset P(V)$ is any projective subvariety and $\delta \in H_{2r}(X, \mathbb{Z})$, then $\mathscr{C}_r(X, \delta)$ becomes a subset of $\mathscr{C}_r(P(V), d)$ where $d \in H_{2r}(P(V), \mathbb{Z}) = \mathbb{Z}$ is the image of δ. The result of Barlet mentioned in (0.1.2) shows, in particular, that (over a field of complex numbers) this subset is Zariski closed and the resulting structure of the algebraic variety on $\mathscr{C}_r(X, \delta)$ does not depend on the projective embedding. (The fact that for a given projective subvariety X the set of $Z \in \mathscr{C}_r(P(V), d)$ lying on X is Zariski closed, is classical.)

In the case of base field of characteristic p, which we do not consider here, the situation is more subtle, see [43, 44].

(0.1.6). Let us return to the situation (0.1.1) of the group H acting on X. We see that for $x \in U$ as in (0.1.1) the subvariety $\overline{H \cdot x}$ is a point of the variety $\mathscr{C}_r(X, \delta)$. Therefore the correspondence $x \mapsto \overline{H \cdot x}$ defines an embedding of the quotient variety U/H into $\mathscr{C}_r(X, \delta)$.

(0.1.7). Definition [31]. The Chow quotient $X//H$ is the closure of U/H in $\mathscr{C}_r(X, \delta)$.

Thus $X//H$ is a projective algebraic variety compactifying U/H. "Infinite" points of $X//H$ are some algebraic cycles in X which are limits (or "degenerations") of generic orbit closures.

(0.1.8). Remarks. a) Definition (0.1.7) does not depend on the freedom in the choice of U since deletion from U of orbits which are already "generic" results in their reappearance as points in the closure.

b) The notion of "genericity" used in Definition (0.1.7) is usually much more restrictive than Mumford's notion of stability [42]. In fact (0.1.7) makes no appeal to stability and is defined entirely in terms of X and the action of H.

(0.2). Torus action on a projective space: secondary polytopes. If H is an algebraic torus acting on a projective variety X, then X may be equivariantly embedded into a projective space with H-action. The case of torus action on a projective space recalled in this subsection will be basic for our study of more general torus actions in this paper.

(0.2.1). Let H be an algebraic torus $(\mathbb{C}^*)^k$. A character of H is the same as a Laurent monomial $t^\omega = t_1^{\omega_1} \cdots t_k^{\omega_k}$, where $\omega = (\omega_1, \ldots, \omega_k) \in \mathbb{Z}^k$ is an integer vector. A collection $A = \{\omega^{(1)}, \ldots, \omega^{(N)}\}$ of vectors from $\mathbb{Z}^k$ defines therefore a diagonal homomorphism from H to $GL(N)$. It is well known that any representation of the torus can be brought into a diagonal form.

(0.2.2). The homomorphism $H \to GL(N)$ constructed from the set A above defines an H-action on the projective space P^{N-1}. The homogeneous coordinates in P^{N-1} are naturally labelled by elements of A. So we shall denote this space by $P(A)$ and the coordinates by $(x_\omega)_{\omega \in A}$ thus dropping the (unnatural) numeration of ω's.

(0.2.3). The Chow quotient $P(A)//H$ was described in [30, 31]. We shall use this description so we recall it here.

(0.2.4). First of all, $P(A)//H$ is a projective toric variety.

To see this, we note that the "big" torus $(\mathbb{C}^*)^A$ acts on $P(A)$ (by dilation of homogeneous coordinates) commuting with H (which is just a subtorus of $(\mathbb{C}^*)^A$). Therefore $P(A)//H$ is the closure, in the Chow variety, of the $(\mathbb{C}^*)^A$-orbit of the variety $X_A = \overline{H \cdot x}$, where $x \in P(A)$ is the point with all coordinates equal to 1.

(0.2.5). It is known that projective toric varieties are classified by lattice polytopes, see [49]. In what follows we shall describe the polytope corresponding to the toric variety $P(A)//H$.

(0.2.6). Let $Q \subset \mathbb{R}^k$ be the convex hull of the set A. A *triangulation* of the pair (Q, A) is a collection of simplices in Q whose vertices lie in A intersecting only along common faces and covering Q. To any such triangulation T we associate its *characteristic function* $\varphi_T \colon A \to \mathbb{Z}$ as follows. By definition, the value of φ_T on $\omega \in A$ is the sum of volumes of all simplices of Q for which ω is a vertex. The volume form is normalized by the condition that the smallest possible volume of a lattice simplex equals 1.

(0.2.7). The *secondary polytope* $\Sigma(A)$ is, by definition, the convex hull of all characteristic functions φ_T in the space $\mathbb{R}^A$.

Secondary polytopes were introduced in [22, 23] in connection with Newton polytopes of multi-dimensional discriminants. It was shown in [22] that the vertices of $\Sigma(A)$ are precisely functions φ_T where the triangulation T is *regular*, i.e., possesses a strictly convex piecewise-linear function.

(0.2.8). THEOREM [30]. *The toric variety $P(A)//H$ corresponds to the convex lattice polytope $\Sigma(A)$.*

(0.2.9). COMPLEMENTS. All the faces of the secondary polytope $\Sigma(A)$ possess a complete description. We shall use in this paper only the case when elements of A are exactly vertices of Q so we shall restrict ourselves to this case, see [22] for the general case. Let us call a *polyhedral decomposition* of Q a collection of convex polytopes in Q whose vertices lie in A, which intersect only along common faces and cover Q. A polyhedral decomposition $\mathcal{D}$ is called *regular*, if it possesses a strictly convex piecewise-linear function. It was shown in [22] that vertices of $\Sigma(A)$ are in bijection with regular polyhedral decompositions of Q. Vertices of the face of $\Sigma(A)$ corresponding to such a decomposition $\mathcal{D}$ are precisely functions φ_T for all regular triangulations T refining $\mathcal{D}$.

(0.2.10). It was shown in [31] that any cycle from $P(A)//H$ is a sum of toric subvarieties (closures of H-orbits), see [31, Proposition 1.1]. In particular, a regular triangulation T represents a 0-dimensional torus orbit in $P(A)//H$, i.e., an algebraic cycle. This cycle has the form $\sum_{\sigma\in T}\mathrm{Vol}(\sigma)\cdot\mathbf{L}(\sigma)$, where $\mathbf{L}(\sigma)$ is the coordinate k-dimensional projective subspace in $P(A)$ spanned by basis vectors corresponding to vertices of σ.

(0.3). Structure of cycles from the Chow quotient.

(0.3.1). THEOREM. *Let H be a reductive group acting on a smooth projective variety X. Suppose that the stationary subgroups H_x, $x\in X$, are trivial for generic x and are never unipotent. Then any component Z_i of any cycle $Z=\sum c_iZ_i\in X//H$ is a closure of a single H-orbit.*

PROOF. For the case when H is a torus, this statement follows from results of [30, 31]. Indeed, we can take an equivariant embedding of X into a projective space P^N with H-action in such a way that the dimension of a generic H-orbit on X is the same as the dimension of a generic H-orbit on P^N. The degeneration of torus orbits on a projective space (and, more generally, on toric varieties) was studied in [30, 31] where it was found that any orbit degenerates in a union of finitely many orbits [31, Proposition 1.1].

Consider now the general case. Let $C(t)$, $t\neq 0$, be a 1-parameter family of closures of generic orbits, $C(0)=\lim\limits_{t\to 0}C(t)$ is their limit in the Chow variety. Let C be any component of $C(0)$. Suppose, contrary to our statement, that C is not a closure of a single orbit. Then, for all points of C the stabilizer H_x has positive dimension. By our assumption, these stabilizers are all nonunipotent. Let x be some fixed generic point of C. Then H_x contains some torus T. Include x in a 1-parameter family of points $x(t)\in C(t)$ such that for $t\neq 0$ the point $x(t)$ lies in the orbit open in $C(t)$. Consider the closures of orbits $\overline{T\cdot x(t)}\subset C(t)$. As $t\to 0$, these closure should degenerate into some cycle Z whose support contains x. But we know that each component of Z is a closure of one T-orbit and so x should lie on the intersection of components of Z. This means that each generic point $x\in C$ lies in the closure of some orbit $H\cdot y$ not coinciding with $H\cdot x$. This is impossible.

(0.3.2). EXAMPLE. The assumption that the stabilizers of points are never unipotent in the formulation of Theorem 0.3.1 cannot be dropped. To construct an example, consider the group $H=SL(2,\mathbb{C})$. It has a standard action on $\mathbb{C}^2$. Let X_0 be the product $\mathbb{C}^2\times\mathbb{C}^2$ on which H acts diagonally. Let $X=P^2\times P^2$ be the natural compactification of X_0 with the obviously extended action of $H=SL(2)$. Generic H-orbits on X are 3-dimensional: two pairs of independent vectors (e_1,e_2) and (f_1,f_2) can be brought to each other by a unique transformation from $SL(2)$ if and only if $\det(e_1,e_2)=\det(f_1,f_2)$. Thus a generic orbit depends on one parameter, namely $\det(e_1,e_2)$. However, when this parameter approaches 0, the orbit degenerates into the 3-dimensional variety of proportional pairs (e_1, e_2). This variety is a union of a 1-parametric family of 2-dimensional orbits $O_\lambda = \{(e_1, e_2): e_1 = \lambda e_2\}$. The stabilizer of each of these orbit is a unipotent subgroup in $H = SL(2)$.

(0.4). Relation to Mumford quotients. It is useful to have a comparison of the Chow quotient with the more standard constructions, namely Mumford's geometric invariant theory quotients [42].

(0.4.1). To define Mumford's quotient, we should choose an H-equivariant projective embedding of X and extend the H-action to the homogeneous coordinate ring $\mathbb{C}[X]$ of X with respect to this embedding. This is equivalent to extending the action to the ample line bundle $\mathcal{L}$ defining the embedding. Such an extension is called *linearization.* Denote the chosen linearization by α. The Mumford's quotient $(X/H)_\alpha$ or $(X/H)_{\mathcal{L},\alpha}$ corresponding to $\mathcal{L}$, α is defined as $\operatorname{Proj} \mathbb{C}[X]^H$, the projective spectrum of the invariant subring [42]. Thus there are two choices in the definition of Mumford quotient: that of an ample line bundle and that of extension of the action to the chosen line bundle.

(0.4.2). By the general theory of [42], points of $(X/H)_{\mathcal{L},\alpha}$ are equivalence classes of α-semistable orbits in X. More precisely, two semistable orbits O, O' are equivalent if any invariant homogeneous function vanishing on O does so on O' and conversely. A Zariski open set in $(X/H)_{\mathcal{L},\alpha}$ is formed by α-stable orbits [42]. They have the property that no two α-stable orbits are equivalent. We shall say that the linearization α is nondegenerate if there are α-stable orbits.

The following result was proved in [31] for the case of a torus acting on a toric variety and (independently and simultaneously) in [26] for torus action on an arbitrary variety.

(0.4.3). THEOREM. *Let H be a reductive group acting on a projective variety X, $\mathcal{L}$ an ample line bundle on X, and α a linearization, i.e., an extension of the H-action on X to $\mathcal{L}$. Suppose that α is nondegenerate. Then there is a regular birational morphism $p_\alpha: X//H \to (X/H)_{\mathcal{L},\alpha}$.*

For any algebraic cycle $Z = \sum n_i Z_i$ in X its support (denoted $\operatorname{supp}(Z)$) is the union $\bigcup Z_i$. The proof of Theorem 0.4.3 consists of three steps:

(0.4.4). For any cycle $Z \in X//H$ as above there is at least one orbit in $\operatorname{supp}(Z)$ which is α-semistable.

(0.4.5). All the α-semistable orbits in $\operatorname{supp}(Z)$ are equivalent, i.e., represent the same point of the Mumford quotient $(X/H)_{\mathcal{L},\alpha}$.

(0.4.6). The map $p_\alpha : X//H \to (X/H)_{\mathcal{L},\alpha}$ that takes $Z \in X//H$ to the point of $(X/H)_{\mathcal{L},\alpha}$ represented by any of the semistable orbits in $\operatorname{supp}(Z)$ is a morphism of algebraic varieties.

(0.4.7). PROOF OF (0.4.4). We use the interpretation of semistability via the moment map [1, 35]. Let H_c be the compact real form of H with

Lie algebra $\mathscr{H}$ and $\mu: X \to \mathscr{H}^*$ be the moment map associated to an H_c-invariant Kähler form on X. Then an orbit O is semistable if and only if $\mu(\bar{O})$ contains the zero element of $\mathscr{H}^*$. Let $O(t)$, $t \neq 0$, be a 1-parameter family of generic orbits and $Z(t)$ be the closure of $O(t)$. Let $Z(0)$ be the limit of $Z(t)$ in the Chow variety. Since $\mu(Z(t))$ is, for $t \neq 0$, a closed set containing 0, the set $\mu(Z(0))$ also contains 0 thus proving that at least one orbit constituting $Z(0)$ is semistable.

(0.4.8). Proof of (0.4.5). Denote by $\mathscr{L}$ the equivariant ample invertible sheaf given by the linearization α. By definition, two semistable orbits O and O' are equivalent in $(X/H)_{\mathscr{L},\alpha}$ if any section of $\mathscr{L}^{\otimes k}$ vanishing at O vanishes also at O'. But the cycle Z is a limit position of closures of single orbits. So our assertion follows by continuity.

(0.4.9). Proof of (0.4.6). Let $X \subset P(V)$ be the equivariant projective embedding given by the linearization. We shall use the approach to the Chow variety via Chow forms, see (0.1.3). Let d be the degree of a generic orbit closure $\overline{H \cdot x}$, $x \in X$. Recall that the Chow form of any cycle Z, $\dim(Z) = r$, $\deg(Z) = d$, is a polynomial $R_Z(l_0, \dots, l_r)$ in the coefficients of $r+1$ indeterminate linear forms $l_i \in V^*$.

(0.4.9.1). Since the property of being a morphism is local, it suffices to prove it in a suitable open covering of $X//H$. More precisely, we are reduced to the following situation.

(0.4.9.2). Let f be an invariant rational function on V homogeneous of degree 0 (so it represents a regular function on some open set of $(X/H)_{\mathscr{L},\alpha}$). We must express the (constant) value of f on a generic orbit O as a rational function of the coefficients of the Chow form R_Z, where $Z = \overline{O}$.

(0.4.9.3). We can write (in characteristic 0 only!)

$$f|_Z = \frac{1}{d} \sum_{\mathbf{x} \in Z \cap L} f(\mathbf{x}),$$

where L is a generic projective subspace in $P(V)$ of codimension r. On the other hand, let $l_1, \dots, l_r$ be equations of L. Then we have the equality of polynomials in $l \in V^*$:

$$R_Z(l, l_1, \dots, l_r) = c \cdot \prod_{\mathbf{x} \in Z \cap L} l(\mathbf{x}),$$

where c is a nonzero number depending on $l_1, \dots, l_r$.

(0.4.9.4). Let $V = \mathbb{C}^{N+1}$ with coordinates $x_0, \dots, x_N$ and let ξ_i be dual coordinates in V^*, so an indeterminate linear form on V is $(\xi, \mathbf{x}) = \sum \xi_i x_i$ for some $\xi_0, \dots, \xi_N$. The Chow form of any 0-cycle $W = \mathbf{x}^{(1)} + \dots + \mathbf{x}^{(d)}$ is the polynomial $\prod(\mathbf{x}^{(i)}, \xi)$. Let us restrict the considerations to the affine chart, say, $\mathbb{C}^N = \{x_0 \neq 0\}$ in $P(V) = P^N$. The coordinate x_0 can then be set to 1 and we can set $\xi_0 = 1$ as well thus obtaining the Chow form of a

0-cycle $W \subset A^N$ as before in the form

$$\Phi_W(\xi) = \prod(1 + x_1^{(i)}\xi_1 + \cdots + x_N^{(i)}\xi_N).$$

The coefficients of this polynomial at various monomials in ξ's are known as elementary symmetric functions in d vector variables $\mathbf{x}^{(1)}, \ldots, \mathbf{x}^{(d)}$, see [28, 39]. By the formula for R_Z in (0.4.9.3), elementary symmetric functions of the d points of intersection $Z \cap L$, for any generic L of codimension r, can be polynomially expressed through the coefficients of R_Z. Therefore we are reduced to the following lemma.

(0.4.9.5). LEMMA. *Let $f(\mathbf{x})$, $\mathbf{x} = (x_1, \ldots, x_N)$, be a rational function in N variables and $d > 0$. Then there is a rational function $U_f = U_f(\Phi)$ in the coefficients of an indeterminate homogeneous polynomial $\Phi(\xi_1, \ldots, \xi_N)$, $\deg(\Phi) = d$, satisfying the following property. If $\mathbf{x}^{(1)}, \ldots, \mathbf{x}^{(d)}$ are points not lying on the polar locus of f then*

$$\sum f(\mathbf{x}^{(i)}) = U_f(\Phi_W),$$

where $\Phi_W = \prod(1 + (\mathbf{x}^{(i)}, \xi))$.

PROOF. It is known since P. A. MacMahon [28, 39] that any symmetric polynomial in $\mathbf{x}^{(i)}$ (in characteristic 0) can be polynomially expressed via elementary symmetric polynomials (in many different ways, if $N > 1$). If $f(\mathbf{x}) = P(\mathbf{x})/Q(\mathbf{x})$, where P, Q are relatively prime polynomials, then

$$\sum f(\mathbf{x}^{(i)}) = \frac{1}{\prod Q(\mathbf{x}^{(i)})} \sum_i P(\mathbf{x}^{(i)}) \prod_{j \neq i} Q(\mathbf{x}^{(j)})$$

is a ratio of two symmetric polynomials and the assertion follows.

The proof of Theorem 0.4.3 is completed.

(0.4.10). REMARK. In [31] it was shown that for the case of torus action on a toric variety, the Chow quotient $X//H$ is, in some sense, the "least common multiple" of all Mumford's quotients corresponding to different linearizations. From the point of view of general reductive groups a more typical case is when the group has 0-dimensional center and hence there is only one linearization. However, we shall see that in this case the Chow quotient still differs drastically from the Mumford one. The reason for this, as we would like to suggest, is that the Chow quotient takes into account not only Mumford quotients corresponding to various linearizations, but also more general symplectic quotients [1] corresponding to coadjoint orbits of H_c. For a torus, a different choice of a coadjoint orbit amounts to a change of a linearization, see [31] so all the symplectic quotients are reduced to Mumford ones. In the general case the symplectic quotients corresponding to nonzero orbits may not have an immediate algebro-geometric interpretation [1]. Nevertheless, their presense is somehow felt in $X//H$.

(0.5). Hilbert quotients (the Bialynicki-Birula–Sommese construction). A different way of speaking about limit positions of generic orbit closures is that of Hilbert schemes. Such a construction was considered by A. Bialynicki-Birula and A. J. Sommese [10] and later by Y. Hu [26].

(0.5.1). Recall [25, 47] that for any projective variety X there is the *Hilbert scheme* $\mathscr{H}_X$ parametrizing all subschemes in X. By definition, a morphism $S \to \mathscr{H}_X$ is a flat family of subschemes in X parametrized by S. The scheme $\mathscr{H}_X$ is of infinite type since no bound on "degrees" of subschemes is imposed. The connected components of $\mathscr{H}_X$ are, nevertheless, finite-dimensional projective schemes.

(0.5.2). Any connected component of the scheme $\mathscr{H}_X$ is canonically mapped into the Chow variety "corresponding" to this component. More precisely, if K is any such connected component then dimensions of subschemes from K are the same and equal, say, r. For any scheme $Z \in K$ we define the algebraic cycle

$$\mathrm{Cyc}(Z) = \sum_{C\,,\ \dim(C)=r} \mathrm{Mult}_C(Z) \cdot C\,, \qquad (0.5.3)$$

where C runs over all r-dimensional irreducible components of the algebraic variety $\mathrm{supp}(Z)$ and $\mathrm{Mult}_C Z$ is the multiplicity given by the scheme structure, see [30, 42].

(0.5.4). In the situation of (0.5.2) it follows from results of [42] that the cycles $\mathrm{Cyc}(Z)$ for all subschemes $Z \subset K$ have the same homology class, say δ_K and the formula (0.5.3) defines a regular morphism

$$K \to \mathscr{C}_r(X\,, \delta_K)\,,$$

see [42].

(0.5.5). Hilbert schemes have the advantage over Chow varieties in that they are defined as objects representing an easily described functor (that of flat families of subschemes, see (0.5.1)). In particular, the Zariski tangent space to the scheme $\mathscr{H}_X$ at a point given by a subscheme Z equals (see [47, Proposition 8.1]):

$$T_Z\mathscr{H}_X = H^0(Z\,, \mathscr{N}_Z)\,, \qquad \text{where} \quad \mathscr{N}_Z = \underline{\mathrm{Hom}}(J_Z/J_Z^2\,, \mathscr{O}_Z).$$

Here J_Z is the sheaf of ideals of the subscheme Z. The sheaf $\mathscr{N}_Z$ is called the normal sheaf of Z. It is locally free if Z is a locally complete intersection. If Z is a smooth variety then $\mathscr{N}_Z$ is the sheaf of sections of the normal bundle of Z.

(0.5.6). Consider the situation of (0.1.1), i.e., an action of an algebraic group H on a projective variety X. Then for a small open H-invariant set $U \subset X$ the orbit closures $\overline{H \cdot x}$ form a flat family. We obtain an embedding $U/H \hookrightarrow \mathscr{H}_X$.

(0.5.7). Definition. The Hilbert quotient $X///H$ is the closure of U/H in the Hilbert scheme $\mathscr{H}_X$.

Thus $X///H$ is a projective algebraic variety compactifying U/H. "Infinite" points of $X///H$ correspond to subschemes in X which are "degenerations" of generic orbit closures.

(0.5.8). The cycle map (0.5.3) provides a canonical regular birational morphism

$$\pi: X///H \longrightarrow X//H \tag{0.5.9}$$

from the Hilbert quotient to the Chow quotient. This morphism may be very nontrivial even in the case when the group H is finite. So $X///H$ provides a still finer compactification.

(0.5.10). In general, the Hilbert quotient is rather hard to describe. For instance, in the case of torus action on the projective space considered in §0.2, the Hilbert quotient is the toric variety corresponding to the so-called *state polytope* of the toric subvariety X_A introduced by D. Bayer, I. Morrison, and M. Stillman [5, 6]. However, its exact description depends not only on the geometry of the set A (as is the case for the secondary polytope) but also on the arithmetic nature of the relation between elements of A.

We shall see later that for the torus action on the Grassmannian Hilbert and Chow quotients coincide, thus allowing us to use the advantages of both approaches.

Chapter 1. Generalized Lie complexes

(1.1). Lie complexes and the Chow quotient of the Grassmannian.

(1.1.1). Let $\mathbb{C}^n$ be the coordinate n-dimensional complex vector space with coordinates $x_1, \dots, x_n$. By $G(k, n)$ we denote the Grassmannian of k-dimensional linear subspaces in $\mathbb{C}^n$. The group $(\mathbb{C}^*)^n$ of diagonal matrices acts on $G(k, n)$. Since homotheties act trivially, we obtain in fact an action of the $(n-1)$-dimensional algebraic torus $H = (\mathbb{C}^*)^n/\mathbb{C}^*$. Our main object of study in this paper will be the Chow quotient $G(k, n)//H$.

(1.1.2). For each subset $I \subset \{1, \dots, n\}$ denote by L_I the coordinate subspace in $\mathbb{C}^n$ defined by equations $x_i = 0$, $i \in I$, and by $\mathbb{C}^I$ the coordinate subspace spanned by basis vectors from I. Thus the codimension of L_I and the dimension of $\mathbb{C}^I$ are equal to $|I|$, the cardinality of I.

A k-dimensional subspace $L \in G(k, n)$ is said to be *generic* if for any $I \subset \{1, \dots, n\}$, $|I| = k$, we have $L \cap L_I = 0$. The space $G^0(k, n)$ of all generic subspaces is an open H-invariant subset in $G(k, n)$. It is called the *generic stratum*. It will serve as the open set U from (0.1).

(1.1.3). The Grassmannian $G(k, n)$ can be seen as the variety of $(k-1)$-dimensional projective subspaces in the projective space P^{n-1}. Using the terminology going back to Plücker, one usually calls $(n-1)$-dimensional families of subspaces in P^{n-1} *complexes.*

(1.1.4). Definition. By a Lie complex we shall mean an algebraic subvariety in $G(k, n)$ which is the closure of the H-orbit $\overline{H \cdot L}$ of some generic subspace $L \in G^0(k, n)$.

(1.1.5). PROPOSITION [19]. *Each Lie complex is an $(n-1)$-dimensional variety containing all the H-fixed points on $G(k, n)$ given by coordinate subspaces $\mathbb{C}^I$, $|I| = k$. These $\binom{n}{k}$ points are the only singular points of a Lie complex. Near each of these points a Lie complex looks like the cone over $P^{k-1} \times P^{n-k-1}$ in the Segre embedding.*

(1.1.6). EXAMPLE. Lie complexes in $G(2, 4)$ were extensively studied in classical literature under the name of *tetrahedral complexes*, see [2, 27] and references in [20]. Let us describe them in more detail. Let $x_1, \dots, x_4$ be homogeneous coordinates in P^3 and L_i be the coordinate plane $\{x_i = 0\}$. The configuration of four planes L_i can be thought of as a tetrahedron. A line $l \in P^3$ lies in the generic stratum $G^0(2, 4)$ if and only if it does not intersect any of the six lines given by the edges of our tetrahedron. For such a line the four points of intersections $l \cap L_i$ are distinct and, as any four distinct points on a projective line, possess the cross-ratio $r(l \cap L_1, \dots, l \cap L_4) \in \mathbb{C} - \{0, 1\}$. Let $\lambda \in \mathbb{C} - \{0, 1\}$ be a fixed number. The tetrahedral complex K_λ is, by definition, the closure of the set of those $l \in G^0(2, 4)$ for which the cross-ratio $r(l \cap L_1, \dots, l \cap L_4)$ equals λ. Its equation in Plücker coordinates is $p_{12}p_{34} + \lambda p_{13}p_{24} = 0$. This can be commented as follows. The classical Plücker relation gives that three quadratic polynomials $p_{12}p_{34}$, $p_{13}p_{24}$, and $p_{14}p_{23}$ on $G(2, 4)$ are linearly dependent, i.e., generate a 1-dimensional linear system (pencil) of hypersurfaces. The tetrahedral complexes are just hypersurfaces from this pencil. They are, therefore, particular cases of quadratic line complexes. As was pointed out in [20], the definition of a tetrahedral complex as the closure of a torus orbit is due to F. Klein and S. Lie.

(1.1.7). Clearly all Lie complexes represent the same class in $(2n-2)$-dimensional homology of the Grassmannian. Denote this class δ. Let us recall an explicit formula for δ found by A. Klyachko [36]. For any Young diagram $\alpha = (\alpha_1 \geqslant \dots \geqslant \alpha_k)$ with no more than k rows and no more than $(n-k)$ columns we shall denote by $|\alpha| = \sum \alpha_i$ the number of cells in α and by σ_α the Schubert class in $H_{2|\alpha|}(G(k, n))$ corresponding to α (see [24] and §3.9 below for details on Schubert cycles). These classes form an integral basis in the homology and the formula of Klyachko gives a decomposition of δ with respect to this basis.

(1.1.8). PROPOSITION [36]. *Let α be a Young diagram with $(n-1)$ cells. The coefficient at σ_α in the decomposition of the fundamental class δ of a Lie complex with respect to Schubert cycles equals*

$$\sum_{i=0}^{k} (-1)^i \binom{n}{i} \dim \Sigma^\alpha(\mathbb{C}^{k-i}),$$

where $\Sigma^\alpha(\mathbb{C}^{k-i})$ is the irreducible representation of $GL(k-i)$ with highest weight α.

(1.1.9). Example. For a Lie complex in $G(2, n)$ the above formula gives

$$\delta = (n-2)\sigma_{n-2,1} + (n-4)\sigma_{n-3,2} + (n-6)\sigma_{n-4,3} + \cdots .$$

In particular, for a Lie complex in $G(2, 4)$ the formula gives $\delta = 2\sigma_{2,1}$ and $\sigma_{2,1}$ is the class of hyperplane sections of $G(2, 4)$ in the Plücker embedding. This agrees with the fact that Lie complexes in $G(2, 4)$ are quadratic complexes.

(1.1.10). The collection of all Lie complexes is naturally identified with $G^0(k, n)/H$, the quotient of the generic stratum (1.1.2). We are interested in the Chow quotient $G(k, n)//H$, which is a projective subvariety in the Chow variety $\mathscr{C}_{n-1}(G(k, n), \delta)$, namely the closure of the set of all Lie complexes.

Any algebraic cycle from $G(k, n)//H$ will be called a *generalized Lie complex*. It is our point of view that generalized Lie complexes are the "right" generalizations of generic torus orbits in the Grassmannian. We shall see later that each generalized Lie complex can be seen as a (possibly reducible) algebraic subvariety in $G(k, n)$.

(1.1.11). Example. The Chow quotient $G(2, 4)//H$ is isomorphic to the projective line P^1. The isomorphism is given by the cross-ratio of four points of intersection $l \cap L_i$ in Example (1.1.6). There are exactly three generalized Lie complexes in $G(2, 4)$ that are not closures of single orbits (i.e., are not genuine Lie complexes). They are limit positions of tetrahedral complexes corresponding to values $0, 1, \infty$ not taken by the cross-ratio. Denote by $Z_{ij} \subset G(2, 4)$ the space of lines intersecting the coordinate line $x_i = x_j = 0$ (the edge of the tetrahedron). This is a linear section of $G(2, 4)$ given by the equation $p_{ij} = 0$. The three limit complexes are

$$Z_{12} + Z_{34}, \quad Z_{13} + Z_{24}, \quad Z_{14} + Z_{23}. \tag{1.1.12}$$

(1.2). Chow strata and matroid decompositions of the hypersimplex.

(1.2.1). Call two k-dimensional linear subspaces $L, L' \subset \mathbb{C}^n$ *equivalent* if $\dim(L \cap L_I) = \dim(L' \cap L_I)$ for any $I \subset \{1, \ldots, n\}$. Corresponding equivalence classes are called *strata*. They are H-invariant subsets in $G(k, n)$. A *base* of a subspace $L \in G(k, n)$ is a k-element subset $I \subset \{1, \ldots, n\}$ such that $L \cap L_I = 0$. It is well known that two subspaces L and L' lie in the same stratum (i.e., are equivalent) if and only if their sets of bases coincide. As a particular case we obtain the generic stratum $G^0(k, n) \subset G(k, n)$ defined as follows. A space L lies in $G^0(k, n)$ if and only if each k-element subset is a base for L.

This stratification was introduced in [18, 19, 21]. The set of bases for any subspace $L \subset G(k, n)$ introduces the structure of a *matroid* of rank k on $\{1, \ldots, n\}$. Because of this, this stratification is often referred to as the *matroid stratification* of the Grassmannian.

(1.2.2). It was remarked in [19, §5.1] that the matroid stratification of the Grassmannian is not a stratification in the sense of Whitney. In particular, the closure of a stratum may fail to be a union of other strata.

(1.2.3). Let $e_1, \dots, e_n$ be standard basis vectors in the coordinate space $\mathbb{R}^n$. We define the convex polytope $\Delta(k, n)$ called the *(k, n)-hypersimplex* to be the convex hull of $\binom{n}{k}$ points $e_{i_1}+\dots+e_{i_k}$, where $1 \leqslant i_1 < \dots < i_k \leqslant n$. All these points are vertices of $\Delta(k, n)$. We shall denote these vertices shortly by $e_I = \sum_{i\in I} e_i$, where $I \subset \{1, \dots, n\}$, $|I| = k$. For any subspace $L \in G(k, n)$ we define its *matroid polytope* $M(L)$ as the convex hull of e_I, where I runs over all bases for L. Thus $\Delta(k, n)$ itself is the matroid polytope for a generic subspace.

The hypersimplex was introduced in [17] and serves as a combinatorial model both for the Grassmannian with torus action and for any Lie complex.

(1.2.4). PROPOSITION [19]. *Let $L \in G(k, n)$. Then:*

a) *Any edge of $M(L)$ is parallel to a vector of the form $e_i - e_j$, $i \neq j$.*

b) *The (complex) dimension of the orbit $H \cdot L$ coincides with the real dimension of the polytope $M(L)$. The closure $\overline{H\cdot L}$ is a projective, normal, toric variety and $M(L)$ is the corresponding polytope (i.e., the fan of $\overline{H\cdot L}$ is the normal fan of $M(L)$). In particular:*

c) *Any Lie complex is a projective toric variety and the corresponding polytope is $\Delta(k, n)$. So p-dimensional H-orbits on any Lie complex are in bijection with p-dimensional faces of $\Delta(k, n)$.*

The following description of faces of $\Delta(k, n)$ was given in [17].

(1.2.5). PROPOSITION. a) *Each face of $\Delta(k, n)$ is itself a hypersimplex.*

b) *Edges of $\Delta(k, n)$ are segments $[e_I, e_J]$ where J differs from I by replacing one element $i \in I$ by another $j \notin I$.*

c) *For $k > 1$ there are exactly $2n$ facets (faces of codimension 1) of $\Delta(k, n)$. They are*

$$\Gamma_i^+ = \mathrm{Conv}\{e_I, i \in I\} \quad \textit{and} \quad \Gamma_i^- = \mathrm{Conv}\{e_I, i \notin I\}$$

for $1 \leqslant i \leqslant n$. Each polytope Γ_i^+ is linearly isomorphic to the hypersimplex $\Delta(k-1, n-1)$, whereas each Γ_i^- is isomorphic to $\Delta(k, n-1)$.

(1.2.6). By a *matroid polytope* in $\Delta(k, n)$ we shall mean any subpolytope $M \subset \Delta(k, n)$ whose vertices are among vertices of $\Delta(k, n)$ and edges have the form described in part a) of Proposition 1.2.4 (i.e., are among edges of $\Delta(k, n)$). According to Proposition 1.2.5 the polytope $M(L)$ for any subspace $L \in G(k, n)$ is a matroid polytope. Matroid polytopes of such form are called *realizable.*

The notion of matroid polytope in $\Delta(k, n)$ was introduced in [19, 21]. It was shown in these papers that such polytopes are in bijection with the structures of rank k matroids on a set $\{1, \dots, n\}$.

(1.2.7). Consider the Plücker embedding of the Grassmannian $G(k, n)$ into the $(\binom{n}{k} - 1)$-dimensional projective space $P(\wedge^k \mathbb{C}^n)$. The homogeneous coordinates in this projective space will be denoted p_I, $I \subset \{1, \dots, n\}$, $|I| = k$. The H-action on $G(k, n)$ extends to the whole $P(\wedge^k \mathbb{C}^n)$. The matroid polytope of a subspace $L \in G(k, n)$ is the image of the orbit closure $\overline{H \cdot L} \subset G(k, n) \subset P(\wedge^k \mathbb{C}^n)$ under the momentum map $\mu\colon P(\wedge^k \mathbb{C}^n) \to \Delta(k, n)$ defined as follows [19, 21]:

$$\mu(x) = \frac{\sum_{|I|=k} |p_I(x)| \cdot e_I}{\sum_{|I|=k} |p_I(x)|}. \tag{1.2.8}$$

(1.2.9). Since $G(k, n)$ is embedded equivariantly into $P(\wedge^k \mathbb{C}^n)$, we obtain the embedding of Chow quotients

$$G(k, n)//H \hookrightarrow P\left(\wedge^k \mathbb{C}^n\right)//H. \tag{1.2.10}$$

The latter quotient is, according to (1.2.7), a toric variety of dimension $\binom{n}{k} - n$ and the corresponding polytope is the secondary polytope of the hypersimplex $\Delta(k, n)$. By comparing the two Chow quotients we deduce from [30, 31] the following proposition.

(1.2.11). PROPOSITION. *Let* $Z = \sum c_i Z_i$ *be a cycle from* $G(k, n)//H$. *Then*

a) *Each component* Z_i *is a closure of some* $(n-1)$*-dimensional* H*-orbit* Z_i^0.
b) *Let* $M(Z_i) = \mu(Z_i)$ *be the matroid polytope of any subspace* $L \in Z_i^0$ *(or, what is the same, the image of* Z_i *under the momentum map). Then the polytopes* $M(Z_i)$ *form a polyhedral decomposition of* $\Delta(k, n)$.

(1.2.12). EXAMPLE. Consider the case $k = 2$, $n = 4$. The hypersimplex $\Delta(2, 4)$ is the 3-dimensional octohedron. Each of the three generalized Lie complexes from (1.1.12) gives a decomposition of this octohedron into a union of two pyramids with a common quadrangular face.

We have the embedding (1.2.10) of $G(2, 4)//H = P^1$ into $P(\wedge^2(\mathbb{C}^4))//H$, a toric variety of dimension 2. This variety is isomorphic to the projective plane P^2. To see this, let us show that the secondary polytope (polygon, in our case) Σ of the octohedron $\Delta(2, 4)$ is in fact a triangle. Indeed, by definition (0.2.7) vertices of Σ are in bijection with regular triangulations of $\Delta(2, 4)$. Each triangulation of $\Delta(2, 4)$ can be obtained as follows. Take any decomposition of $\Delta(2, 4)$ into two pyramids as above and then decompose each of these pyramids into two tetrahedra in a compatible way:

(1.2.13)

Thus there are three triangulations that correspond to vertices of Σ and three pyramidal decompositions that correspond to edges of Σ and hence Σ is a triangle.

Since the symmetry group of the octohedron acts on Σ, it is a regular triangle. Hence the toric variety corresponding to Σ has P^2 as its normalization. The fact that this variety is normal can be established by direct computation of vertices of Σ as points of the integer lattice $\mathbb{Z}^6$ (according to (0.2.6)). We leave this computation to the reader.

So the toric variety $P(\wedge^2(\mathbb{C}^4))//H$ is a projective plane. The subvariety $G(2, 4)//H = P^1$ is a conic in this projective plane inscribed into the coordinate triangle:

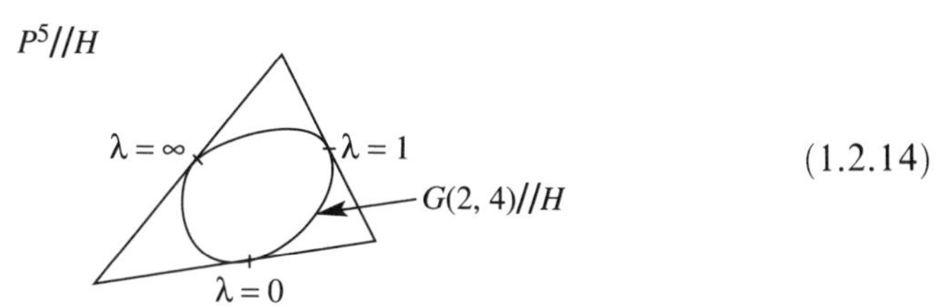

(1.2.14)

(1.2.15). PROPOSITION. *Let* $Z = \sum c_i Z_i$ *be a cycle from* $G(k, n)//H$. *Then all the multiplicities* c_i *equal* 1 (*or* 0).

PROOF. The recipe for the calculation of c_i given in [30, 31] is the following. We consider the affine $\mathbb{Z}$-lattice Ξ_i generated by the vertices of the polytope $M(Z_i)$ which is imbedded into the affine lattice Ξ generated by all the vertices of $\Delta(k, n)$. Then $c_i = [\Xi : \Xi_i]$. Let us show that in fact $\Xi = \Xi_i$. Choose some vertex e_I of $M(Z_i)$. Subtracting it from points of Ξ we identify Ξ with $\{(a_1, \dots, a_n) \in \mathbb{Z}^n : \sum a_i = 0\}$. Consider now all edges of $M(Z_i)$ containing e_I. By Proposition 1.2.4 b), they all have the form $e_j - e_l$, where $j \in I$, $l \notin I$. Since $M(Z_i)$ has full dimension, there are at least $n - 1$ independent edges. However, any $n - 1$ independent vectors of the form $e_j - e_l$ generate the lattice Ξ.

By the above proposition, generalized Lie complexes (="infinite" points of the Chow quotient $G(k, n)//H$) can be thought of as usual reducible subvarieties (instead of cycles) in the Grassmannian, which further justifies their name. We will therefore denote these complexes by $Z = \bigcup Z_i$ to emphasize that they are varieties.

(1.2.16). DEFINITION. Two cycles $Z, Z' \in G(k, n)//H$ are called equivalent if the corresponding polyhedral decompositions of $\Delta(k, n)$ coincide. Equivalence classes under this relation are called Chow strata.

By considering again the Plücker embedding we see that our stratification of $G(k, n)//H$ is induced from the stratification of the toric variety $P(\wedge^k \mathbb{C}^n)//H$ given by the torus orbits. Each Chow stratum can be specified by a finite list of usual strata corresponding to individual matroid polytopes from the polyhedral decomposition.

(1.2.17). DEFINITION. A polyhedral decomposition $\mathcal{P}$ of the hypersimplex $\Delta(k, n)$ is called a matroid decomposition if all the polytopes from $\mathcal{P}$ are matroid polytopes (1.2.7). A matroid decomposition is called realizable if it comes from a generalized Lie complex (1.2.11).

Thus matroid decompositions of $\Delta(k, n)$ are precisely the labels by which Chow strata are labelled. The notion of a matroid polytope being equivalent to that of matroid (1.2.1), a matroid decomposition represents a new kind of combinatorial structure: a collection of usual matroids with certain properties (that the corresponding polytopes form a decomposition of the hypersimplex).

(1.3). An example: matroid decompositions of the hypersimplex $\Delta(2, n)$**.** In this section we give a complete description of matroid decompositions of the hypersimplex $\Delta(2, n)$. The structure involved will turn out to be identical to that in the description of stable n-pointed curves of Grothendieck [12] and Knudsen [37].

Recall that vertices of $\Delta(2, n)$ are of the form $e_{ij} := e_i + e_j$, $i \neq j$, $1 \leqslant i$, $j \leqslant n$, where e_i are the standard basis vectors of $\mathbb{R}^n$.

(1.3.1). PROPOSITION. a) *Matroid polytopes in* $\Delta(2, n)$ *are in bijection with pairs* (J, R), *where* $J \subset \{1, \dots, n\}$ *is a nonempty subset and* R *is an equivalence relation on* J *with at least two equivalence classes. The matroid polytope* $M(J, R)$ *corresponding to* (J, R) *above has vertices* e_{ij}, *where* $i, j \in J$ *are such that* iRj *does not hold.*

b) *The dimension of* $M(J, R)$ *equals* $|J| - 1$ *if* R *has* $\geqslant 3$ *equivalence classes and equals* $|J| - 2$ *if* R *has exactly two equivalence classes.*

PROOF. This follows from [21, Example 1.10 and Proposition 4, §2].

(1.3.2). COROLLARY. *Matroid polytopes in* $\Delta(2, n)$ *which have full dimension* $(n-1)$, *are in bijection with equivalence relations on* $\{1, \dots, n\}$ *with* $\geqslant 3$ *equivalence classes.*

Thus matroid decompositions of $\Delta(2, n)$ are certain "compatible" systems of equivalence relations on the same set $\{1, \dots, n\}$. We are going to describe them.

(1.3.3). By a graph we mean a finite 1-dimensional simplicial complex. So a graph Γ is defined by its set of vertices Γ_0 and the set of edges Γ_1 together with the incidence relation connecting these sets. If v is a vertex of a graph Γ, the *valency* of v is, by definition, the number of edges containing v.

By a *tree* we mean a connected graph T without loops such that every vertex of T has the valency either 1 or $\geqslant 3$. The vertices of valency 1 will be called *endpoints* of T. For any two vertices v, w of a tree T there is a unique edge path without repetitions joining these vertices. This path will be denoted $[v, w]$.

Let $A_1, \dots, A_n$ be formal symbols. By a tree bounding the endpoints $A_1, \dots, A_n$, we mean a tree T with exactly n endpoints that are put into bijection (or just identified) with symbols A_i. Two such trees T, T' are called isomorphic if there is an isomorphism of graphs $T \to T'$ preserving A_i.

(1.3.4). Let T be a tree bounding endpoints $A_1, \dots, A_n$. Any vertex of T which is not an endpoint will be called interior. Let $v \in T_0$ be an interior vertex. We define an equivalence relation $\cong_v$ on $\{1, \dots, n\}$ by setting $i \cong_v j$ if the edge path $[A_i, A_j]$ does not contain the vertex v.

In other words, the deletion of the vertex v splits the tree into several connected components and $i \cong_v j$ if the endpoints A_i, A_j are situated in the same component. The equivalence classes under $\cong_v$ are in bijection with edges of T containing v.

(1.3.5). Proposition (1.2.5) implies that $\Delta(2, n)$ has n facets (faces of codimension 1) $\Gamma_i^+ = \mathrm{Conv}\{e_{ij}, j \neq i\}$ which are $(n-2)$-dimensional simplices. It is clear that in any polyhedral decomposition of $\Delta(2, n)$ each Γ_i^+ is a facet of exactly one polytope from decomposition.

(1.3.6). **Theorem.** *Matroid decompositions of the hypersimplex $\Delta(2, n)$ are in one-to-one correspondence with isomorphism classes of trees bounding endpoints $A_1, \dots, A_n$.*

- *Explicitly, if T is such a tree, the corresponding decomposition $\mathcal{P}(T)$ consists of matroid polytopes $M(\cong_v)$ (Proposition (1.3.1)) for all interior vertices v of T.*
- *Conversely, the tree T can be recovered from the corresponding matroid decomposition $\mathcal{P}$ as follows. Internal vertices of T are barycenters of polytopes (of maximal dimension) from $\mathcal{P}$. Endpoints of T are barycenters of facets Γ_i^+. The barycenter of each Γ_i^+ is joined to the barycenter of the unique polytope from $\mathcal{P}$ containing Γ_i; the barycenters of two polytopes from $\mathcal{P}$ are joined if and only if these polytopes have a common facet.*

(1.3.7). **Remark.** Let v be an interior vertex of the tree T. The vertices of the polytope $M(\cong_v)$ are those vertices $e_{ij} = e_i + e_j$ of $\Delta(2, n)$ for which the edge path $[A_i, A_j]$ does contain v.

Proof of (1.3.6). Let T be any tree bounding $A_1, \dots, A_n$. Let us show that the collection of polytopes $M(\cong_v)$ forms a polyhedral decomposition of $\Delta(2, n)$. By definition, this means that the following two properties hold:

(1.3.8). Intersection of any two polytopes $M(\cong_v)$, $M(\cong_w)$ is a common face of both of them.

(1.3.9). The union of the polytopes $M(\cong_v)$ is the entire hypersimplex $\Delta(2, n)$.

(1.3.10). **Proof of** (1.3.8). We shall prove a somewhat stronger statement that $M(\cong_v) \cap M(\cong_w)$ is the convex hull of vertices common to

$M(\cong_v)$, $M(\cong_w)$. By (1.3.7), a vertex e_{ij} is common to $M(\cong_v)$, $M(\cong_w)$ if the edge path $[A_i, A_j]$ contains $[v, w]$ as a subpath. Let us subdivide the set $\{1, \ldots, n\}$ into three parts: X_+, X_-, X_0. We set $i \in X_+$ if the edge path $[A_i, v]$ does not contain points on $[v, w]$ other than v. We set $i \in X_-$ if the edge path $[A_i, w]$ does not contain points of $[v, w]$ other than w. We set $i \in X_0$ in all other cases:

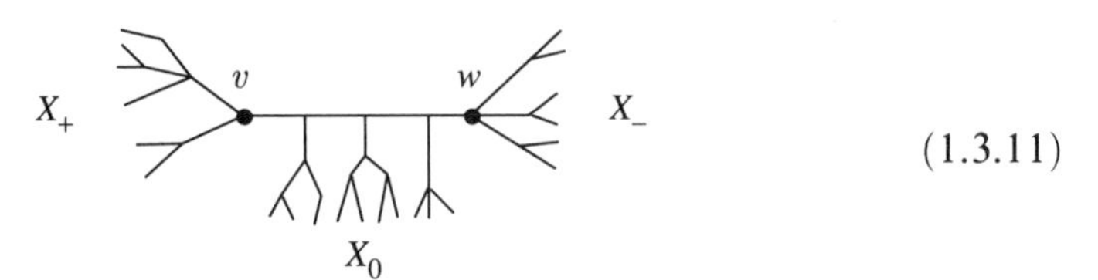

(1.3.11)

Recall that $\Delta(2, n)$ lies in $\mathbb{R}^n$ as the convex hull of sums $e_{ij} = e_i + e_j$ of two distinct basis vectors. Consider the linear function f on $\mathbb{R}^n$ such that $f(e_i) = +1, -1$, or 0 if $i \in X_+, X_-$, or X_0 respectively. Then f is nonnegative on all vertices of $M(\cong_v)$ and nonpositive on all vertices of $M(\cong_w)$. The only vertices of $M(\cong_v)$, $M(\cong_w)$ for which $f = 0$ are the vertices common to both these polytopes. Assertion (1.3.8) is proved.

(1.3.12). PROOF OF (1.3.9). It suffices to show that any face of codimension 1 of any $M(\cong_v)$ lies either on the boundary of $\Delta(2, n)$ or is a face of another polytope $M(\cong_w)$. The following description of facets (=faces of codimension 1) of matroid polytopes follows from [21, §2, Theorem 5].

(1.3.12.1). PROPOSITION. *Let $J = \{1, \ldots, n\}$ and $M = M(J, R)$ be the matroid polytope of full dimension corresponding to an equivalence relation R on J. Its facets are the following:*

(1) *Facets $\Gamma_j^+(M)$ defined for any j unless $\{j\}$ is an equivalence class in itself and the total number of classes is three. Such a facet is the matroid polytope $M(J', R')$, where $J' = \{1, \ldots, n\} - \{j\}$ and R' is the equivalence relation induced by R. It lies entirely in the boundary of $\Delta(2, n)$.*
(2) *Facets $\Gamma_K^-(M)$ defined for any equivalence class $K \in J/R$. This is the matroid polytope $M(J, R'')$, where R'' is the equivalence relation with only two classes, one being K and the other being formed by all elements not in K.*

The notation $\Gamma^\pm$ is compatible with the notation for the facet of the full hypersimplex $\Delta(2, n)$ introduced in Proposition 1.2.5.

(1.3.12.2). COROLLARY. *Let T be a tree bounding $A_1, \ldots, A_n$ and $v \in T$ be an interior vertex. The facets of the matroid polytope $M(\cong_v)$ not lying in the boundary of $\Delta(2, n)$ are in bijection with edges of T containing v whose second end is also an interior vertex. The facet corresponding to such an edge e is of the form Γ_K^-, where K is the $\cong_v$-equivalence class corresponding to e.*

Now the assertion (1.3.9) follows from Corollary 1.3.12.2 since every edge of the tree T joins two vertices and the two matroid polytopes corresponding to these vertices have a common facet.

(1.3.13). We have proved that any tree bounding $A_1, \dots, A_n$ yields a matroid decomposition of $\Delta(2, n)$. Conversely, let $\mathcal{P}$ be any matroid decomposition. By taking barycenters of polytopes from $\mathcal{P}$ and joining them as described in Theorem 1.3.6, we obtain a certain graph T. Let us show that T is a tree which generates the decomposition $\mathcal{P}$. The fact that T is a tree follows from the next lemma.

(1.3.14). LEMMA. *Let $M \subset \Delta(2, n)$ be a matroid polytope of full dimension and $\Gamma \subset M$ be a facet not lying on the boundary of $\Delta(2, n)$. Then Γ is equal to the intersection of the entire $\Delta(2, n)$ with a hyperplane.*

PROOF. According to Proposition 1.3.12.1, the polytope Γ has the form $M(J, R)$, where R is an equivalence relation on $J = \{1, \dots, n\}$ with only two equivalence classes, say A and B. Define a linear function g on $\mathbb{R}^n$ whose value on the basis vector e_i equals 1 if $i \in A$ and equals (-1) if $i \in B$. Then Γ is the intersection of $\Delta(2, n)$ with the kernel of g. Lemma (1.3.14) is proved.

(1.3.15). To complete the proof of Theorem 1.3.6, it remains to show that the tree obtained from the matroid decomposition $\mathcal{P}$ in (1.3.13), generates $\mathcal{P}$. This is left to the reader.

(1.3.16). EXAMPLE. There are four matroid decompositions of the octohedron $\Delta(2, 4)$: one consists of $\Delta(2, 4)$ itself and each of the others decomposes the octohedron into two pyramids (Example 1.2.12). These decompositions correspond to the following trees bounding endpoints $A_1, \dots, A_4$:

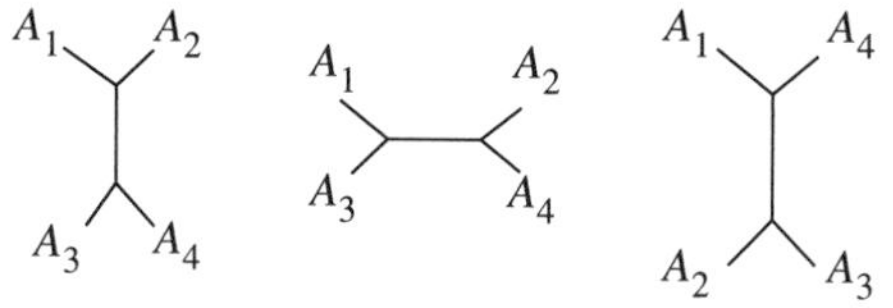

(1.3.17)

(1.3.18). We shall show in §1.4 that all matroid decompositions of $\Delta(2, n)$ are realizable.

(1.4). Relation to the secondary variety for the product of two simplices. In this section we shall compare the Chow quotient $G(k, n)//H$ with a toric variety of the same dimension. This toric variety will correspond to the convex polytope that is the secondary polytope for the product of two simplices.

(1.4.1). Let $P \subset \mathbb{R}^m$ be any convex polytope and $x \in P$ be any vertex. Denote by $N_x P$ the union of all half-lines drawn from x through all the points of P. This is an affine cone, which we call the *normal cone* to P at x. The base of this cone, i.e., a transversal section of $N_x P$ by an affine hyperplane, will be called the *vertex figure* of P at x. Thus vertices of the vertex figure correspond to edges of P containing x.

(1.4.2). Let A be any finite set. Denote by Δ^A the simplex (of dimension $|A|-1$) whose set of vertices is A. By definition, Δ^A is the subset in the space $\mathbb{R}^A$ of functions $A \to R$ consisting of functions $f(a)$ such that $f(a) \geqslant 0, \forall a$, and $\sum f(a) = 1$. To any $a \in A$ there corresponds a vertex δ_a of Δ^A. This is the function $A \to \mathbb{R}$ taking a to 1 and other elements to 0.

(1.4.3). Let e_I be a vertex of the hypersimplex $\Delta(k, n)$. The corresponding vertex figure is the product of two simplices $\Delta^{k-1} \times \Delta^{n-k-1}$ or, in the more invariant notation of (1.4.2), $\Delta^I \times \Delta^{\bar{I}}$, where $\bar{I}$ is the complement to I.

Indeed, edges of $\Delta(k, n)$ containing e_I are $[e_I, e_I+e_j-e_i]$, where $i \in I$, $j \notin I$. The required isomorphism takes such an edge to the vertex (δ_i, δ_j) of $\Delta^I \times \Delta^{\bar{I}}$.

The toric variety associated to the polytope $\Delta^{k-1} \times \Delta^{n-k-1}$ is the product of projective spaces $P^{k-1} \times P^{n-k-1}$ and the structure of $\Delta(k, n)$ near a vertex corresponds to the structure of a Lie complex near its singular point (Proposition 1.1.5).

(1.4.4). PROPOSITION-DEFINITION. *Let $\mathscr{P}$ be a matroid decomposition of $\Delta(k, n)$ and $e_I \in \Delta(k, n)$ a vertex. Let $\mathscr{P}_I$ be the induced polyhedral decomposition of the vertex figure $\Delta^I \times \Delta^{\bar{I}}$. Then all the vertices of polytopes constituting $\mathscr{P}_I$ lie among the vertices of $\Delta^I \times \Delta^{\bar{I}}$. If $\mathscr{P}$ is a realizable matroid decomposition then $\mathscr{P}_I$ is a regular polyhedral subdivision of $\Delta^I \times \Delta^{\bar{I}}$.*

Recall [22] that a polyhedral subdivision is called regular if it admits a strictly convex piecewise-linear function.

PROOF. Vertices of polytopes from $\mathscr{P}_I$ correspond to edges of polytopes from $\mathscr{P}$ containing e_I. Since all these polytopes are matroid polytopes, the edges in question correspond to vertices of $\Delta^I \times \Delta^{\bar{I}}$. If $\mathscr{P}$ is realizable then it is regular as a polyhedral subdivision of $\Delta(k, n)$ and so is $\mathscr{P}_I$.

(1.4.5). Let $I \subset \{1, \dots, n\}$ be a k-element subset. The coordinate subspace $\mathbb{C}^I \in G(k, n)$ is a fixed point under the action of the torus H (1.1.1). Therefore we have the action of H on the tangent space $T_I := T_{\mathbb{C}^I} G(k, n)$.

The tangent space to $G(k, n)$ at any point L is canonically identified, see [47], with $\mathrm{Hom}(L, \mathbb{C}^n/L)$. Therefore we have the isomorphism of H-modules

$$T_I = \mathrm{Hom}(\mathbb{C}^I, \mathbb{C}^{\bar{I}}). \tag{1.4.6}$$

In other words, T_I is decomposed into $k(n-k)$ one-dimensional weight subspaces V_{ij}, $i \in I$, $j \notin I$, such that for any $t = (t_1, \dots, t_n) \in H$ and any $v \in V_{ij}$ one has $(t_1, \dots, t_n) \cdot v = (t_i/t_j)v$.

(1.4.7). The character lattice of the torus H is the sublattice in $\mathbb{Z}^n$ consisting of vectors with the sum of coordinates equal to 0. The character corresponding to the subspace V_{ij} is the vector $e_i - e_j \in \mathbb{Z}^n$. The collection of all vectors $e_i - e_j$, $i \in I$, $j \notin I$, forms the set of vertices of the simplex $\Delta^I \times \Delta^{\overline{I}}$.

(1.4.8). Call a point v of the tangent space T_I (and the corresponding point of the projectivization $P(T_I)$) *generic* if all the weight components of v are nonzero. By a generic H-orbit in T_I or $P(T_I)$ we mean the orbit of a generic point.

(1.4.9). For the torus orbits in the Grassmannian $G(k, n)$ we also have a notion of genericity introduced in (1.1.2). Closures of generic orbits in $G(k, n)$ were called Lie complexes. Let $Z = \overline{H \cdot L}$, $L \in G^0(k, n)$, be any Lie complex and $TC_I Z := TC_{\mathbb{C}^I} Z \subset T_I$ its tangent cone at the point $\mathbb{C}^I$. It follows from (1.1.5) that $TC_I Z$ is the closure of a generic H-orbit in T_I. So we obtain the following proposition.

(1.4.10). PROPOSITION. *Let us identify the quotient $G^0(k, n)/H$ (i.e., the set of generic (1.1.2) H-orbits on $G(k, n)$) with the set of Lie complexes. Then the correspondence $Z \mapsto$ (the projectivization of $TC_{\mathbb{C}^I} Z$) defines an open embedding of $G^0(k, n)/H$ into the set of generic H-orbits in $P(T_I)$.*

(1.4.11). We are going to compare the Chow quotient $G(k, n)//H$ with $P(T_I)//H$. The latter variety is, according to results recalled in §0.2, a projective toric variety of the same dimension $k(n-k) - n + 1$ as $G(k, n)//H$. Due to (1.4.7) and Theorem 0.2.8, the toric variety $P(T_I)//H$ corresponds to the secondary polytope of $\Delta^I \times \Delta^{\overline{I}}$. We shall therefore call this variety the *secondary variety* of $\Delta^I \times \Delta^{\overline{I}}$.

(1.4.12). PROPOSITION. *The open embedding from (1.4.10) extends to a regular birational (in particular, surjective) morphism*

$$f_I \colon G(k, n)//H \to P(T_I)//H. \tag{1.4.13}$$

This morphism takes any generalized Lie complex $\mathscr{L}$ to the projectivization of its tangent cone at $\mathbb{C}^I$ (which is regarded as an algebraic cycle with multiplicities 0 or 1 in $P(T_I)$).

PROOF. Let $U_I \subset G(k, n)$ be the affine chart consisting of the k-dimensional subspace L such that $L \oplus \mathbb{C}^{\overline{I}} = \mathbb{C}^n$. Each such subspace can be regarded as the graph of a linear operator from $\mathbb{C}^I$ to $\mathbb{C}^{\overline{I}}$. This correspondence identifies U_I with the space of $k \times (n-k)$-matrices thus introducing coordinates z_{ij}, $i \in I$, $j \notin I$, in U_I. The tangent vector space T_I becomes identified with U_I, cf. (1.4.6). Consider the action of $\mathbb{C}^*$ on U_I by homotheties (simultaneous multiplication of the coordinates z_{ij} by a scalar). Since this action is a part of the torus action on $G(k, n)$, we deduce that for any Lie complex Z the intersection $Z \cap U_I$ will be a conic (i.e., $\mathbb{C}^*$-invariant) subvariety of U_I. The same will hold, by continuity, for any generalized Lie complex. Therefore the map f_I takes any generalized Lie complex Z (which is a subvariety in $G(k, n)$) into the subvariety in $P(T_I) = P(U_I)$ represented by the conic subvariety $Z \cap U_I$ in U_I. We need to show that f_I is a regular morphism of algebraic varieties.

Since both $G(k, n)//H$ and $P(U_I)//H$ are projective, it suffices, by Serre's GAGA theorem [24], to show that f_I is a holomorphic map of complex analytic spaces corresponding to these varieties. However, this follows from the description of the Chow varieties given by D. Barlet [3]. More specifically, Barlet gave a condition for a family of p-dimensional cycles $Z(s) \subset X$ parametrized by a reduced analytic space S to be analytic near a point $s_0 \in S$. This condition is essentially that for any p and any codimension p analytic subvariety $Y \subset X$ that intersects $Z(s_0)$ properly, the 0-cycle $Z(s) \cap Y$ depends analytically on s near s_0. To prove that $f_I(Z)$ depends analytically on $Z \in G(k, n)//H$ we note that any analytic subvariety Y of codimension $(n-1)$ in $P(U_I)$ can be lifted to U_I (by setting one of the coordinates to be 1) to a subvariety $\tilde{Y}$. If Y intersects some $f_I(Z)$ properly then so does $\tilde{Y}$ with respect to Z due to the conic property. Analytic dependence of $\tilde{Y} \cap Z$ on Z implies the analytic dependence of $Y \cap f_I(Z)$ which is just the image of $\tilde{Y} \cap Z$ in the projectivization. Proposition (1.4.12) is proved.

Later we shall make use of morphisms f_I to construct "coordinate charts" on the Chow quotient. The following fact is an immediate consequence of (1.4.12).

(1.4.14). Corollary. *Each regular polyhedral decomposition (in particular, triangulation) of the product of simplices $\Delta^I \times \Delta^{\bar{I}}$ has the form $\mathcal{P}_I$ for some realizable matroid decomposition of $\Delta(k, n)$.*

Thus the problem of classification of all realizable matroid decompositions of the hypersimplex contains the classification problem for triangulations of the product of two simplices.

(1.5). Relation to Hilbert quotients.

(1.5.1). The Hilbert quotients were defined in §0.5. We want to compare the Chow quotient $G(k, n)//H$ with the Hilbert quotient $G(k, n)///H$. Recall that there exists a regular birational morphism $\pi \colon G(k, n)///H \to G(k, n)//H$ to the Chow quotient, see (0.5.8).

(1.5.2). Theorem. *The morphism π is an isomorphism.*

Proof. By definition, points of $G(k, n)///H$ are subschemes that are limit positions of Lie complexes. By Proposition 1.2.15, every such subscheme Z is reduced at a generic point of every one of its component.

(1.5.3). LEMMA. *Any subscheme Z from $G(k, n)///H$ is reduced.*

PROOF. Consider the intersection of Z with some coordinate Schubert chart U_I. It suffices to prove, for every I, that $Z \cap U_I$ is reduced. The action of H in U_I is a linear one. This is the action corresponding to the products of simplices and it follows from the result of B. Sturmfels [48, Theorem 6.1] that any limit position of generic H-orbits in U_I is reduced.

Lemma (1.5.3) implies that the morphism π is bijective on $\mathbb{C}$-points. To show that it is an isomorphism of algebraic varieties, it suffices to show that for any $Z \in G(k, n)//H$ and any nonzero Zariski tangent vector ξ to $G(k, n)///H$ at Z the vector $d\pi(\xi)$, the image of ξ under the differential of π, is nonzero. The tangent space $T_Z\mathscr{H}_G$ to the whole Hilbert scheme at Z equals $H^0(Z, \mathscr{N})$, where $\mathscr{N}_Z = \underline{\mathrm{Hom}}_{\mathscr{O}_G}(J_Z/J_Z^2, \mathscr{O}_Z)$ is the normal sheaf of Z (see §0.5). Let Z_{reg} be the smooth part of Z. The restriction of $\mathscr{N}_Z$ to Z_{reg} is the sheaf of section of the normal bundle of Z_{reg} in the usual sense. Hence our vector $\xi \subset T_Z\mathscr{H}_G$ gives a normal vector field on Z_{reg}. Since Z is reduced, this field is nonzero if ξ is non-zero, so the assertion follows. Theorem (1.5.2) is proved.

(1.6). The (hyper-)simplicial structure on the collection of $G(k, n)//H$.

(1.6.1). Recall (1.1.2) that $G^0(k, n)$ denotes the generic stratum in the Grassmannian $G(k, n)$. For any $i \in \{1, \ldots, n\}$ there are the *intersection and projection maps*

$$G^0(k, n-1) \xleftarrow{B_i} G^0(k, n) \xrightarrow{A_i} G^0(k-1, n-1) \tag{1.6.2}$$

defined as follows. Let the embedding $J_i\colon \mathbb{C}^{n-1} \hookrightarrow \mathbb{C}^n$ take $(x_1, \ldots, x_{n-1})$ to $(x_1, \ldots, x_{i-1}, 0, x_i, \ldots, x_n)$. The intersection map A_i sends a k-dimensional subspace $L \subset \mathbb{C}^n$ to $J_i^{-1}L$. The projection map B_i is induced by the projection $\mathbb{C}^n \to \mathbb{C}^{n-1}$ forgetting the ith coordinate.

The formal structure of these maps is analogous to that of faces of the hypersimplex $\Delta(k, n)$ (Proposition 1.2.4). The existence of such a system of "face" maps was the original reason for introducing hypersimplices and then Grassmannians into the problem of combinatorial calculation of characteristic classes [17]. More recently, these maps were used by A. A. Beilinson, R. D. MacPherson, and V. V. Schectman in [7] to give a "constructible" approximation to K-theory.

(1.6.3). As was noted in [7], the maps (1.6.2) descend to maps of the quotients

$$G^0(k, n-1)/(\mathbb{C}^*)^{n-1} \xleftarrow{b_i} G^0(k, n)/(\mathbb{C}^*)^n \xrightarrow{a_i} G^0(k-1, n-1)/(\mathbb{C}^*)^{n-1} \tag{1.6.4}$$

of the generic strata by their respective tori.

(1.6.5). Clearly there is no way to extend maps (1.6.2) to the entire Grassmannians: if the subspace L is contained in the hyperplane $\{x_i = 0\}$ then $A_i(L)$ will have wrong dimension and similarly for B_i. However, it turns out that for Chow quotients the situation is different.

(1.6.6). THEOREM. *The maps* a_i, b_i *in* (1.6.4) *can be extended to regular morphisms of projective algebraic varieties*

$$G(k, n-1)//(\mathbb{C}^*)^{n-1} \overset{\tilde{b}_i}{\longleftarrow} G(k, n)//(\mathbb{C}^*)^n \overset{\tilde{a}_i}{\longrightarrow} G(k-1, n-1)//(\mathbb{C}^*)^{n-1}. \tag{1.6.7}$$

The proof of Theorem (1.6.6) will occupy the rest of this section.

(1.6.8). Let $e_1, \dots, e_n \in \mathbb{C}^n$ be the standard basis vectors and $\mathbb{C}_i^{n-1}$ be the coordinate hyperplane spanned by e_j, $j \neq i$. For any $i \subset \{1, \dots, n\}$ we consider the varieties

$$G_i^+ = \{L \in G(k, n)\colon e_i \in L\}, \qquad G_i^- = \{L \in G(k, n)\colon L \subset \mathbb{C}_i^{n-1}\}. \tag{1.6.9}$$

They are the analogs of the family of coordinate hyperplanes in P^{n-1}. The next proposition is immediate.

(1.6.10). PROPOSITION. a) *As abstract varieties,* G_i^+ *are isomorphic to* $G(k-1, n-1)$ *and* G_i^- *to* $G(k, n-1)$.

b) *Both* G_i^+ *and* G_i^- *are linear sections of* $G(k, n)$ *in the Plücker embedding. More precisely, let* Π_i^+ *and* Π_i^- *be projective subspaces in* $P(\wedge^k \mathbb{C}^n)$ *given by vanishing of the Plücker coordinates* p_I, $i \notin I$, *or, respectively,* p_I, $i \in I$. *Then* $G_i^\pm = G(k, n) \cap \Pi_i^\pm$.

c) *The image of the subvarieties* $G_i^\pm$ *under the moment map* μ *from* (1.1) *is the facet* $\Gamma_i^\pm$ *of* $\Delta(k, n)$ *introduced in* (1.2.5). *Moreover, we have* $G_i^\pm = \mu^{-1}(\Gamma_i^\pm)$.

(1.6.11). The isomorphisms in part a) of (1.6.10) are as follows. The isomorphism $u_i\colon G(k-1, n-1) \to G_i^+$ takes a $(k-1)$-dimensional $\Lambda \subset \mathbb{C}^{n-1}$ to $J_i(\Lambda) \oplus \mathbb{C}e_i$, where the embedding $J_i\colon \mathbb{C}^{n-1} \hookrightarrow \mathbb{C}^n$ was defined in (1.6.1). The isomorphism $v_i\colon G(k, n-1) \to G_i^-$ takes a k-dimensional $M \subset \mathbb{C}^{n-1}$ into $J_i(M)$.

(1.6.12). Let us turn to the construction of $\tilde{a}_i$, $\tilde{b}_i$. Note that the maps a_i, b_i of quotients of generic strata have the following transparent description in terms of Lie complexes (closures of generic orbits).

(1.6.13). LEMMA. *Let* $Z = \overline{H \cdot L}$ *be a Lie complex in* $G(k, n)$. *Then the Lie complex* $a_i(Z)$ *in* $G(k-1, n-1)$, *representing the orbit of* $A_i(L)$, *is equal to* $u_i^{-1}(Z \cap G_i^+) = u_i^{-1}(Z \cap \Pi_i^+)$. *Similarly, the Lie complex* $b_i(Z)$ *in* $G(k, n-1)$, *representing the orbit of* $B_i(L)$, *is equal to* $v_i^{-1}(Z \cap G_i^-) = v_i^{-1}(Z \cap \Pi_i^-)$.

PROOF. Denote by $g_i(t)$ the diagonal matrix $(1, \dots, 1, t, 1, \dots, 1) \in (\mathbb{C}^*)^n$ (where t is on the ith place). Let $\mathrm{pr}_i\colon \mathbb{C}^n \to \mathbb{C}_i^{n-1}$ be the coordinate projection. Let $L \in G^0(k, n)$ be a generic subspace. Then we have

$$(L \cap \mathbb{C}_i^{n-1}) \oplus \mathbb{C}e_i = \lim_{t\to\infty} g_i(t) \cdot L, \qquad \mathrm{pr}_i(L) = \lim_{t\to 0} g_i(t)L. \tag{1.6.14}$$

This shows that $Z \cap G_i^+$ (resp. $Z \cap G_i^-$) contains the orbit of $A_i(L)$ (resp. $B_i(L)$).

On the other hand, the vanishing of Plücker coordinates p_I for which $e_I \notin \Gamma_i^{\pm}$ forms a system of equations for $Z \cap G_i^+$, as follows from the general theory of toric varieties [49]. Lemma (1.6.13) is proved.

The lemma we have just proved shows that a_i and b_i are given by intersecting Lie complexes with projective subspaces $\Pi_i^{\pm}$. However, $\dim(Z \cap \Pi_i^{\pm}) = \dim(Z) - 1$, whereas $\Pi_i^{\pm}$ have high codimension. Thus to prove that the intersection gives a regular morphism of Chow quotients, extra work is needed.

(1.6.15). PROOF THAT $\tilde{a}_i, \tilde{b}_i$ ARE REGULAR MORPHISMS. For any face $\Gamma \subset \Delta(k, n)$ and any Lie complex Z we shall denote by $Z(\Gamma)$ the closure of the H-orbit in Z corresponding to Γ. In particular, the codimension 1 faces of $\Delta(k, n)$ are $\Gamma_i^{\pm}$ from Proposition 1.2.5 and the corresponding orbit closures $Z(\Gamma_i^{\pm}) = Z \cap G_i^{\pm} = Z \cap \Pi_i^{\pm}$ are the varieties we are studying. Our aim is to show that the Chow form of $Z(\Gamma_i^{\pm})$ can be polynomially expressed via that of Z.

(1.6.15.1). LEMMA. *Let $\{p_I = 0\}$ be a coordinate hyperplane in $P(\wedge^k \mathbb{C}^n)$ defined by the condition that the Plücker coordinate p_I vanishes. Then we have the equality of cycles*

$$Z \cap \{p_I = 0\} = \sum_{\Gamma\,:\, e_I \notin \Gamma} Z(\Gamma),$$

where Γ runs over codimension 1 faces of $\Delta(k, n)$ (i.e., over $\Gamma_i^{\pm}$) not containing the vertex e_I.

PROOF. The lemma says that the order of vanishing of p_I on $Z(\Gamma)$ equals 0 if $e_I \in \Gamma$ and 1 if $e_I \notin \Gamma$. According to the general rule (valid for any toric variety in an equivariant projective embedding [49]) this order equals the distance from e_I to the affine hyperplane spanned by Γ, the distance being measured in natural integer units induced by the lattice. In our case, if $e_I \notin \Gamma$, then the said distance equals one.

(1.6.15.2). COROLLARY. *Let $R_Z(l_1, \dots, l_n)$, $l_i \in \wedge^k(\mathbb{C}^n)^*$, be the Chow form of a Lie complex Z. Let π_I be the coordinate projection to the coordinate hyperplane $\mathrm{Ker}(p_I) \subset \wedge^k(\mathbb{C}^n)$. Then for any linear functionals*

$\lambda_1, \dots, \lambda_{n-1} \in (\operatorname{Ker} p_I)^*$ *we have*

$$R_Z(p_I, \pi_I^*\lambda_1, \dots, \pi_I^*\lambda_{n-1}) = \prod_{\Gamma :\, e_I \notin \Gamma} R_{Z(\Gamma)}(\lambda_1, \dots, \lambda_{n-1})$$

where in the right-hand side we have Chow forms of subvarieties $Z(\Gamma) \subset P(\operatorname{Ker} p_I)$.

(1.6.15.3). END OF THE PROOF THAT $\tilde{a}_i$, $\tilde{b}_i$ ARE REGULAR. Consider some facet of $\Delta(k, n)$, say, Γ_i^+, and let $e_I \notin \Gamma_i^+$ be some vertex (i.e., $i \in I$). Consider the coordinate projection $\operatorname{Ker} p_I \to \Pi_i^+$ of coordinate subspaces in $\wedge^k(\mathbb{C}^n)$. Here Π_i^+ is defined, as above, by the vanishing of all p_I with $i \in I$. The projection of $\bigcup_{\Gamma :\, e_i \notin \Gamma} Z(\Gamma)$ to Π_i^+ is the component $Z(\Gamma_i^+)$ we are interested in, plus the union of some coordinate $(n-2)$-dimensional subspaces (which are images of other components). Let $\mu_1, \dots, \mu_{n-1} \in (\Pi_i^+)^*$ be linear forms. Denote by $\tilde{\mu}_i$ the extension of μ_i to the entire $\operatorname{Ker}(p_I)$ by means of the coordinate projection $\operatorname{Ker} p_I \to \Pi_i^+$. We obtain the equality

$$\prod_{\Gamma :\, e_I \notin \Gamma} R_{Z(\Gamma)}(\tilde{\mu}_1, \dots, \tilde{\mu}_{n-1})$$
$$= R_{Z(\Gamma_i^+)}(\mu_1, \dots, \mu_{m-1}) \cdot \prod_S R_S(\mu_1, \dots, \mu_{n-1})^{\nu_S},$$

where S runs over coordinate $(k-1)$-dimensional subspaces in π_i^+, R_S is the Chow form of the subspace S, and ν_S are some exponents. From this equality we obtain that $R_{Z(\Gamma_i^+)}$ can be obtained from the right-hand side by division to a fixed polynomial. Since the left-hand side itself depends polynomially on R_Z, Theorem (1.6.6) is completely proved.

Chapter 2. Projective configurations and the Gelfand-MacPherson isomorphism

I. M. Gelfand and R. W. MacPherson have established in [20, 40] an important correspondence between torus orbits in $G(k, n)$ and projective configurations, i.e., $GL(k)$-orbits on $(P^{k-1})^n$. In this section we shall show that this correspondence extends to an isomorphism of Chow quotients.

(2.1). Projective configurations and their Chow quotient.

(2.1.1). Consider the $(k-1)$-dimensional projective space P^{k-1}. By a *configuration* we shall mean an ordered collection $M = (x_1, \dots, x_n)$ of n points in P^{k-1}. The general linear group $GL(k)$ acts on P^{k-1} by projective transformations. This induces an action on the space $(P^{k-1})^n$ of configurations.

The study of orbits of this action is a classical problem of projective geometry. See [14] for investigations from the standpoint of Mumford's geometric invariant theory.

(2.1.2). The elements of a configuration M can also be visualized as hyperplanes (in the dual projective space). This point of view will be useful later. In this subsection we shall just consider elements of M as points.

(2.1.3). We will be interested in the Chow quotient $(P^{k-1})^n//GL(k)$. To apply Definition 0.1.7 it is first desirable to know which configurations are "generic enough". The answer, of course, is the following.

(2.1.4). A configuration $M = (x_1, \dots, x_n)$ of points in P^{k-1} will be said to be *in general position* if any i of these points, $i \leqslant k$, span a projective subspace of dimension exactly $i-1$. The set of all such configurations will be denoted by $(P^{k-1})^n_{\mathrm{gen}}$. Orbits of configurations in general position will be referred to as *generic orbits* in $(P^{k-1})^n$.

(2.1.5). Generic $GL(k)$-orbits on $(P^{k-1})^n$ depend on continuous parameters only when $n \geqslant k+2$. We shall assume in the sequel that this condition holds. In this case generic orbits have dimension k^2-1 since the stabilizer of a generic configuration consists only of homotheties.

(2.1.6). For any $0 \leqslant m \leqslant k-1$ denote by $[m]$ the $2m$-dimensional homology class of P^{k-1} represented by P^m. By the Künneth formula, the graded homology space of $(P^{k-1})^n$ is the n-fold tensor power of the graded homology space $H_*(P^{k-1})$. Therefore, the basis for the $2p$th homology group $H_{2p}((P^{k-1})^n)$ is given by tensor products $[m_1]\otimes\cdots\otimes[m_n]$, $\sum m_i = p$.

(2.1.7). Proposition. *The closure of any generic $GL(k)$-orbit in $(P^{k-1})^n$, $n \geqslant k+2$, is a variety of dimension k^2-1 and of homology class*

$$\delta = \sum_{\substack{m_1+\cdots+m_n=k^2-1 \\ m_i \leqslant k-1}} [m_1]\otimes\cdots\otimes[m_n].$$

The set of closures of all generic orbits forms a subvariety in the Chow variety $\mathscr{C}_{k^2-1}((P^{k-1})^n, \delta)$ that is isomorphic to the quotient $(P^{k-1})^n_{\mathrm{gen}}/GL(k)$.

Proof. Let $Z = \overline{GL(k)\cdot M}$ be the closure of any (k^2-1)-dimensional orbit and $\delta \in H_{2(k^2-1)}((P^{k-1})^n, \mathbb{Z})$ its homology class. The coefficient in δ at $[m_1]\otimes\cdots\otimes[m_n]$ can be calculated as follows. Take generic projective subspaces $L_i \subset P^{k-1}$ of codimension m_i. Our coefficient is just the intersection number of Z with $L_1\times\cdots\times L_n$. In other words, this is the number of projective transformations that take each point x_i of our configuration $M = (x_1, \dots, x_n)$ inside L_i. The condition $g(x_i) \subset L_i$ is a linear condition on matrix elements of a matrix $g \in GL(k)$ of codimension m_i. Taking into account all L_i, we obtain a system of k^2-1 linear equations on matrix elements of g. By Bertini's theorem applied to Z, if $(L_1, \dots, L_n)$ are generic enough, the intersection $Z \cap (L_1\times\cdots\times L_n)$ consists of finitely many points. For our linear system this implies that for generic L_j just one of the following two cases holds:

(2.1.7.1). The space of solutions of the system is 1-dimensional and consists of multiples of a nondegenerate matrix.

(2.1.7.2). The space of solutions is contained in the variety of degenerate matrices.

In the first case the coefficient equals one, in the second it equals 0. We need to show that for $x_1, \dots, x_n$ in general position, the case a) always holds.

(2.1.7.3). Consider the product of Grassmannians $\Pi = \prod G(k - m_i, k)$, i.e., the variety of all tuples $(L_1, \dots, L_n)$ as above. Let $\Pi_Y \subset \Pi$ be the subvariety of those tuples for which $x_i \subset L_i$ for all i.

(2.1.7.4). Lemma. *Let $Y = (x_1, \dots, x_n)$ be any configuration with $(k^2 - 1)$-dimensional orbit. Then the truth of the case* (2.1.7.2) *above* (*or, equivalently, the vanishing of the coefficient at $[m_1] \otimes \cdots \otimes [m_n]$*) *is equivalent to the following fact: For a generic tuple of subspaces $(L_1, \dots, L_n) \subset \Pi_Y$ its stabilizer in $PGL(k-1)$ has positive dimension.*

Proof. Case (2.1.7.2) means that the union of $GL(k)$-orbits of points from Π_Y is not dense in Π. The codimension of Π_Y in Π equals $k^2 - 1$. Therefore case (2.1.7.2) means that for any $Y \in \Pi_Y$ its orbit has dimension smaller than $k^2 - 1$.

Now it is clear that if $x_1, \dots, x_n$ are in general position then for any $m_1, \dots, m_n$ with the sum $k^2 - 1$ it is possible to choose codimension m_i subspaces L_i through x_i such that the whole collection $(L_1, \dots, L_n)$ has trivial stabilizer in $PGL(k)$. Proposition 2.1.7 is proved.

(2.1.8). Example. Consider the action of $GL(2)$ in $(P^1)^4$. The closure of the orbit of a 4-tuple of distinct points is a 3-dimensional variety. It contains four 2-dimenional orbits W_i, where W_i is the set of points $(x_1, \dots, x_4)$ such that all three x_j with $j \neq i$ coincide with each other but differ from x_i. The closure of each W_i is isomorphic to $P^1 \times P^1$. These closures intersect along the 1-dimensional orbit, which is the set of coinciding tuples.

(2.1.9). We are interested in the Chow quotient $(P^{k-1})^n // GL(k)$. By definitions, points of this quotient are certain algebraic cycles in $(P^{k-1})^n$ of dimension $k^2 - 1$ and homology class δ given by Proposition 2.1.7. Moreover, since the stabilizer of a configuration cannot be a unipotent subgroup in $PGL(k)$, we can apply Theorem 0.3.1 to conclude that components of any cycle from $(P^{k-1})^n // GL(k)$ are closures of $(k^2 - 1)$-dimensional orbits.

(2.1.10). Example. For four distinct points on P^1 the only invariant is the cross-ratio, which identifies $(P^1)^4_{\text{gen}} / GL(2)$ with $P^1 - \{0, 1, \infty\}$. Denote by Z_λ the closure of the orbit given by 4-tuples with cross-ratio λ. When $\lambda \to 0, 1, \infty$, the variety Z_λ degenerates into one of three cycles in $(P^1)^4$. Namely, let Δ_{ij} to be the subset in $(P^1)^4$ given by $\{x_i = x_j\}$. Then the three cycles in question are

$$\Delta_{12} + \Delta_{34}, \quad \Delta_{13} + \Delta_{24}, \quad \Delta_{14} + \Delta_{23}.$$

For example, suppose that our four points x_i depend on a parameter t and degenerate in such a way that $x_1(0) = x_2(0)$ but $x_3(0)$ and $x_4(0)$ are different from them (see (2.1.11)). Let $Z(t)$, $t \neq 0$, be the closure of the orbit of $M(t) = (x_i(t)_{i=1,\dots,4})$. Then, of course, the orbit of the limit position $(x_i(0))$, i.e., Z_{12}, will be a part of the cycle $Z(0) = \lim_{t\to 0} Z(t)$, but not the only part! Indeed, we can perform, for each t, a projective transformation $g(t)$ which stretches $x_1(t)$ and $x_2(t)$ back to some fixed distance. This transformation shrinks the remaining points $x_3(t)$ and $x_4(t)$ close to each other. The limit of the point $g(t)Y$ will lie on the second component Z_{34}.

$x_1(t)$ $\quad$ $t \to 0$ $\quad$ $x_1 = x_2$ $\quad$ x_1
$x_2(t)$ $\quad$ x_3 $\quad$ + $\quad$ x_2 $\quad$ (2.1.11)
$x_3(t)$
$x_4(t)$ $\quad$ x_4 $\quad$ $x_3 = x_4$

(2.1.12). Similarly, if we have a degeneration $M(t) = (x_1(t), \dots, x_n(t))$ of a family of n points on P^1 such that just two points merge, e.g., $x_1(0) = x_2(0)$ and all the other $x_i(0)$ remain distinct, then $\lim_{t\to 0} \overline{GL(2)M(t)}$ consists of two components. The first is the orbit of the limit configuration $(x_1(0) = x_2(0), x_3(0), \dots, x_n(0))$. The second component is the set of $(x_1, \dots, x_n)$ such that $x_3 = \cdots = x_n$ and x_1, x_2 are arbitrary. We shall see later (Chapter 4) that this phenomenon exactly corresponds to the degeneration of $(P^1, x_1(t), \dots, x_n(t))$ in the Knudsen's moduli space $\overline{M_{0,n}}$ of stable n-punctured curves of genus 0.

(2.2). The Gelfand-MacPherson correspondence.

(2.2.0). Let us recall the original idea of [20, 40] which explains how to construct a configuration from a point in the Grassmannian. It will be more convenient for us to speak in this section about configurations of *hyperplanes* instead of points.

(2.2.1). Let $L \subset \mathbb{C}^n$ be a k-dimensional subspace not lying in any coordinate hyperplane $H_i = \{x_i = 0\}$. Then $(L \cap H_i)$ forms a configuration of hyperplanes in L, i.e., a point in $(P(L^*))^n$. If a subspace L' is obtained from L by the action of a torus element, we shall obtain a projectively isomorphic configuration of hyperplanes in L'. A class of projective isomorphisms of configurations of n hyperplanes in a $(k-1)$-dimensional projective space is the same as a $GL(k)$-orbit in the Cartesian power of a fixed projective space $(P^{k-1})^n$. Note that not every configuration of hyperplanes can be obtained, up to an isomorphism, from L as above. To make the assertion precise, denote by $G_{\max}(k, n) \subset G(k, n)$ the set of L such that $\dim(H \cdot L) = n - 1$. Similarly denote by $((P^{k-1})^n)_{\max}$ the set of configurations $\Pi = (\Pi_1, \dots, \Pi_n)$ such that $\dim(GL(k) \cdot \Pi) = k^2 - 1$. The Gelfand-MacPherson correspondence induces the bijection of orbit sets

$$\Lambda\colon G_{\max}(k, n)/H \to ((P^{k-1})^n)_{\max}/GL(k), \tag{2.2.2}$$

see [20].

(2.2.3). Note that in general sets in both sides of (2.2.2) are not algebraic varieties since $G_{\max}(k, n)$ and $(P^{k-1})^n)_{\max}$ contain unstable points. For comparison of Mumford's quotients of both sides in (2.2.2) see section (2.4) below.

The main result of this section is the following theorem.

(2.2.4). THEOREM. *The Gelfand-MacPherson correspondence* (2.2.2) *extends to an isomorphism of Chow quotients*

$$\Lambda\colon G(k, n)//H \longrightarrow (P^{k-1})^n//GL(k). \tag{2.2.5}$$

This fact permits one to apply the information about behavior of $(n-1)$-dimensional torus orbits (which may be obtained by techniques of toric varieties and A-resultants [30]), to the study of $(k^2 - 1)$-dimensional orbits of $GL(k)$, which are at first glance harder to understand.

(2.2.6). COROLLARY. *Every cycle in* $(P^{k-1})^n//GL(k)$ *is a sum of closures of some* $(k^2 - 1)$*-dimensional orbits with multiplicities* 0 *or* 1.

Before starting to prove Theorem 2.2.4, let us give a simple matrix interpretation of the correspondence in question.

(2.2.7). Let $M(k, n)$ be the vector space of all complex k by n matrices, $M_0(k, n) \subset M(k, n)$ the space of matrices of rank k, and $M'(k, n) \subset M(k, n)$ the space of matrices whose every row is a nonzero vector in $\mathbb{C}^k$. The group $GL(k)$ acts on $M(k, n)$ from the right and the group $(\mathbb{C}^*)^n \subset GL(n)$ acts from the left, and we have the identifications

$$GL(k)\backslash M_0(k, n) = G(k, n), \qquad M'(k, n)/(\mathbb{C}^*)^n = (P^{k-1})^n. \tag{2.2.8}$$

The Gelfand-MacPherson correspondence comes from consideration of both types of orbits in (2.2.5) as double $(GL(k), (\mathbb{C}^*)^n)$-orbits in $\mathrm{Mat}(k, n)$.

(2.2.9). Let us carry on these considerations for Chow quotients. Note that each $(GL(k), (\mathbb{C}^*)^n)$-orbit in the vector space $M(k, n)$ is invariant under multiplications by scalars (in this vector space) and thus may be identified with a subvariety in the projectivization $P(M(k, n))$. Instead of double orbits we can speak about left orbits of the product $GL(k) \times (\mathbb{C}^*)^n$. Consider the Chow quotient $P(M(k, n))//GL(k) \times (\mathbb{C}^*)^n$. To prove Theorem 2.2.4 it suffices to construct isomorphisms

$$G(k, n)//(\mathbb{C}^*)^n \xrightarrow{\alpha} P(M(k, n))//GL(k)\times(\mathbb{C}^*)^n \xleftarrow{\beta} (P^{k-1})^n//GL(k). \tag{2.2.10}$$

(2.2.11). The existence of these morphisms does not present any problem.

The morphism α associates to any cycle $Z = \sum c_i Z_i$ in $G(k, n)//(\mathbb{C}^*)^n$ the cycle $\sum \overline{p^{-1}(Z_i)}$ where $p: P(M_0(k, n)) \to G(k, n)$ is the projection from (2.2.8). (The multiplicities c_i all are equal to 1 by Proposition 1.2.15.) Similarly, β associates to any cycle $W = \sum m_i W_i$ in $(P^{k-1})^n//GL(k)$ the cycle $\sum m_i \overline{q^{-1}(W_i)}$, where $q: P(M'(k, n)) \to (P^{k-1})^n$ is the other projection arising from (2.3). To show that α and β are regular maps, it suffices to apply Barlet's criterion of analytic dependence of a cycle on a parameter [3]. Since α and β are both given by inverse images in fibrations, this criterion is trivially applicable.

(2.2.12). Let us show that α is an isomorphism. To do this, note that any generic $(GL(k) \times (\mathbb{C}^*)^n)$-orbit in $P(M(k, n)$ has dimension $(n-1)(k^2-1)$. Each component of a cycle Z from $P(M(k, n))//GL(k) \times (\mathbb{C}^*)^n$ is the closure of a single orbit, which therefore should be the inverse image of an orbit of maximal dimension in $G(k, n)$. The algebraic cycle W formed by these orbits lies clearly in the Chow quotient $G(k, n)//(\mathbb{C}^*)^n$ and this is the unique element of this Chow quotient such that $\alpha(W) = Z$. This proves that α is bijective on $\mathbb{C}$-points. Denote by α^{-1} the inverse map. To prove that α is indeed an isomorphism of algebraic varieties we need to prove that α^{-1} is regular too (which need not necessarily be the case if the varieties involved are not normal). However, this again follows from Barlet's criterion similarly to the proof of Proposition 1.4.12.

Similarly we prove that β is an isomorphism. Theorem 2.2.4 is proved.

(2.3). Duality (or association). It is known classically (since A. B. Coble [11]) that projective equivalence classes of configurations of n ordered points in P^{k-1} are in bijection with projective equivalence classes of configurations of n points in P^{n-k-1}. This correspondence is known as the *association* [11, 14] and was used in the context of matroid theory (see [21, §2.3]).

The most transparent way to define the association is via the Gelfand-MacPherson correspondence.

(2.3.1). Let us identify the dual subspace to the coordinate space $\mathbb{C}^n$ with $\mathbb{C}^n$ by means of the standard pairing. By considering orthogonal complements to k-dimensional subspaces we obtain an isomorphism $G(k, n) \cong G(n-k, n)$. The torus $H = (\mathbb{C}^*)^n$ acts in both Grassmannians and the above isomorphism is H-equivariant. Hence it induces the isomorphism of coset spaces $G(k, n)/H \to G(n-k, n)/H$. Taking into account the Gelfand-MacPherson isomorphism (2.2.2), we obtain the following isomorphism

$$A_{k,n}: (P^{k-1})^n_{\max}/GL(k) \to (P^{n-k-1})^n_{\max}/GL(n-k), \tag{2.3.2}$$

where the subscript " max " means the set of points whose orbits have the maximal dimension. The isomorphism (2.3.2) will be called the *association isomorphism*. By construction, this system of isomorphisms is involutive, i.e., $A_{k,n} \circ A_{n-k,n} = \mathrm{Id}$.

(2.3.3). If $(x_1, \dots, x_n)$ and $(y_1, \dots, y_n)$ are n-tuples of points in P^{k-1} and P^{n-k-1} respectively, then we say that (y_i) is associated to (x_i) (and vice versa) if their orbits under projective transformations have maximal dimensions and are taken into each other by the association isomorphism.

(2.3.4). Explicitly, the configuration associated to (x_i) can be calculated as follows. Let $\mathbf{x}_i \in \mathbb{C}^k$ be vectors whose projectivizations are x_i. By definition, we have to find a k-dimensional subspace $L \subset \mathbb{C}^n$ and an isomorphism $\mathbb{C}^k \to L^*$ which takes $\mathbf{x}_i$ into the restriction of the ith coordinate function to L. Then the vectors $\mathbf{y}_i \in \mathbb{C}^n/L$, defined as the projections of the standard basis vectors, will represent the associated configuration. In other words, we have to find a complete $(n-k)$-dimensional system of linear relations between $\mathbf{x}_i$, namely $y_{j1}\mathbf{x}_1 + \cdots + y_{jn}\mathbf{x}_n = 0$, $j = 1, \dots, n-k$. Vectors $\mathbf{y}_i$ representing the associated configuration (y_i) are given by columns of the matrix $\|y_{ji}\|$. This gives the following criterion.

(2.3.5). Proposition. *Let* $\mathbf{x}_i \in \mathbb{C}^k$, $\mathbf{y}_i \in \mathbb{C}^{n-k}$ *be n-tuples of vectors such that the corresponding configurations of points* $x_i \in P^{k-1}$, $y_i \in P^{n-k-1}$ *have orbits of maximal dimension. Then* (y_i) *is associated to* (x_i) *if and only if there is a unique, up to a constant, linear relation in* $\mathbb{C}^k \otimes \mathbb{C}^{n-k}$:

$$\sum_i \lambda_i(\mathbf{x}_i \otimes \mathbf{y}_i) = 0 \tag{2.3.6}$$

such that all $\lambda_i \neq 0$.

This can be reformulated as follows.

(2.3.7). Let P be some projective space and $C \subset P$ be some finite subset. We say that C is a *circuit* (in the sense of matroid theory, see [21]) if C is projectively dependent but any of its proper subsets is projectively independent.

(2.3.8). Reformulation. *Let* $x = (x_1, \dots, x_n) \in (P^{k-1})^n$ *and* $y = (y_1, \dots, y_n) \in (P^{n-k-1})^n$ *be two n-tuples whose orbits with respect to projective transformations have maximal dimensions (i.e.,* k^2-1 *and* $(n-k)^2-1$*). Consider the Segre embedding* $P^{k-1} \times P^{n-k-1} \hookrightarrow P^{k(n-k)-1}$. *Then* y *is associated to* x *if and only if the points* $(x_i, y_i) \in P^{k(n-k)-1}$ *form a circuit.*

"Normally" one would expect that points (x_i, y_i) are projectively independent.

(2.3.9). For the case of $2k$ points in P^{k-1} the source and the target of the association isomorphism are the same, so it is possible to speak about a configuration being self-associated. The following characterization of self-associated configurations due to A. Coble [11] is a corollary of Reformulation 2.3.8.

(2.3.10). Corollary. *Let* $x = (x_1, \dots, x_{2k}) \in (P^{k-1})^{2k}$ *be a* $2k$*-tuple. Consider the* 2*-fold Veronese embedding* $P^{k-1} \hookrightarrow P(S^2\mathbb{C}^k)$. *Then the configuration* x *is self-associated if and only if the images of* x_i *in the Veronese embedding form a circuit.*

(2.3.11). Example. Let $k = 2$. The association induces an isomorphism between the set of projective equivalence classes of n-tuples of distinct points on P^1 and the set of projective equivalence classes of n-tuples of distinct points in P^{n-3}. This correspondence can be seen geometrically as follows ([14, Chapter III, §2, Proposition 2]).

Given n points $y_1, \dots, y_n \in P^{n-2}$ in general position, there is a unique rational normal curve (Veronese curve, for short) in P^{n-3} through these points. This curve is isomorphic to P^1 and hence y_i represent on it a configuration of n points in P^1, which is the configuration associated to that of y_i in P^{n-3}. Conversely, given n distinct points on P^1, we consider the $(n-3)$-fold Veronese embedding of P^1. It identifies P^1 with a Veronese curve in P^{n-3}. The images y_i of x_i in this embedding are in general position as it may be seen by calculating the Vandermonde determinant. These points represent the configuration associated to that of x_i on P^1.

(2.3.12). Example. Let $x_1, \dots, x_6$ be a configuration of six points in P^2 in general position. Corollary 2.3.10 means in this case that the configuration (x_i) is self-associated if and only if six points x_i lie on a conic. Further examples can be found in [11, 14].

(2.3.13). Theorem 2.2.4 implies that the association isomorphism extends to the Chow quotients of the spaces of projective configurations. In other words, we have the following fact.

(2.3.14). Corollary. *There is an isomorphism of Chow quotients*

$$(P^{k-1})^n // GL(k) \to (P^{n-k-1})^n // GL(n-k).$$

(2.4). Gelfand-MacPherson correspondence and Mumford quotients.

For completeness we include here the comparison of Mumford's quotients of $G(k, n)$ modulo torus and of P^{k-1} modulo projective transformations.

(2.4.1). First of all, the theory of Mumford is sensitive not only to the structure of orbits, but also to the choice of groups generating these orbits. In order that things behave well, we should consider the action of the subgroup $H_1 = \{(t_1, \dots, t_n) \in (\mathbb{C}^*)^n : \prod t_i = 1\}$ on $G(k, n)$ and the action of the subgroup $SL(k) \subset GL(k)$ on $(P^{k-1})^n$.

(2.4.2). Recall (see §0.4) that to define Mumford's quotient by any group G acting on any variety X we should fix two things: an ample line bundle $\mathcal{L}$ on X and a linearization, i.e., an extension α of the G-action to $\mathcal{L}$.

(2.4.3). First consider the H_1-action on the Grassmannian $G(k, n)$. The Picard group of $G(k, n)$ is generated by the sheaf $\mathcal{O}(1)$ in the Plücker embedding. So there is essentially no freedom in choosing $\mathscr{L}$. We set $\mathscr{L} = \mathscr{O}(1)$. For this choice of $\mathscr{L}$ a linearization is given by an integral vector $a = (a_1, \ldots, a_n)$ defined modulo multiples of $(1, \ldots, 1)$. Denote by $t^a = t_1^{a_1} \cdots t_n^{a_n}$ the character of H_1 corresponding to a. The H_1-action on $\mathbb{C}^n$ corresponding to a has the form

$$(t_1, \ldots, t_n) \mapsto \operatorname{diag}(t^a \cdot t_1, \ldots, t^a \cdot t_n).$$

This action induces an H_1-action on $\wedge^k \mathbb{C}^n$, which is the linearization corresponding to a.

(2.4.4). Denote by $A(k, n)$ the coordinate ring of $G(k, n)$ in the Plücker embedding. It is well known [14] that $A(k, n)$ can be identified with the ring of polynomials $\Phi(M)$ in entries of an indeterminate $(k \times n)$-matrix $M = \|v_{ij}\|$ which satisfy the condition $\Phi(gM) = \Phi(M)$ for any $g \in SL(k)$. In particular, the Plücker coordinate p_I corresponds to the polynomial in v_{ij} given by the $(k \times k)$-minor of M on columns from I.

The Mumford quotient $(G(k, n)/H_1)_{\mathscr{O}(1), a}$ is, by definition, the projective spectrum $\operatorname{Proj}(A(k, n)^{H_1})$ of the invariant subring in $A(k, n)$.

(2.4.5). Consider now the $SL(k)$-action on $(P^{k-1})^n$. For any integral vector $a = (a_1, \ldots, a_k)$ denote by $\mathscr{O}(a) = \mathscr{O}(a_1, \ldots, a_n)$ the line bundle on $(P^{k-1})^n$ whose local sections are multihomogeneous functions of multidegree $(a_1, \ldots, a_k)$. It is well known [24, 25] that bundles $\mathscr{O}(a)$ exhaust the Picard group of $(P^{k-1})^n$. Since the center of $SL(k)$ has dimension 0, for any $a \in \mathbb{Z}^n$ the bundle $\mathscr{O}(a)$ has exactly one $SL(k)$-linearization. This linearization will be denoted by λ.

(2.4.6). The bundle $\mathscr{O}(a)$ is ample if and only if all $a_i > 0$. Assuming that this is the case, let $B(k, n, a) = \bigoplus_d B(k, n, a)_d$ be the homogeneous coordinate ring of $(P^{k-1})^n$ in the projective embedding given by $\mathscr{O}(a)$. The degree d homogeneous component $B(k, n, a)_d$ of this ring consists of polynomials $F(w_1, \ldots, w_n)$ in coordinates on n vectors $w_i \in \mathbb{C}^k$ such that $F(t_1 w_1, \ldots, t_n w_n) = t^{da} F(w_1, \ldots, w_n)$ for any $t_i \in \mathbb{C}^*$. Writing the vectors in coordinate form as columns $w_i = (v_{1i}, \ldots, v_{ki})^t$, we realize elements of $B(k, n, a)_d$ as polynomials $F(M)$ in entries of an indeterminate $(k \times n)$-matrix $M = \|v_{ij}\|$ such that $F(M \cdot t) = t^{da} F(M)$. The Mumford's quotient $((P^{k-1})^n / SL(k))_{\mathscr{O}(a), \lambda}$ is, by definition, the projective spectrum $\operatorname{Proj}(B(k, n, a)^{SL(k)})$.

(2.4.7). **Theorem.** *Let $a = (a_1, \ldots, a_n) \in \mathbb{Z}^n$ be a vector of integers. If at least one $a_i \leqslant 0$ then the Grassmannian $G(k, n)$ does not contain a-stable orbits. If all a_i are positive then we have an isomorphism of Mumford quotients*

$$(G(k, n)/H_1)_{\mathscr{O}(1), a)} \cong ((P^{k-1})^n / SL(k))_{\mathscr{O}(a), \lambda}.$$

PROOF. Both varieties are projective spectra of the same ring $R = \bigoplus R_d$, where R_d consists of polynomials $\Phi(M)$, $M \in \mathrm{Mat}(k \times n)$, such that

$$\Phi(M \cdot t) = t^{da}\Phi(M), \quad t \in (\mathbb{C}^*)^n, \qquad \Phi(gM) = \Phi(M), \quad g \in SL(k).$$

(2.4.8). REMARK. The algebra generated by $(k \times k)$-minors of an indeterminate $k \times n$ matrix is known as the bracket algebra. Traditionally, it makes its appearance in two seemingly different contexts. The first appearance is as the coordinate ring of the Grassmannian $G(k, n)$ in the Plücker embedding. The other is in the study of (semi-)invariants in the system of vectors by the symbolic method (see [14]). However, the idea of serious use of Grassmannians for the study of projective configurations appeared only fairly recently in the papers of Gelfand and MacPherson.

Chapter 3. Visible contours of (generalized) Lie complexes and Veronese varieties

The Grassmannian point of view on projective configurations (i.e., the Gelfand-MacPherson isomorphism, see §3.2) simplifies considerably the study of the Chow quotient. Indeed, instead of working with $(k^2 - 1)$-dimensional subvarieties on $(P^{k-1})^n$ that are closures of $PGL(k)$-orbits, we have to deal with Lie complexes in $G(k, n)$ that are $(n - 1)$-dimensional toric varieties.

(3.1). Visible contours and the logarithmic Gauss map.

(3.1.1). There is a classical method (see, e.g., [2, 24, 27]) to analyze any complex of projective subspaces in P^{n-1}, i.e., an $(n - 1)$-dimensional subvariety $Z \subset G(k, n)$. Namely, take any point $p \in P^{n-1}$ and consider the subvariety

$$Z_p = \{L \in Z : p \in L\}$$

of subspaces in Z containing p. This subvariety will be called the *visible contour* of Z at p.

Let $G(k-1, n-1)_p \subset G(k, n)$ be the variety of all $(k - 1)$-dimensional projective subspaces containing p. It is clear that $Z_p = Z \cap G(k-1, n-1)_p$.

(3.1.2). Still another step towards a visualization of the complex Z at a point p is done as follows. Let P_p^{n-1} be the space of lines in P^{n-1} through p. Then $G(k-1, n-1)_p$ is identified with the variety of all $(k-2)$-dimensional projective subspaces in P_p^{n-2}. We define the *visible sweep* of Z at p to be the subvariety $\mathrm{Sw}_p(Z) \subset P_p^{n-2}$ that is the union of all projective subspaces corresponding to elements of Z_p.

(3.1.3). REMARKS. a) If $k = 2$, then Z consists of lines in P^{n-1}. Lines belonging to the complex Z can be thought of as rays of light piercing the space, so Z_p is the contour, which is seen by an observer at a point p. In this case the visible contour is the same as the visible sweep.

b) Although the consideration of the locus Z_p is classical, there seems to be no good name in the literature for it. The term "complex cone" which is used sometimes [27] for the union of subspaces from Z_p (i.e., the cone over the visible sweep, in our terminology) is obviously unsuitable for modern usage.

c) Dually, one can take any hyperplane $\Pi \subset P^{n-1}$ and consider the locus of subspaces from Z that lie in Π.

(3.1.4). Since $\operatorname{codim} G(k-1, n-1)_p = n-k$, we find that for any complex $Z \subset G(k, n)$ and a generic p the variety Z_p has dimension $p-1$. Thus Z_p is a curve if Z consists of lines, a surface if Z consists of planes, etc.

(3.1.5). We shall use the approach of visible contours to study Lie complexes and, more generally, closures of arbitrary $(n-1)$-dimensional torus orbits in $G(k, n)$ (such closures can be components of generalized Lie complexes). Visible sweeps of Lie complexes will be studied in §3.6.

(3.1.6). Let us realize our torus $H = (\mathbb{C}^*)^n/\mathbb{C}^*$ as an open subset in P^{n-1} consisting of points with all homogeneous coordinates nonzero. The point $e = (1 : \cdots : 1) \in P^{n-1}$ becomes the unit in H. Denote by $\mathbf{h}$ the Lie algebra of H. It is identified with the tangent space to P^{n-1} at e. Explicitly, $\mathbf{h} = \mathbb{C}^n/\{(a, \ldots, a)\}$. For any $x \in H$ let $\mu_x : H \to H$ be the operator of multiplication by X.

Any subvariety $L \subset P^{n-1}$ not lying inside a coordinate hyperplane gives a subvariety $L \cap H$ in the algebraic group H.

(3.1.7). Definition. Let $X \subset H$ be a p-dimensional algebraic subvariety. The logarithmic Gauss map of X is the (rational) map $\gamma_X : X \to G(p, \mathbf{h})$ that takes a smooth point $x \in X$ to the p-dimensional subspace $d(\mu_x^{-1})(T_x X) \subset T_e H = \mathbf{h}$, the translation to the unity of the tangent space $T_x X \subset T_x H$.

The name "logarithmic" comes from the fact that the explicit formula for γ_X involves logarithmic derivatives (see below).

(3.1.8). Let $L \subset P^{n-1}$ be a $(k-1)$-dimensional projective subspace not lying in a coordinate hyperplane. The orbit closure $\overline{H \cdot L}$ has dimension $n-1$, i.e., it is a complex. Since this complex is H-invariant, its visible contour $(\overline{H \cdot L})_p$ at any point $p \in P^{n-1}$ with all coordinates nonzero is isomorphic to the visible contour at the point $e = (1 : \cdots : 1)$.

Before stating the next proposition let us note that the Grassmannian $G(k-1, n-1)_e$, where visible contours lie, is canonically identified with $G(k-1, \mathbf{h})$. (Correspondingly, the space P_e^{n-2}, where visible sweeps lie, is $P(\mathbf{h})$.)

(3.1.9). Proposition. a) *If the subspace L does not lie in a coordinate hyperplane, then the visible contour $(\overline{H \cdot L})_e$ coincides with the closure of the image of $L \cap H$ under the logarithmic Gauss map. In particular, this visible contour is a rational variety.*

b) *The intersection of* $\overline{H \cdot L}$ *with the sub-Grassmannian* $G(k-1, n-1)_e$ *is proper and transversal at its generic point.*

PROOF. a) Neither the complex $\overline{H \cdot L}$ nor the image of $L \cap H$ under the logarithmic Gauss map will change if we translate L by the H-action. So we can (and will) assume that L contains the point $e = (1 : \cdots : 1)$. For $h \in H$ the translated subspace $h^{-1}L$ contains e if and only if $h \in L$. Thus the variety $(H \cdot L)_e = (H \cdot L) \cap G(k-1, n-1)_e$ consists of subspaces $h^{-1} \cdot L$, $h \in L$. In other words, $(H \cdot L)_e$ is the image of the map $L \cap H \to G(k-1, n-1)_e$ taking $h \in L \cap H$ to the subspace $h^{-1} \cdot L$. This map clearly coincides with the logarithmic Gauss map.

b) If Z is any complex in $G(k, n)$ then the assertion will be true for the intersection of Z with $G(k-1, n-1)_p$, where $p \in P^{n-1}$ is a generic point. In our case, due to the invariance under the torus action, the situation at any $p = (p_1 : \cdots : p_n)$ with all $p_i \neq 0$ is the same as at e.

(3.1.10). THEOREM. *Suppose that* $L \subset P^{n-1}$ *is a* $(k-1)$*-dimensional subvariety belonging to the generic stratum* $G^0(k, n)$. *Then the logarithmic Gauss map* $\gamma_{L \cap H}$ *extends to a regular embedding* $L \hookrightarrow G(k-1, \mathbf{h}) = G(k-1, n-1)_e$. *In other words, the visible contour* $(\overline{H \cdot L})_e$ *is identified with* L *itself.*

PROOF. First let us show that the logarithmic Gauss map γ extends to all of L as a regular map. This will be done by calculation in coordinates which we shall also use on other occasions.

(3.1.11). Let $x_1, \ldots, x_n$ be homogeneous coordinates in P^{n-1}. Let $p = (y_1 : \cdots : y_n) \in L$ be any point. Since L lies in the generic stratum, there are $1 \leqslant i_1 < \cdots < i_{k-1} \leqslant n$ such that $y_j \neq 0$ for $j \notin \{i_1, \ldots, i_{k-1}\}$. After renumbering variables we can (and will) assume that $y_k, \ldots, y_n$ are nonzero. Consider the affine space $L - \{x_n = 0\}$, which contains our point p. Introduce in this space affine coordinates $z_1, \ldots, z_{k-1}$, where $z_i = x_i/x_n$. We can set x_n to be 1 on $L - \{x_n = 0\}$ and express all the other coordinates as affine-linear functions in z_i, i.e.,

$$x_i = z_i, \qquad i = 1, \ldots, k-1,$$

$$x_i = f_i(z) = \sum_{\nu=1}^{k-1} a_{i\nu} z_\nu + a_k, \qquad i = k, \ldots, n-1.$$

We also set $f_i(z) = z_i$ for $i = 1, \ldots, k-1$.

(3.1.12). We identify the torus $H = (\mathbb{C}^*)^n/\mathbb{C}^* \subset P^{n-1}$ with the set $\{(t_1, \ldots, t_{n-1}, 1) \in (\mathbb{C}^*)^n\}$, i.e., with $(\mathbb{C}^*)^{n-1}$. Its Lie algebra is therefore identified with $\mathbb{C}^{n-1}$. In this notation the map γ takes a point $z = (z_1, \ldots, z_{k-1})$ to the $(k-1)$-dimensional subspace in $\mathbb{C}^{n-1}$ spanned by the rows of a $(k-1)$ by $(n-1)$ matrix $\|\partial \log f_i / \partial z_j\|$, $i = 1, \ldots, n-1$,

$j = 1, \ldots, k-1$. We can multiply the jth row by z_j without changing this subspace. After this the matrix takes the form

$$\begin{pmatrix} 1 & 0 & \ldots & 0 & \frac{a_{k1}z_1}{f_k(z)} & \ldots & \ldots & \frac{a_{n-1,1}z_1}{f_{n-1}(z)} \\ 0 & 1 & \ldots & 0 & \frac{a_{k2}z_2}{f_k(z)} & \ldots & \ldots & \frac{a_{n-1,2}z_2}{f_{n-1}(z)} \\ \ldots & \ldots & \ldots & \ldots & \ldots & \ldots & \ldots & \ldots \\ \ldots & \ldots & \ldots & \ldots & \ldots & \ldots & \ldots & \ldots \\ 0 & 0 & \ldots & 1 & \frac{a_{k,k-1}z_{k-1}}{f_k(z)} & \ldots & \ldots & \frac{a_{n-1,k-1}z_{k-1}}{f_{n-1}(z)} \end{pmatrix}. \qquad (3.1.13)$$

This matrix is clearly regular near our point p, since $f_k(p), \ldots, f_{n-1}(p)$ are nonzero. The rank of the matrix (3.1.13) being equal to k, we deduce that γ is regular at p. We have proved that γ extends to a regular morphism $L \to G(k-1, n-1)$.

(3.1.14). Let us complete the proof of Theorem 3.1.10 by showing that the logarithmic Gauss map γ is an embedding. Consider the set of all $(k-1)$ by $(n-1)$ matrices of which the first $(k-1)$ columns form the unit $(k-1)$ by $(k-1)$ matrix. The entries of the remaining $(n-k)$ columns are independent affine coordinates in the open Schubert cell $\mathbb{C}^{(k-1)(n-k)} \subset G(k-1, n-1)$. Let us show that entries of any given column of (3.1.13), whose number is greater that k, alone suffice to separate all points of L. Indeed, consider, say, the pth column, $p > k$, and regard its entries as defining a transformation

$$(z_1, \ldots, z_{k-1}) \mapsto (s_1, \ldots, s_{k-1}), \qquad \text{where} \quad s_i = \frac{a_{pi}z_i}{\sum_{\nu=1}^{k-1} a_{p\nu}z_\nu}.$$

This is a projective transformation corresponding to the k by k matrix

$$T_p = \begin{pmatrix} a_{p1} & 0 & 0 & \ldots & 0 \\ 0 & a_{p2} & 0 & \ldots & 0 \\ \ldots & \ldots & \ldots & \ldots & \\ a_{p1} & a_{p2} & a_{p3} & \ldots & a_{pk} \end{pmatrix}. \qquad (3.1.15)$$

Since our subspace $L \subset P^{n-1}$ belongs to the generic stratum, every entry of the matrix a_{ij}, $i = 1, \ldots, k-1$, $j = 1, \ldots, n-k$, is nonzero. Hence the matrix (3.1.15) defines a nondegenerate projective transformation and separates the points (as well as the tangent vectors). Theorem 3.1.10 is proved.

(3.2). Bundles of logarithmic forms on P^{k-1} and visible contours.

(3.2.1). It is well known [24] that maps from any projective variety X to Grassmannians of the form $G(r, V)$, $\dim(V) > r$, are in correspondence with rank r vector bundles on X. More precisely, given such a bundle E we consider the vector space $V = H^0(X, E)^*$. Suppose that E is generated by global sections and let N be the dimension of V. Define a map $\varphi_E \colon X \to G(r, V) = G(N-r, H^0(X, E))$ as follows. For a point $x \in X$ the value $\varphi_E(x)$ is the codimension r subspace in $H^0(X, E)$ consisting of all the sections vanishing at x. Conversely, suppose we are given a map $\varphi \colon X \to$

$G(r, V)$. Let S be the tautological rank r bundle on $G(r, V)$ (whose fiber at a subspace L is L itself). Associate to φ the bundle $\varphi^* S^*$ on X.

The bundles on P^{k-1} corresponding to visible contours of Lie complexes have the following description.

(3.2.2). Let $M = (M_1, \dots, M_n)$ be a configuration of n hyperplanes in P^{k-1} in general position. Then M is a divisor with normal crossings and we can define the sheaf $\Omega^1_{P^{k-1}}(\log M)$ of differential 1-forms on P^{k-1} with logarithmic poles along M, see [13]. By definition, the space of sections of this sheaf near a point $x \in P^{k-1}$ is generated (over $\mathcal{O}_{P^{k-1}, x}$) by 1-forms regular at x and also by forms $d \log f_i$, where f_i are local equations of hyperplanes from M containing x.

An important property of the sheaf $\Omega^1_{P^{k-1}}(\log M)$ is that it is locally free, i.e., can be seen as a rank $(k-1)$ vector bundle over P^{k-1}.

(3.2.3). Proposition. *Let $M = (M_1, \dots, M_n)$ be a configuration of n hyperplanes in the projective space $L = P^{k-1}$ in general position and let f_i be a linear form defining M_i. Then:*

a) *The space $W = H^0(L, \Omega^1_L(\log M))$ has dimension $n-1$ and consists of forms*

$$\sum_i \alpha_i d \log f_i = d \log \prod_i f_i^{\alpha_i}, \qquad \alpha_i \in \mathbb{C}, \quad \sum \alpha_i = 0.$$

Higher cohomology groups of $\Omega^1_L(\log M)$ vanish.

b) *The vector bundle $E = \Omega^1_L(\log M)$ defines a regular embedding $\varphi_E : L \hookrightarrow G(k-1, W^*)$.*

c) *Suppose that L is realized as a subspace in the coordinate P^{n-1} so that M_i is given by the vanishing of the ith coordinate. Then φ_E coincides with the (extension of) the logarithmic Gauss map $\gamma_{L \cap H}$, and the image $\varphi_E(L)$ coincides with the visible contour of the Lie complex $\overline{H \cdot L}$.*

(3.2.4). Proof of (3.2.3) a). The sheaf Ω^1_L of regular 1-forms is obviously a subsheaf of $\Omega^1_L(\log M)$. To describe the quotient denote, following P. Deligne [13], by $\tilde{M}$ the disjoint union of hyperplanes in M and let $\epsilon : \tilde{M} \to L$ be the natural map. Then we have the exact sequence

$$0 \to \Omega^1_L \to \Omega^1_L(\log M) \xrightarrow{\text{Res}} \epsilon_* \mathcal{O}_{\tilde{M}} \to 0 \tag{3.2.5}$$

where Res is the Poincaré residue morphism, see [13]. Consider the corresponding long exact sequence of cohomology. The equality $H^0(L, \epsilon_* \mathcal{O}_{\tilde{M}}) = \mathbb{C}^n$ means that the residue of a global logarithmic form along each M_i is constant. The sum of the residues given by the boundary map $H^0(L, \epsilon_* \mathcal{O}_{\tilde{M}}) \to H^1(L, \Omega^1) = \mathbb{C}$ should be zero. Since the forms defined in the formulation are indeed global sections of our sheaf, we obtain the statement about

H^0. The vanishing of higher H^i follows from known information about the cohomology of the sheaf $\mathcal{O}$ on P^{k-2} and of the sheaf Ω^1 on P^{k-1}.

(3.2.6). PROOF OF (3.2.3) b) AND c). We can assume that M is given by the intersection of an embedded $L = P^{k-1} \subset P^{N-1}$ with coordinate hyperplanes $\{x_i = 0\}$. By a), the basis of $H^0(L, \Omega^1_L(\log M))$ is given by 1-forms $d\log(x_1/x_n)$, $i = 1, \dots, n-1$. We identify the space of section with $\mathbb{C}^{n-1}$ by using this basis. Now looking at explicit formula (3.1.12), we find that the map $\varphi_E: L \to G(k-1, n-1)$ is defined by the formula identical to that of the logarithmic Gauss map.

(3.2.7). PROPOSITION. *The Chern classes of* $E = \Omega^1_{P^{k-1}}(\log(M_1 + \dots + M_n))$ *have the form*

$$c_i(E) = \binom{n-k+i}{i} \in H^{2i}(P^{k-1}, \mathbb{Z}) = \mathbb{Z}.$$

In particular, the determinant (= *top exterior power*) *of* E *is isomorphic to* $\mathcal{O}_{P^{k-1}}(n-k)$.

PROOF. This follows at once from the exact sequence (3.2.5).

(3.2.8). EXAMPLE. Consider a Lie complex in $G(2, n)$, the Grassmannian of lines in P^{n-1}. Let this complex have the form $Z = \overline{H \cdot l}$, where l is a line belonging to the generic stratum. The visible contour Z_e lies in the projective space P^{n-2}_e of all lines in P^{n-1} through the point e. Proposition 3.2.3 means that Z_e is a rational normal curve (Veronese curve, for short) in P^{n-2}_e. More precisely, it is the embedding of $l = P^1$ defined by the invertible sheaf $\Omega^1_l(\log(m_1 + \dots + m_n)) \cong \mathcal{O}_l(n-2)$. Here $m_i \in l$ is the point of intersection of l with the coordinate hyperplane $\{x_i = 0\}$.

(3.3). The visible contour as a Veronese variety in the Grassmannian.

(3.3.1). Recall [24, 25] the d-fold *Veronese embedding*

$$P^{k-1} = P(\mathbb{C}^k) \hookrightarrow P(S^d\mathbb{C}^k), \qquad x \mapsto x^d \tag{3.3.2}$$

of P^{k-1} into the projectivization of $(S^d\mathbb{C}^k)$, the space of homogeneous degree d polynomials in k variables. This is the embedding corresponding to the line bundle $\mathcal{O}(d)$. We say that a $(k-1)$-dimensional subvariety $X \subset P^N$ is a d-fold *Veronese variety* if there is a projective equivalence $P^N \cong P(S^d\mathbb{C}^k)$ taking X into the image of (3.3.2). A Veronese curve in P^N is the same as a rational normal curve of degree N.

(3.3.3). The dimension of the projective space $P(S^d\mathbb{C}^k)$ of the d-fold Veronese embedding (3.3.2) equals $N = \binom{d+k-1}{d-1} - 1$. Note that the same dimension is attained by the projective space of the Plücker embedding of the Grassmannian $G(k-1, d+k-1)$. Therefore it makes sense to look for those Veronese subvarieties in $P^{\binom{d+k-1}{k-1}-1} = P(\wedge^{k-1}\mathbb{C}^{d+k-1})$ that lie on the Grassmannian.

(3.3.4). We say that a $(k-1)$-dimensional subvariety

$$X \subset G(k-1, d+k-1)$$

is a d-fold Veronese variety if it becomes one after the Plücker embedding of $G(k-1, d+k-1)$.

(3.3.5). PROPOSITION. *Let* $M = (M_1, \dots, M_n)$ *be a configuration of hyperplanes in* P^{k-1} *in general position,* $E = \Omega^1_{P^{k-1}}(\log M)$ *the corresponding logarithmic bundle, and* $\varphi_E\colon P^{k-1} \to G(k-1, n-1)$ *the embedding corresponding to* E. *Then* $\varphi_E(P^{k-1})$ *is an* $(n-k)$*-fold Veronese variety in* $G(k-1, n-1)$.

PROOF. Let us construct an isomorphism of linear spaces $\wedge^{k-1}\mathbb{C}^{n-1} \to S^{n-k}\mathbb{C}^k$ taking $\varphi_E(P^{k-1})$ into the standard Veronese variety. Let S be the tautological rank $(k-1)$ bundle on $G(k-1, n-1)$. Then $E = \varphi_E^*(S^*)$ and hence

$$\varphi_E^*\left(\wedge^{k-1} S^*\right) = \wedge^{k-1} E = \mathscr{O}_{P^{k-1}}(n-k)$$

by Proposition 3.2.7. Thus we obtain a linear map of restriction

$$\begin{aligned}\wedge^{k-1}\mathbb{C}^{n-1} &= H^0\left(G(k-1, n-1), \wedge^{k-1}S^*\right) \xrightarrow{r} H^0\left(P^{k-1}, \varphi_E^*(\wedge^{k-1}S^*)\right) \\ &\cong H^0(P^{k-1}, \mathscr{O}(n-k)) = S^{n-k}\mathbb{C}^k.\end{aligned} \tag{3.3.6}$$

(3.3.7). Let us show that the restriction map r in (3.3.6) is an isomorphism.

Since spaces in both sides have the same dimension, it suffices to show that r is injective, i.e., that the variety $X = \varphi_E(P^{k-1})$ does not lie in any hyperplane in $P(\wedge^{k-1}\mathbb{C}^{n-1})$. Take an affine chart in P^{k-1} in which the last hyperplane M_n is the infinite one. All the other hyperplanes M_i are then defined by vanishing of affine-linear functions f_i, $i = 1, \dots, n-1$, on $\mathbb{C}^{k-1} = P^{k-1} - M_n$. The fact that the variety X lies in a hyperplane means that there is a collection of numbers $a_{i_1, \dots, i_{k-1}}$, not all of them zero, such that the meromorphic $(k-1)$-form

$$\Omega = \sum_{1 \leqslant i_1 < \cdots < i_{k-1} \leqslant n-1} a_{i_1, \dots, i_{k-1}} d\log f_{i_1} \wedge \cdots \wedge d\log f_{i_{k-1}}$$

on $\mathbb{C}^{k-1}$ vanishes identically. However, the coefficient $a_{i_1, \dots, i_{k-1}}$ can be read off Ω as the residue at the intersection point $M_{i_1} \cap \cdots \cap M_{i_{k-1}}$ so all the coefficients should be zero. Proposition 3.3.5 is proved.

(3.3.8). Let $\mathbf{h}$ be the quotient of $\mathbb{C}^n$ by the subspace of $(a, \dots, a)$, $a \in \mathbb{C}$. Note that this subspace is canonically identified with the Lie algebra of the torus H and with the tangent space to P^{n-1} at the point $e = (1, \dots, 1)$. We shall denote, therefore, by $G(k-1, n-1)_e$ the Grassmannian of $(k-1)$-dimensional subspaces in $\mathbf{h}$.

(3.3.9). DEFINITION. By a special Veronese subvariety in $G(k-1, n-1)_e$ we mean a subvariety of the form $\varphi_E(P^{k-1})$, where:

a) $E = \Omega^1_{P^{k-1}}(\log M)$ is the logarithmic bundle corresponding to some configuration $M = (M_1, \dots, M_n)$ of hyperplanes in general position;

b) The space $H^0(E)$ is identified with $\{(a_1, \dots, a_n) \in \mathbb{C}^n : \sum a_i = 0\}$ as in Proposition 3.2.3, and its dual with $\mathbf{h}$.

Thus the notion of a special Veronese variety makes an explicit appeal to a choice of coordinate system.

(3.3.10). Note that by Proposition 3.2.3 special Veronese varieties are precisely the visible contours of Lie complexes in $G(k, n)$. In particular, around a generic point of $G(k-1, n-1)_e$ these varieties define a foliation with $(k-1)$-dimensional fibers, which is just the intersection of $G(k-1, n-1)_e$ with the foliation given by the orbits of H. Let us note also the following corollary.

(3.3.11). COROLLARY. *The set $G^0(k, n)/H = (P^{k-1})^n_{\text{gen}}/GL(k)$ of projective equivalence classes of configuration of n hyperplanes in P^{k-1} in general position is in one-to-one correspondence with the set of special Veronese varieties in $G(k-1, n-1)_e$. This correspondence takes a configuration $M = (M_1, \dots, M_n)$ into the subvariety $\varphi_E(P^{k-1})$, where $E = \Omega^1(\log M)$.*

(3.3.12). Clearly all special Veronese varieties in $G(k-1, n-1)_e$ represent the same homology class $\Delta \in H_{2k-2}(G(k-1, n-1), \mathbb{Z})$. A precise determination of Δ will be given in §3.9 below. By Corollary 3.3.11 we obtain an embedding of $G^0(k, n)/H$ into the Chow variety $\mathscr{C}_{k-1}(G(k-1, n-1), \Delta)$. Denote by V the closure of $G^0(k, n)/H$ in this Chow variety. So it is the variety of cycles in $\mathscr{C}_{k-1}(G(k-1, n-1), \Delta)$ that are limit positions of special Veronese varieties.

(3.3.13). Similarly, all special Veronese varieties in $G(k-1, n-1)_e$ form a flat family. Let $\mathscr{H}$ be the Hilbert scheme parametrizing all subschemes in $G(k-1, n-1)$, cf. (0.5.1). Define the variety W to be the closure of $G^0(k, n)/H$ in the Hilbert scheme $\mathscr{H}$. So it is the variety of subschemes in $\mathscr{C}_{k-1}(G(k-1, n-1), \Delta)$ which are limit positions of special Veronese varieties.

Our next result shows that all the information about the Chow quotient $G(k, n)//H$ is contained in visible contours.

(3.3.14). THEOREM. *The correspondence $Z \mapsto Z_e$ extends to an isomorphism of the variety $G(k, n)//H$ with V and W.*

PROOF. Since, by Proposition 3.1.9 b), every orbit closure $\overline{H \cdot L}$ intersects the variety $G(k-1, n-1)_e$ properly, we can conclude, by using the result of Barlet [4], that the map $Z \mapsto Z_e = Z \cap G(k-1, n-1)_e$ defines a regular morphism $\psi: G(k, n)//H \to V$. Proposition 3.1.9 implies that ψ is a set-theoretical bijection. To show that it is an isomorphism of algebraic varieties it suffices to apply once again the arguments involving normal vector fields on a generalized Lie complex Z similar to those used in the proof of Theorem 1.5.2. Similarly for the Hilbert scheme compactification.

(3.3.15). REMARK. A natural problem would be to study all Veronese subvarieties in Grassmannians. In general, not every such variety is projectively equivalent to a special one. This is because there are rank $(k-1)$ vector bundles on P^{k-1} that have the same Chern classes as $\Omega^1(\log M)$, but not having this form. A study of bundles $\Omega^1(\log M)$ from the point of view of stable vector bundles on projective spaces will be undertaken in a subsequent paper of I. Dolgachev and the author.

(3.4). Properties of special Veronese varieties.

(3.4.1). As has been recalled in Chapter 2, Lie complexes in $G(k, n)$ correspond to projective equivalence classes of configurations of n hyperplanes in P^{k-1} in general position. We have seen in the previous subsection that the space P^{k-1} can be recovered from the corresponding Lie complex Z as its visible contour Z_e. Let us recover the configuration too.

(3.4.2). For any points $x_1, \dots, x_m \in P^{n-1}$ let $\langle x_1, \dots, x_m\rangle$ denote their projective span. Define also $G_{\langle x_1, \dots, x_m\rangle}$ as the subvariety in $G(k, n)$ formed by the P^{k-1}'s containing $\langle x_1, \dots, x_m\rangle$. As an abstract variety, it is isomorphic to $G(k-p, n-p)$, where $p = \dim\langle x_1, \dots, x_m\rangle + 1$. Let $e_i \in P^{n-1}$ be the images of the standard basis vectors in $\mathbb{C}^n$.

(3.4.3). PROPOSITION. *Let $M = (M_1, \dots, M_n)$ be a configuration of n hyperplanes in P^{k-1} in general position, $E = \Omega^1_{P^{k-1}}(\log M)$, and $X = \varphi_E(P^{k-1}) \subset G(k-1, n-1)_e$ the corresponding special Veronese variety (i.e., the visible contour of the Lie complex corresponding to M). Then $\varphi_E(M_j) = X \cap G_{\langle e, e_j\rangle}$.*

PROOF. Let us give a coordinate description of φ_E which, unlike the description given in (3.1.11), is symmetric with respect to permutation of hyperplanes.

(3.4.4). Let $z_1, \dots, z_k$ be homogeneous coordinates in P^{k-1} and let $g_j(z) = \sum_i a_{ij} z_j$ be the linear equations of M_j, $j = 1, \dots, n$. Let $\mathbf{h}$ denote, as before, the quotient $\mathbb{C}^n/\{(a, \dots, a)\}$. The map $\varphi_E: P^{k-1} \to G(k-1, \mathbf{h}) = G(k-1, n-1)_e$ is defined as follows.

(3.4.5). Let $z = (z_1 : \cdots : z_k) \in P^{k-1}$ be generic. Consider the Jacobian $(k \times n)$-matrix $N(z) = \|\partial \log f_i/\partial z_j\|$. Due to the identity $\sum_j z_j \frac{\partial \log f_i}{\partial z_j} = 1$, $\forall i$, the k-dimensional subspace spanned by rows of this matrix contains the vector $(1, \dots, 1)$ and hence defines a $(k-1)$-dimensional subspace in $\mathbf{h}$

which is precisely $\varphi_E(z)$. For any subset $I \subset \{1, \dots, n\}$, $|I| = k$, denote by $p_I(N(z))$ the $(k \times k)$-minor of $N(z)$ on columns from I.

(3.4.6). We can (and will) assume, by renumbering the coordinates, that the number j in the formulation of Proposition 3.4.3 equals 1. The subspace generated by rows of $N(z)$ lies in the sub-Grassmannian $G_{\langle e, e_1\rangle}$ if and only if all minors $p_I(N(z))$, $1 \notin I$, vanish. To evaluate the limit of this subspace for $x \to M_i$, multiply $N(z)$ by $\mathrm{diag}(f_1(z), \dots, f_k(z))$ from the left. Then any minor p_I of $\mathrm{diag}(f_1(z), \dots, f_k(z)) \cdot N(z)$ with $1 \notin I$, will contain a row vanishing on M_i and hence will vanish on M_i itself. On the other hand, the minor $p_{1,2,\dots,k}$ of $\mathrm{diag}(f_1(z), \dots, f_k(z)) \cdot N(z)$ is constant. Since we can write instead of $(f_1, \dots, f_k)$ any $(f_1, f_{i_2}, \dots, f_{i_k})$, $1 < i_2 < \dots < i_k \leqslant n$, this proves that $\varphi_E(M_1) = \varphi_E(P^{k-1}) \cap G_{\langle e, e_1\rangle}$. Proposition 3.4.3 is proved.

(3.4.7). Corollary. *Any special Veronese variety contains* $\binom{n}{k-1}$ *points*

$$\langle e, e_{i_1}, \dots, e_{i_{k-1}}\rangle \in G(k-1, n-1)_e, \qquad 1 \leqslant i_1 < \dots < i_{k-1} \leqslant n.$$

(3.4.8). Proposition. *The intersection* $\varphi_E(P^{k-1}) \cap G_{\langle e, e_j\rangle}$ *is itself a special Veronese variety corresponding to the projective space* M_j *and the configuration of hyperplanes* $(M_i \cap M_j)$, $i \neq j$.

Proof. Straightforward. Left to the reader.

(3.4.9). Example. Consider the case $k = 2$, when $G(k, n)$ consists of lines in P^{n-1}. By §1.2, the variety of Lie complexes in $G(2, n)$ is the same as the quotient $((P^1)^n - \bigcup\{x_i = x_j\})/GL(2)$, i.e., the set of projective equivalence classes of n-tuples of distinct points on P^1. As we have seen in Example 3.2.8, the visible contour of any Lie complex in $G(2, n)$ is a Veronese curve in P_e^{n-2}, the variety of lines in P^{n-1} throught e. Corollary 3.4.7 means that every special Veronese curve in P_e^{n-2} contains n points $\langle e, e_i\rangle$ in general position.

It is a classical fact that for any points $p_1, \dots, p_n \in P^{n-2}$ in general position the set $V_0(p_1, \dots, p_n)$ of all Veronese curves through p_i is in bijection with $(P^1)^n_{\mathrm{gen}}/GL(2)$ (see [14, Chapter III, §2, Proposition 3]).

(3.4.10). Example (continued). As a transparent particular case, consider the case of four points $p_1, \dots, p_4$ in P^2. Veronese curves in P^2 are just smooth conics. Conics through $p_1, \dots, p_4$ form a 1-dimensional pencil $\mathscr{L} = P^1$. There are exactly three degenerate conics in this pencil, namely unions of lines

$$\langle p_1, p_2\rangle \cup \langle p_3, p_4\rangle, \quad \langle p_1, p_3\rangle \cup \langle p_2, p_4\rangle, \quad \langle p_1, p_4\rangle \cup \langle p_2, p_3\rangle.$$

The set of cross-ratios of $p_1, \dots, p_4$ regarded on conics from $\mathscr{L}$ is in bijection with the set of nondegenerate conics from $\mathscr{L}$, i.e., with P^1 minus three points.

(3.5). Steiner construction of Veronese varieties in the Grassmannian. Veronese curves in projective spaces possess a lot of remarkable properties, see [2, 14, 46]. Most of these properties do not generalize to higher-dimensional Veronese varieties. It is our opinion that the "right" class of ambient spaces for p-dimensional Veronese varieties is formed not by projective spaces but by Grassmannians of the form $G(p, V)$. In this section we show that (special) Veronese varieties admit a "synthetic" construction in the spirit of Steiner.

(3.5.1). Consider some projective space P^{m-1}. Let $L \subset P^{m-1}$ be a projective subspace of codimension d. Denote by $]L[$ the space of all hyperplanes in P^{m-1} containing L. We shall call it the *star* of L. This is a projective space of dimension $d-1$.

(3.5.2). We may like to have a *parametrization* of the star $]L[$, i.e., an identification $f: P^{d-1} \to]L[$ of $]L[$ with the standard P^{d-1}. Such an identification is the same as a linear operator $E: \mathbb{C}^m \to \mathbb{C}^d$ whose kernel is $\mathbf{L}$, the linear subspace corresponding to L. Indeed, given such an E, we obtain a bijection $\Pi \mapsto E^{-1}(\Pi)$ between hyperplanes in $\mathbb{C}^d$ and hyperplanes in $\mathbb{C}^m$ containing $\mathbf{L}$, i.e., hyperplanes from $]L[$.

In coordinate notation, we write E as a row of linear functions $g_i: \mathbb{C}^m \to \mathbb{C}$ and to any $(\lambda_1 : \cdots : \lambda_d) \in P^{d-1}$ associate the hyperplane $\mathrm{Ker}(\sum \lambda_i g_i) \in]L[$.

(3.5.3). It will be convenient for us to view a parametrization f above as a linear form $\sum \lambda_i g_i$ on $\mathbb{C}^m$ whose entries are linear forms in $\lambda_1, \ldots, \lambda_k$. This is tantamount to viewing a linear operator $E: \mathbb{C}^m \to \mathbb{C}^d$ as an element of $\mathbb{C}^d \otimes (\mathbb{C}^m)^*$.

(3.5.4). Recall Steiner's construction of Veronese curves in P^m [24]. Take m projective subspaces of codimension 2, $L_1, \ldots, L_m \subset P^m$. The star $]L_i[$ of each L_i is just a pencil of hyperplanes, i.e., it is isomorphic to the projective line P^1. Let us identify these pencils with each other, e.g., by choosing projective equivalences $f_i: P^1 \to]L_i[$. Consider the curve in P^m that is the image of P^1 under the map

$$t \mapsto (f_1(t) \cap \cdots \cap f_m(t)).$$

This is a Veronese curve. It depends on the choice of subspaces L_i and of identifications f_i.

In classical terminology, one would say that a Veronese curve can be obtained as the locus of intersections of corresponding hyperplanes from m pencils in correspondence.

(3.5.5). Construction (The Grassmannian Steiner construction). *Take $n-k$ projective subspaces in P^{n-2}, say $L_1, \ldots, L_{n-k}$, of codimension k. Put the stars $]L_i[$ into 1–1 correspondence with each other, e.g., by choosing projective isomorphisms $f_i: P^{k-1} \to]L_i[$. Then consider the subvariety in*

$G(k-1, n-1)$ given by the parametrization

$$t \mapsto (f_1(t) \cap \cdots \cap f_{n-k}(t)), \qquad t \in P^{k-1}.$$

This is a direct generalization of the construction in (3.5.4). Using the fact (3.5.3) that parametrized stars are the same as linear forms with coefficients linearly depending on parameters, we get the following reformulation of the construction.

(3.5.6). REFORMULATION. *Let $A: \mathbb{C}^k \to \mathrm{Hom}(\mathbb{C}^{n-1}, \mathbb{C}^{n-k})$ be a linear operator such that for any nonzero $z \in \mathbb{C}^k$ the operator $A(z): \mathbb{C}^{n-1} \to \mathbb{C}^{n-k}$ is surjective. The Grassmannian Steiner construction is the subvariety in $G(k-1, n-1)$ consisting of points* $\mathrm{Ker}\, A(z)$, *$z \in \mathbb{C}^k - \{0\}$.*

(3.5.7). THEOREM. *Any special $(k-1)$-dimensional Veronese variety in $G(k-1, n-1)$ can be obtained by the Grassmannian Steiner construction.*

PROOF. Let X be a special Veronese variety coming from a configuration $(M_1, \ldots, M_n)$ of hyperplanes in P^{k-1} in general position. Similarly to (3.1.11), we can assume that M_n is the infinite hyperplane and choose affine coordinates $z_1, \ldots, z_{k-1}$ in $\mathbb{C}^{k-1} = P^{k-1} - M_n$ such that M_i is given by the equation $z_i = 0$ for $i = 1, \ldots, k-1$.

(3.5.8). Consider the coordinate space $\mathbb{C}^{n-1}$ with coordinates $y_1, \ldots, y_n$ and basis vectors $e_1, \ldots, e_{n-1}$. Decompose it into the direct sum $\mathbb{C}^{k-1} \oplus \mathbb{C}^{n-k}$, where $\mathbb{C}^{k-1}$ is spanned by $e_1, \ldots, e_{k-1}$ and $\mathbb{C}^{n-k}$ by $e_k, \ldots, e_{n-1}$.

(3.5.9). By definition, the variety $X \subset G(k-1, n-1)$ has the rational parametrization $z \mapsto \gamma(z)$, where $z \subset P^{k-1}$ and γ is the logarithmic Gauss map. Explicit formula (3.1.13) gives that for generic $z \in \mathbb{C}^{k-1}$ the subspace $\gamma(z)$ is the graph of the linear operator $\mathbb{C}^{k-1} \to \mathbb{C}^{n-k}$ given by the matrix

$$B(z) = \begin{pmatrix} \frac{a_{k1}z_1}{f_k(z)} & \frac{a_{k2}z_2}{f_k(z)} & \cdots & \frac{a_{k,k-1}z_{k-1}}{f_k(z)} \\ \frac{a_{k+1,1}z_1}{f_k(z)} & \frac{a_{k+1,2}z_2}{f_k(z)} & \cdots & \frac{a_{k+1,k-1}z_{k-1}}{f_k(z)} \\ \cdots & \cdots & \cdots & \cdots \\ \frac{a_{n-1,1}z_1}{f_{n-1}(z)} & \frac{a_{n-1,2}z_2}{f_{n-1}(z)} & \cdots & \frac{a_{n-1,k-1}z_{k-1}}{f_{n-1}(z)} \end{pmatrix} \tag{3.5.10}$$

where $f_j(z) = \sum_{\nu=1}^{k-1} a_{j\nu} z_\nu + a_{jk}$ is the equation of the hyperplane M_j, $j = k, \ldots, n-1$. In other words, the subspace $\gamma(z)$ is spanned by the $(k-1)$ vectors $e_i + \sum_{j=k}^{n-1} \frac{a_{ji}z_i}{f_j(z)} e_j$. It is immediate to see that $\gamma(z)$ is the intersection of $(n-k)$ hyperplanes given, in the standard coordinates $y_1, \ldots, y_{n-1}$, by linear equations

$$f_j(z)y_j - (a_{j1}z_1)y_1 - \cdots - (a_{j,k-1}z_{k-1})y_{k-1} = 0, \quad j = 1, \ldots, n-k. \tag{3.5.11}$$

The linear functions in (3.5.11), considered together, define a linear operator $a(z): \mathbb{C}^{n-1} \to \mathbb{C}^{n-k}$ whose matrix elements are affine functions of $z_1, \ldots, z_{k-1}$.

Let us complete the affine coordinates $z_1, \dots, z_{k-1}$ in $\mathbb{C}^{k-1}$ to homogeneous coordinates $z_1, \dots, z_k$ in P^{k-1} so that the vanishing of z_k defines the infinite hyperplane. Then affine-linear functions $f_j(z_1, \dots, z_{k-1})$ will become linear functions $F_j(z) = F_j(z_1, \dots, z_k) = \sum_{\nu=1}^{k} a_{j\nu} z_\nu$. The $(n-k)$ linear functions in (3.5.11) give rise to a family of linear operators $A(z_1, \dots, z_k)\colon \mathbb{C}^{n-1} \to \mathbb{C}^{n-k}$ given by the matrix

$$\begin{pmatrix} -a_{k1}z_1 & \dots & -a_{k,k-1}z_{k-1} & F_k(z) & 0 & \dots & 0 \\ -a_{k+1,1}z_1 & \dots & -a_{k+1,k-1}z_{k-1} & 0 & F_{k+1}(z) & \dots & 0 \\ \dots & \dots & \dots & \dots & \dots & \dots & \dots \\ -a_{n-1,1}z_1 & \dots & -a_{n-1,k-1}z_{k-1} & 0 & 0 & \dots & F_{n-1}(z) \end{pmatrix} \tag{3.5.12}$$

whose entries are linear functions in $z_1, \dots, z_k$. Theorem 3.5.7 is proved.

(3.5.13). REMARK. It is immediate to extract from the formula (3.5.11) the $(n-k)$ subspaces $L_i \subset \mathbb{C}^{n-1}$ whose stars $]L_i[$ are identified (in the synthetic version (3.5.5) of the Grassmannian Steiner construction). Namely, L_i, $i = k, k+1, \dots, n-1$, is the span of the vectors $e_1, \dots, e_{k-1}, e_i$.

We observe that the position of these subspaces is rather special. The identifications of the stars are also very special. The extremely interesting question of possibility of Steiner construction of more general Veronese varieties in Grassmannians will be treated elsewhere.

(3.5.14). Varieties in *projective spaces* defined by various generalizations of Steiner's construction (3.5.4) were studied in detail in the book [46] by T. G. Room. To obtain such a generalization, one takes r subspaces $L_1, \dots, L_r \subset P^{m-1}$ of codimension d, identifies all the stars $]L_i[$ with each other, and considers the codimension r subspaces in P^{m-1} that are the intersections of corresponding hyperplanes from these stars. If $d < r$, then the union of these subspaces is a proper subvariety in P^{m-1}, which will be called a *projectively generated variety* [46]. The fundamental remark of Room is that any projectively generated variety in P^{m-1} can be given by a system of equations where each equation has the form of the determinant of a matrix consisting of linear forms on P^{m-1}.

We shall use this idea in the next section to get a better hold of Veronese varieties in Grassmannians.

(3.6). The sweep of a special Veronese variety.

(3.6.1). Let X be any subvariety in the Grassmannian $G(k-1, n-1)$, i.e., a family of $(k-2)$-dimensional projective subspaces in P^{n-2}. The *sweep* of X is, by definition, the subvariety $\mathrm{Sw}(X) \subset P^{n-2}$ defined as the union of the subspaces from X.

(3.6.2). We shall be mostly interested in the case when $X \subset G(k-1, n-1)$ is a $((k-1)$-dimensional) special Veronese variety, see (3.3.9). In other words (3.3.10), X is the visible contour of a Lie complex $Z = \overline{H \cdot L}$, i.e., the locus of those subspaces P^{k-1} in P^{n-1} that belong to the complex Z and contain the chosen point $e = (1, \ldots, 1)$. The sweep of X is what was called in (3.1.2) the visible sweep of the complex Z at e. So it is the projectivization of the cone in P^{n-1} with vertex e given by the union of all P^{k-1}'s from the complex Z which contain e.

(3.6.3). Let $\mathbf{h} = \mathbb{C}^n/\{(a, \ldots, a)\}$ be the Lie algebra of the maximal torus $H \subset PGL(n)$. Recall (3.1.8) that the Grassmannian $G(k-1, n-1)$ in which the visible contours (and hence special Veronese varieties) lie, is in fact $G(k-1, \mathbf{h})$. Therefore the sweep of any special Veronese variety lies naturally in the projective space $P(\mathbf{h})$.

(3.6.4). Let $t_1, \ldots, t_n$ be standard coordinate functions on $\mathbb{C}^n$. A linear form $\sum c_i t_i$ descends to a linear form on $\mathbf{h}$ if $\sum c_i = 0$. In particular, the *roots,* i.e., the linear forms $t_i - t_j$, are forms on $\mathbf{h}$.

(3.6.5). The possibility of defining X by the Grassmannian Steiner construction (3.5.5) implies that $\mathrm{Sw}(X)$ is always a projectively generated variety in the sense of (3.5.14).

(3.6.6). THEOREM. *Suppose that* $k \leqslant n-k$. *Let* $z_1, \ldots, z_k$ *be homogeneous coordinates in* P^{k-1}. *Suppose that a configuration* $M = (M_1, \ldots, M_n)$ *of hyperplanes in* P^{k-1} *consists of* k *coordinate hyperplanes* $M_i = \{z_i = 0\}$, $i = 1, \ldots, k$, *and* $(n-k)$ *other hyperplanes* $M_j = \{\sum_{i=1}^k a_{ji} z_j = 0\}$. *Then the sweep of the Veronese variety in* $P^{n-2} = P(\mathbf{h})$ *corresponding to* M *is given by vanishing of all* $(k \times k)$*-minors of the following* $(k \times (n-k))$*-matrix of linear forms on* $\mathbf{h}$:

$$A^\dagger(t_1, \ldots, t_n) = \|a_{ji}(t_j - t_i)\|, \qquad i = 1, \ldots, k, \quad j = k+1, \ldots, n. \tag{3.6.7}$$

PROOF. Let $A: \mathbb{C}^k \to \mathrm{Hom}(\mathbb{C}^{n-1}, \mathbb{C}^{n-k})$ be the linear system of linear operators such that X consists of kernels of $A(z)$, $z \in \mathbb{C}^k - \{0\}$. An explicit formula for A is given in (3.5.12). Using partial dualization, let us associate to A a linear operator

$$A^\dagger : \mathbb{C}^{n-1} \to \mathrm{Hom}(\mathbb{C}^k, \mathbb{C}^{n-k}).$$

A point $t \in \mathbb{C}^{n-1}$ lies in the kernel of $A(z)$ for some nonzero $z \in \mathbb{C}^k$ if and only if the linear operator $A^\dagger(t): \mathbb{C}^k \to \mathbb{C}^{n-k}$ has nontrivial kernel, i.e., the rank of $A^\dagger(t)$ is less than k. Thus the sweep $\mathrm{Sw}(X)$ is defined by the condition that all $k \times k$ minors of the matrix $A^\dagger(t)$ of linear forms on P^{n-2} vanish.

To see that $A^\dagger$ has the claimed form, we use the formula (3.5.12). This formula was written with respect to the nonsymmetric system of coordinates $y_1, \ldots, y_{n-1}$ in $\mathbf{h}$. In the language of (3.6.4) we have $y_i = t_i - t_n$. Substituting this into (3.5.12) and transposing the matrix, we arrive at the formula (3.6.7). Theorem 3.6.6 is proved.

(3.6.8). Let us describe a more geometric construction for the sweep of the Veronese variety corresponding to a projective configuration.

Let $\mathrm{Mat}(k, n-k)$ be the vector space of all k by $(n-k)$ matrices and $P(\mathrm{Mat}(k, n-k))$ be the projectivization of this space. The projectivization of the set of matrices of rank 1 is just the Segre embedding $P^{k-1} \times P^{n-k-1} \subset P(\mathrm{Mat}(k, n-k))$. Let $\nabla \subset P(\mathrm{Mat}(k, n-k))$ be the projectivization of the space of matrices of rank $< k$. This is an algebraic subvariety of codimension $n-2k+1$.

(3.6.9). Let now $(x_1, \dots, x_n)$ be a configuration of points in P^{k-1} in general position. (Recall that modulo projective isomorphism, configurations of points give the same orbit space as configurations of hyperplanes.) Let $(y_1, \dots, y_n)$ be the configuration of points in P^{n-k-1} associated to $x_1, \dots, x_n$ (see §2.3 about association). By Reformulation (2.3.8), the points $z_i = (x_i, y_i) \subset P^{k-1} \times P^{n-k-1} \subset P(\mathrm{Mat}(k, n))$ form a circuit, i.e., span a projective space, say L, whose dimension is $n-2$ and are in general position as points of L. The space $P_e^{n-2} = P(\mathbf{h})$ also comes with a circuit given by points $\overline{e}_i$, projectivizations of images of the basis vectors $e_i \in \mathbb{C}^n$ in $\mathbf{h} = \mathbb{C}^n/\mathbb{C}$. Hence there is a unique projective transformation $\varphi: L \to P(\mathbf{h})$ taking z_i to $\overline{e}_i$. We shall be interested in the intersection $\nabla \cap L \subset L$, where ∇ is the determinantal variety in (3.6.8).

(3.6.10). PROPOSITION. *The map φ identifies the subvariety $\nabla \cap L \subset L$ with the sweep $\mathrm{Sw}(X(x_1, \dots, x_n)) \subset P(\mathbf{h})$ of the Veronese variety corresponding to the configuration $(x_1, \dots, x_n)$.*

PROOF. Denote the sweep $\mathrm{Sw}(X(x_1, \dots, x_n))$ by S. Let $M_1, \dots, M_n$ be the hyperplanes in the dual space P^{k-1} corresponding to x_i. After choosing suitable homogeneous coordinates we can apply Theorem (3.6.6) which gives a representation of S as the inverse image of ∇ under the linear embedding $A^\dagger: \mathbf{h} \to \mathrm{Mat}(k, n-k)$. We regard $A^\dagger$ as a map $\mathbb{C}^n \to \mathrm{Mat}(k, n-k)$ using the isomorphism $\mathbf{h} = \mathbb{C}^n/\mathbb{C}$. Let $e_i \in \mathbf{C}^n$ be the standard basis vectors. Proposition (3.6.10) is a consequence of the following statements:

(3.6.11). The matrices $A^\dagger(e_i)$ lie in the Segre embedding

$$P^{k-1} \times P^{n-k-1} \subset P(\mathrm{Mat}(k, n-k)), \tag{3.6.12}$$

i.e., $\mathrm{rank}\, A^\dagger(a_i) = 1$.

(3.6.13). The configuration of hyperplanes $\mathrm{Ker}(A^\dagger(e_i)) \subset P^{k-1}$ is projectively isomorphic to the configuration $(M_1, \dots, M_n)$ and the configuration of points $\mathrm{Im}(A^\dagger(e_i)) \subset P^{n-k-1}$ is associated to $(M_1, \dots, M_n)$.

Both these statements are immediate from the explicit form (3.6.7) of the matrix $A^\dagger$.

(3.6.14). COROLLARY. *Let $n = 2k$. Then Veronese varieties corresponding to a configuration $(M_1, \dots, M_{2k}) \subset P^{k-1}$ and to the associated configuration have the same sweep.*

(3.6.15). Any determinantal variety, i.e., variety defined by vanishing of minors of a matrix of linear forms, bears two canonical families of projective subspaces, so-called α- and β-families [46]. Let us recall their construction and explain their relevance to our situation.

Let $k \leqslant n-k$ and $\nabla \subset P(\mathrm{Mat}(k, n-k))$ denote, as before, the projectivization of the space of matrices of rank $< k$. For any 1-dimensional subspace $\lambda \subset \mathbb{C}^k$ set

$$\Pi_\alpha(\lambda) = P\Big(\{M\colon \mathbb{C}^k \to \mathbb{C}^{n-k}\colon M(\lambda) = 0\}\Big) \subset \nabla. \tag{3.6.16}$$

This is a projective subspace in ∇ of codimension $k-1$. Thus we get a family of projective subspaces in ∇ (called α-subspaces) of codimension $k-1$, parametrized by $P^{k-1} = P(\mathbb{C}^k)$.

Similarly, for any hyperplane $\Lambda \subset \mathbb{C}^{n-k}$ set

$$\Pi_\beta(\Lambda) = P\Big(\{M : \mathbb{C}^k \to \mathbb{C}^{n-k} : \mathrm{Im}(M) \subset \Lambda\}\Big) \subset \nabla. \tag{3.6.17}$$

This is a projective subspace in ∇ of codimension $n-k-1$. We get a family of projective subspaces in ∇ (called β-subspaces) of codimension $n-k-1$ parametrized by $P^{n-k-1} = P((\mathbb{C}^{n-k})^*)$.

(3.6.18). Let $L \subset P(\mathrm{Mat}(k, n-k))$ be a projective subspace of dimension $n-2$. Consider the variety $S = L \cap \nabla$. It contains projective subspaces $\Pi_\alpha(\lambda) \cap L$ whose dimension is at least $n-k-1$ (they will be called the α-subspaces in S) and subspaces $\Pi_\beta(\Lambda) \cap L$ whose dimension is at least $k-1$ (they will be called β-subspaces in S). The role of these subspaces in our situation is as follows.

(3.6.19). Proposition. *Let $(x_1, \dots, x_n)$ be a configuration of points in P^{k-1} in general position and $(y_1, \dots, y_n)$ the associated configuration in P^{n-k-1}. Let $X \subset G(k-1, n-1)$ and $X' \subset G(n-k, n-1)$ be the Veronese varieties corresponding to (x_i) and (y_i) (their dimensions equal, respectively, $k-1$ and $n-k-1$). Let $S, S' \subset P_e^{n-2} = P(\mathbf{h})$ be the sweeps of these varieties. Then $S = S'$ and the subspaces from X (resp. from X') lying on $S = S'$ are precisely the β- (resp. α-) subspaces on S defined in* (3.6.18).

Proof. This is a reformulation of Theorem 3.5.7 about the Steiner construction of X.

(3.7). An example: Visible contours and sweeps of Lie complexes in $G(3, 6)$. In this section we study in detail the construction of §3.6 in the particular case corresponding to the configuration of six points on P^2. In other words, we consider the case $k = 3$, $n = 6$.

(3.7.1). To any sextuple $(x_1, \dots, x_6) \in (P^2)^6_{\mathrm{gen}}$ we have associated the Veronese surface $X(x_1, \dots, x_6) \subset G(2, 5)$. Its sweep, denoted $S(x_1, \dots, x_6) \subset P^4$, is a cubic hypersurface since by Theorem 3.6.6 it is given by vanishing of the determinant of a (3×3)-matrix of linear forms. We shall study such hypersurfaces.

(3.7.2). We are interested in configurations modulo projective isomorphism. So we can consider equally well the sextuple of lines $M_i \subset \check{P}^2$ dual to x_i. This sextuple represents the same element of $(P^2)^6/GL(3)$.

(3.7.3). We can always assume that lines M_i have the particular form considered in Theorem 3.6.6: for $i = 1, 2, 3$ the line M_i is given by the equation $z_i = 0$ and M_j for $j = 4, 5, 6$ is given by the equation $\sum a_{ij} z_j = 0$. The 3×3 matrix $\|a_{ij}\|$ is defined by a projective isomorphism class of $(M_1, \dots, M_6)$ not uniquely but only up to multiplication of rows and columns by nonzero scalars. Generic position of lines L_ν implies that all $a_{ij} \neq 0$. Hence by multiplication of rows and columns by scalars we can take $\|a_{ij}\|$ into a unique matrix of the form

$$\begin{pmatrix} 1 & 1 & 1 \\ 1 & a & b \\ 1 & c & d \end{pmatrix}. \tag{3.7.4}$$

In this way the quotient $(P^2)^6_{\text{gen}}/GL(3)$ becomes identified with the space of (a, b, c, d) such that all the minors of the matrix (3.7.4) are nonzero.

(3.7.5). PROPOSITION. *The points $x_1, \dots, x_6$ lie on a conic (or, equivalently, the lines $M_1, \dots, M_6$ are tangent to a conic) if and only if the matrix elements a, b, c, d of (3.7.4) satisfy the equation $\Psi(a, b, c, d) = 0$, where*

$$\begin{aligned} \Psi(a, b, c, d) &= \det \begin{pmatrix} a(1-c) & b(1-d) \\ c(1-a) & d(1-b) \end{pmatrix} \\ &= ad - bc + abc + bcd - acd - abd. \end{aligned} \tag{3.7.6}$$

PROOF. This is Proposition 2.13.1 of [41].

(3.7.7). The Veronese variety $X = X(x_1, \dots, x_6)$ lies in $G(2, 5)$, the space of lines in $P^4 = P(\mathbf{h})$, $\mathbf{h} = \mathbb{C}^6/\mathbb{C}$. Let $p_i \in P^4$, $i = 1, \dots, 6$, be the point corresponding to the standard basis vector $e_i \in \mathbb{C}^6$ (in the realization of P^4 as the space of lines in P^5 through $e = (1, \dots, 1)$, the point p_i corresponds to the line $\langle e, e_i \rangle$). By Corollary 3.4.7, the variety X contains all the lines $\langle p_i, p_j \rangle$. This implies that points p_i are singular points of the sweep $S = S(x_1, \dots, x_6)$ of X. Indeed, the tangent directions at any p_i to lines $\langle p_i, p_i \rangle$, $j \neq i$, span the whole tangent space $T_{p_i} P^4$ which will therefore coincide with the Zariski tangent space of S at p_i.

More detailed information about the singularities of S is given in the next proposition which is the main result of this section.

(3.7.8). PROPOSITION. a) *If $x_1, \dots, x_6 \in P^2$ are in general position and do not lie on a conic then the sweep $S(x_1, \dots, x_6)$ has only six singular points, namely p_i, and these points are simple quadratic singularities.*

b) *If* $x_1, \dots, x_6$ *are in general position and lie on a conic* $K \subset P^2$, *then* $S(x_1, \dots, x_6)$ *has a curve* C *of singular points, which is a Veronese curve in* P^4 *containing* $p_1, \dots, p_6$. *In this case the configuration of* x_i *on* $K \cong P^1$ *is isomorphic to that of* p_i *on* C. *The variety* $S(x_1, \dots, x_6)$ *is the union of all chords of* C.

PROOF. Consider the varieties

$$P^2 \times P^2 \subset \nabla \subset P^8 = P(\mathrm{Mat}(3, 3)), \tag{3.7.9}$$

where ∇ consists of degenerate matrices and $P^2 \times P^2$ of matrices of rank 1. It is well known that $P^2 \times P^2 = \mathrm{Sing}(\nabla)$.

Assume that the configuration $(M_1, \dots, M_6)$ of lines dual to x_i has the form specified in Theorem 3.6.6 with the 3×3 matrix $\|a_{ij}\|$ given in the normal form (3.7.4). Let $A^\dagger \colon P^4 = P(\mathbf{h}) \to P(\mathrm{Mat}(3, 3))$ be the embedding given by formula (3.6.7). In other words (taking into account the normal form of $\|a_{ij}\|$) we have

$$A^\dagger(t_1, \dots, t_6) = \begin{pmatrix} t_1 - t_4 & t_1 - t_5 & t_1 - t_6 \\ t_2 - t_4 & a(t_2 - t_5) & b(t_2 - t_6) \\ t_3 - t_4 & c(t_3 - t_5) & d(t_3 - t_6) \end{pmatrix}. \tag{3.7.10}$$

Theorem 3.6.6 implies that our sweep S equals $(A^\dagger)^{-1}(\nabla)$.

(3.7.11). Let $L \subset P(\mathrm{Mat}(3, 3))$ be the image of $A^\dagger$. It is immediate to check that the degree of the Segre variety $P^2 \times P^2 \subset P^8$ equals 6. On the other hand, the subspace $L \subset P^4$ already intersects $P^2 \times P^2$ in six points $q_i = A^\dagger(e_i)$, $i = 1, \dots, 6$. Hence there remains one of two possibilities:

Case 1. L intersects $P^2 \times P^2$ transversally at six points $q_i = A^\dagger(e_i)$.

Case 2. The intersection $L \cap (P^2 \times P^2)$ contains a component of positive dimension.

Part a) of Proposition 3.7.8 will follow from the next two lemmas.

(3.7.12). LEMMA. *If* $x_1, \dots, x_6$ *are in general position and do not lie on a conic then for* $L = A^\dagger(P(\mathbf{h}))$ *the Case* 1 *holds.*

(3.7.13). LEMMA. *If for* L *the Case* 1 *holds then* $L \cap \nabla$ *has* q_i *as the only singular points and the singularities at* q_i *are simple quadratic.*

(3.7.14). PROOF OF LEMMA 3.7.13. The only possibility which we have to exclude is that there is a point $q \in L \cap \nabla$ that is smooth in ∇ (i.e., $\mathrm{rank}(q) = 2$) and such that $T_q L \subset T_q \nabla$. To rule out this possibility, let $\lambda \in P^2$ be the point corresponding to $\mathrm{Ker}(q)$ and let $\Pi = \Pi_\alpha(\lambda)$ be the corresponding α-subspace, i.e., the projectivization of the space of all 3×3 matrices annihilating λ. Then $\dim(\Pi) = 5$. Since L is connected in the embedded tangent space to ∇ at q (which is 7-dimensional), we have $\dim(L \cap \Pi) \geqslant 2$. However, by Proposition 3.6.19, the intersection of L with all α-subspaces should be 1-dimensional.

(3.7.15). PROOF OF LEMMA 3.7.12. It suffices to show that each q_i is an isolated singular point of the intersection $L \cap (P^2 \times P^2)$. To do this, we prove that the tangent spaces $T_{q_i}L$, $T_{q_i}(P^2 \times P^2) \subset T_{q_i}P^8$ intersect only at 0. Since the roles of q_i are symmetric, it is enough to consider $i = 1$. Using the normal form (3.7.4) of $A^\dagger$, one sees that the point $q_1 = A^\dagger(e_1)$ is given by the matrix

$$\begin{pmatrix} 1 & 1 & 1 \\ 0 & 0 & 0 \\ 0 & 0 & 0 \end{pmatrix}.$$

The tangent space at q_1 to the locus of rank 1 matrices is easily seen to consist of matrices of the form

$$\begin{pmatrix} \lambda_1 & \lambda_2 & \lambda_3 \\ \lambda_4 & \lambda_4 & \lambda_4 \\ \lambda_5 & \lambda_5 & \lambda_5 \end{pmatrix}. \tag{3.7.16}$$

Therefore the intersection $T_{q_1}L \cap T_{q_1}(P^2 \times P^2)$ is obtained from the space of solutions $(t_1, \dots, t_6)$ of the linear system

$$t_2 - t_4 = a(t_2 - t_5) = b(t_2 - t_6), \quad t_3 - t_4 = c(t_3 - t_5) = d(t_3 - t_6) \tag{3.7.17}$$

by factorization over the 2-dimensional subspace spanned by the vectors $e_1 = (1, 0, \dots, 0)$ and $(1, 1, \dots, 1)$. Hence the proof is reduced to the following statement:

(3.7.18). LEMMA. *If x_i do not lie on a conic (i.e., if the polynomial $\Psi(a, b, c, d)$ given by* (3.7.6) *does not vanish) then the linear system* (3.7.17) *has a 2-dimensional space of solutions.*

PROOF. This is a system of four equations on six variables whose matrix of coefficients has the form

$$\begin{pmatrix} 0 & 1-a & 0 & -1 & a & 0 \\ 0 & 1-b & 0 & -1 & 0 & b \\ 0 & 0 & 1-c & -1 & c & 0 \\ 0 & 0 & 1-d & -1 & 0 & d \end{pmatrix}.$$

Let us delete the first column, then move the column with (-1)'s to the left and then subtract the first row from all the other rows. We obtain the matrix

$$\begin{pmatrix} -1 & 1-a & 0 & a & 0 \\ 0 & a-b & 0 & -a & b \\ 0 & -1+a & 1-c & c-a & 0 \\ 0 & -1+a & 1-d & -a & d \end{pmatrix}.$$

It is immediate to see that all 4×4 minors of this matrix are of the form $\pm\Psi(a, b, c, d)$.

We have proved part a) of Proposition 3.7.8.

(3.7.19). Let us prove part b) of Proposition 3.7.8. So assume that $x_1, \dots, x_6 \in P^2$ are points in general position lying on a conic K. Consider the 2-fold Veronese embedding

$$v_2: P^2 \hookrightarrow P^5 = P(S^2(\mathbb{C}^3)) \subset P(\mathrm{Mat}(3, 3)), \tag{3.7.20}$$

where $P(S^2(\mathbb{C}^3))$ is embedded into $P(\mathrm{Mat}(3, 3))$ as the space of symmetric matrices. Let $L \subset P(S^2(\mathbb{C}^3))$ be the projective envelope of $v_2(x_i)$. Let also $\nabla_{\mathrm{sym}} \subset P(S^2(\mathbb{C}^3))$ be the space of degenerate quadratic forms, i.e., $\nabla_{\mathrm{sym}} = \nabla \cap P(S^2(\mathbb{C}^3))$. Note that $P^2 = \mathrm{Sing}(\nabla_{\mathrm{sym}})$ has codimension 2 in ∇_{sym}.

Since x_i lie on a conic, their configuration is self-associated (see Example 2.3.12). Now the interpretation of $S(x_1, \dots, x_6)$ given in (3.6.9), (3.6.10) implies that $S(x_1, \dots, x_6) = L \cap \nabla_{\mathrm{sym}}$.

The conic K is equal to $v_2^{-1}(L \cap v_2(P^2))$. Since x_i are in general position, K is smooth. Now since $P^2 = \mathrm{Sing}(\nabla_{\mathrm{sym}})$, we find that $C = v_2(K) = \mathrm{Sing}(L \cap \nabla_{\mathrm{sym}})$ is the singular curve of $L \cap \nabla_{\mathrm{sym}} = S(x_1, \dots, x_6)$. This is clearly a Veronese curve in $L = P^4$. The Veronese embedding v_2 identifies k with C and points $x_i \in K$ with our distinguished points $p_i \in C$. Finally, ∇_{sym} is the union of chords of $v_2(P^2)$ (every degenerate quadratic form in three variables is a sum of two quadratic forms of rank 1). Hence $S(x_1, \dots, x_6)$ contains the union of chords of C. Since the latter is also a 3-dimensional variety, the two varieties in question coincide.

Proposition 3.7.8 is completely proved.

(3.7.21). Fix six points $p_1, \dots, p_6 \in P^4$ in general position. Any family of points p_i can be transformed into any other family by a unique projective isomorphism. Let $\mathscr{L} \subset P(S^3(\mathbb{C}^5))$ be the linear system of all cubic hypersurfaces in P^4 that contain p_i as singular points. It is clear by dimension count that $\mathscr{L}$ has dimension 4. On the other hand, taking p_i to be the standard points (images of e_i in $P(\mathbf{h})$) we have constructed a 4-dimensional family of sweeps $S(x_1, \dots, x_6)$ that all belong to $\mathscr{L}$. Hence we have the following corollary.

(3.7.22) Corollary. *Let $p_1, \dots, p_6 \in P^4$ be points in general position. A generic cubic hypersurface $S \subset P^4$ for which p_i are singular points is projectively equivalent to the visible sweep of some Lie complex in $G(3, 6)$. In particular, S can be realized as $P^4 \cap \nabla$ for a suitable embedding $P^4 \subset P(\mathrm{Mat}(3, 3))$. The variety S contains two families of lines (α- and β-families, in the determinantal realization), whose parameter spaces P, P' are isomorphic to P^2. These families give rise to two Veronese surfaces in $G(2, 5)$, which correspond to a pair of associated configurations $(M_1, \dots, M_6) \subset P$, $(M_1', \dots, M_6') \subset P'$ of lines. Explicitly, $M_i \subset P$ (resp. $M_i' \subset P'$) is the locus of all lines from the first (resp. second) family that contain the point p_i.*

(3.7.23). Remarks. a) Any generic intersection $P^4 \cap \nabla \subset P(\mathrm{Mat}(3, 3))$ intersects the Segre variety $P^2 \times P^2 = \mathrm{Sing}(\nabla)$ in six points (since $6 = \deg(P^2 \times P^2)$) and hence is a cubic hypersurface of the form studied in the above corollary.

b) The correspondence

$$(x_1, \dots, x_6) \bmod PGL(3) \longmapsto S(x_1, \dots, x_6)$$

is two-to-one. Therefore, this correspondence defines a double covering of fourfolds $\pi: (P^2)^6_{\mathrm{gen}}/GL(3) \to \mathscr{L}$. This covering is well known classically, see [14, 41]. It extends to a map of the Mumford quotient $((P^2)^6/GL(3))_{\mathrm{Mumf}} \to \mathscr{L}$, which is a double cover ramified along a hypersurface $W \subset \mathscr{L}$ of degree 2. This hypersurface is called the modular variety of level 2 (see [14]). The projective dual $\check{W} \subset \check{\mathscr{L}}$ is a so-called Segre cubic threefold, i.e., a cubic hypersurface with ten ordinary singular points. (It is known that all such threefolds are projectively isomorphic.)

(3.8). Chordal varieties of Veronese curves.

(3.8.1). Let $C \subset P^r$ be a Veronese curve. An $(s-1)$-dimensional projective subspace $L \subset P^r$ is called *chordal* to C if it intersects C in s points (counted with multiplicities). Denote by $Ch_{s-1}(C)$ the variety of all chordal $(s-1)$-dimensional subspaces of C. Clearly $Ch_{s-1}(C)$ is isomorphic to the s-fold symmetric power of $C \cong P^1$, i.e., to the projective space P^s. We obtain therefore a special class of embeddings $P^s \subset G(s, r+1)$. It turns out that these embeddings give particular cases of Veronese varieties in Grassmannians considered in §3.3 above.

(3.8.2). Proposition. *The chordal variety of any Veronese curve is a Veronese variety in the Grassmannian.*

This proposition is classical and due to L. M. Brown [9]. In modern language it is a consequence (or, rather, a reformulation) of the following fact.

(3.8.3). Proposition. *There is a unique isomorphism of $GL(2)$-modules*

$$\xi: \wedge^k (S^n(\mathbb{C}^2)) \longrightarrow S^{n-k-1}(S^k(\mathbb{C}^2)) \otimes \left(\wedge^2 (\mathbb{C}^2)\right)^{\otimes k(k-1)/2}$$

such that for any $l_1, \dots, l_k \in \mathbb{C}^2$,

$$\xi(l_1^n \wedge \cdots \wedge l_k^n) = (l_1 \cdots l_k)^{n-k-1} \otimes \prod_{i<j} (l_i \wedge l_j).$$

(3.8.4). Put now $s = k-1$, $r = n-2$ so that we obtain Veronese varieties in $G(k-1, n-1)$, i.e., find ourselves in the setting of §§ 3.3 and 3.4. Recall that special Veronese varieties in $G(k-1, n-1)_e$ (i.e., in the space of P^{k-1}'s in P^{n-1} through $e = (1, \dots, 1)$) are in bijection with the set of projective equivalence classes of n-tuples of points in P^{k-1} in general position. Let us clarify the place in this picture of chordal varieties of Veronese curves.

Suppose that n distinct points $x_1, \dots, x_n \in P^{k-1}$ happen to lie on a Veronese curve D (of degree $k-1$) in P^{k-1}. Then they are in general position in P^{k-1}, as it follows from the calculation of the Vandermonde determinant. Thus they represent an element of $(P^{k-1})^n_{\text{gen}}/GL(k)$. Such an element, by Corollary 3.3.11, is represented by a unique special Veronese variety $X(x_1, \dots, x_n) \subset G(k-1, n-1)_e$ of dimension $k-1$. On the other hand, the curve D being isomorphic to P^1, the points x_i represent an element from $(P^1)^n_{\text{gen}}/GL(2)$. The latter set, as we have seen in Example 3.4.9, is identified with the set of Veronese curves in $P^{n-2}_e = G(1, n-1)_e$ through n points $\langle e, e_i \rangle$. Let $C(x_1, \dots, x_n)$ be the special Veronese curve representing the configuration of x_i on D.

(3.8.5). THEOREM. *The special Veronese variety $X(x_1, \dots, x_n)$ coincides with the chordal variety of the Veronese curve $C(x_1, \dots, x_n)$.*

Using the language of hyperplane configurations, this can be reformulated as follows.

(3.8.6). REFORMULATION. *Let D be a Veronese curve in P^{k-1}, $M = (M_1, \dots, M_n) \subset P^{k-1}$ a configuration of hyperplanes which are osculating to D (i.e., each M_i intersects D in just one point x_i with multiplicity $(k-1)$). Then the embedding $\varphi_E \colon P^{k-1} \hookrightarrow G(k-1, n-1)$ defined by the vector bundle $E = \Omega^1_{P^{k-1}}(\log M)$ maps P^{k-1} isomorphically to the chordal variety of some other Veronese curve $C \subset P^{n-1}$. The curve C is the image of D in the projective embedding defined by the line bundle $\Omega^1_D(\log(x_1 + \dots + x_n))$.*

PROOF. An easy calculation in coordinates shows that the restriction of 1-forms defines an isomorphism

$$H^0(P^{k-1}, \Omega^1_{P^{k-1}}(\log M)) \longrightarrow H^0(D, \Omega^1_D(\log(x_1 + \dots + x_n))). \qquad (3.8.7)$$

Denote for short the bundle $\Omega^1_{P^{k-1}}(\log M)$ on P^{k-1} by E and the bundle $\Omega^1_D(\log(x_1 + \dots + x_n))$ on D by F. Let φ_E and φ_F be the corresponding maps to the Grassmannian and the projective space respectively. We shall show that under the identification (3.8.7) the image of φ_E is the chordal variety of the image of φ_F. By definition of φ_E, φ_F (see section (3.2) above) this is equivalent to part a) of the following statement:

(3.8.8). LEMMA. *Let $p \in P^{k-1}$ be a generic point.*

a) *There are $(k-1)$ points $y_1(p), \dots, y_{k-1}(p) \in D$ such that a form $\omega \in H^0(P^{k-1}, E)$ vanishes at p (as a section of E) if and only if the restriction of ω to D (as a 1-form) vanishes at all y_i.*

b) *Explicitly, points* $y_i(p)$ *are precisely the points of osculation of the* $k-1$ *osculating hyperplanes to* D *passing through* p.

Note that through any generic point $p \in P^{k-1}$ there pass exactly $k-1$ osculating hyperplanes to D. (Since osculating hyperplanes to D form a Veronese curve $\hat{D}$ in the dual projective space, this just means that the degree of $\hat{D}$ is also $k-1$.)

We shall verify Lemma 3.8.8 in coordinates.

(3.8.9). Let us regard the affine space $\mathbb{C}^k$ as the space of polynomials $f(t) = \sum_{i=0}^{k-1} a_i t^i$ of degree $\leqslant k-1$ in one variable. The hyperplanes M_i have the form $M_i = \{f\colon f(x_i) = 0\}$, where $x_i \in \mathbb{C}$ are distinct numbers. The equation of M_i is, therefore, the evaluation map $f \mapsto f(x_i)$. Any point $p \in P^{k-1}$ is represented as a polynomial $f(t)$. If α_i are the roots of $f(t)$ then the points $y_i(p)$ are the polynomials $(t-\alpha_i)^{k-1}$. We can normalize any polynomial f to have the form $f(t) = \prod(t-\alpha_i)$.

A section of the bundle E is given as the logarithmic differential of the function

$$f \mapsto \prod_{j=1}^{n} f(x_j)^{\lambda_j} = \prod_{i=1}^{k-1}\prod_{j=1}^{n}(x_j-\alpha_i)^{\lambda_j}, \qquad \sum \lambda_i = 0.$$

Our assertion means that this function has a critical point at a given f if and only if the function $\prod(t-x_j)^{\lambda_j}$ has a critical value at each $t=\alpha_i$. But this follows from the equality

$$\frac{\partial}{\partial t}\log\prod_j (t-x_j)^{\lambda_j}\Big|_{t=\alpha_i} = -\frac{\partial}{\partial \alpha_r}\log\prod_j (x_j-\alpha_i)^{\lambda_j} = \sum_j \frac{\lambda_j}{\alpha_i - t_j}.$$

Lemma 3.8.8 and hence Theorem 3.8.5 are proved.

(3.8.10). Examples. a) Consider the case of five points in P^2, i.e., $k=3$, $n=5$. Such configurations lead to Veronese surfaces in $G(2,4) \subset P^5$. Since every five points in P^2 lie on a unique conic, any special Veronese surface in $G(2,4)$ is the chordal variety of a Veronese curve (twisted cubic) in P^3.

b) Similarly, the case of n points in P^{n-3} leads to Veronese $(n-3)$-folds in $G(n-3, n-1)$, which are chordal varieties of Veronese curves in P^{n-2}. Note that $G^0(k,n)/H = G^0(n-k,n)/H$ by duality and hence the case of n points in P^{n-3} is equivalent to that of n points on P^1 (see section (2.3)). Any n points in P^{n-3} in general position lie on a unique Veronese curve, which provides the dual configuration.

c) In the case $k=3$, $n=6$ (six points in P^2) we associate to sextuples $(x_1,\dots,x_6) \in (P^2)^6_{\text{gen}}$ Veronese surfaces $X(x_1,\dots,x_6)$ in $G(2,5)$, the space of lines in P^4. When $x_1,\dots,x_6$ lie on a conic, the surface $X(x_1,\dots,x_6)$ is the chordal surface of a Veronese curve in P^4. We have seen this is Proposition 3.7.8.

(3.8.11). REMARK. It is a remarkable fact that chordal varieties of Veronese curves (regarded as subvarieties in Grassmannians) possess deformations that do not come from Veronese curves at all (and represent general projective configurations).

(3.9). The homology class of a Veronese variety in the Grassmannian and dimensions of representations of $GL(n-k)$. To each isomorphism class of generic configurations of n points in P^{k-1} we have associated a certain embedding of P^{k-1} into the Grassmannian $G(k-1, n-1)$, namely the Veronese variety. For instance, configurations of points on P^1 correspond to Veronese curves in P^{n-2} through a fixed set of n generic points. In this section we calculate the homology class Δ represented by these Veronese varieties in $G(k-1, n-1)$. It turns out that the coefficients of the expansion of Δ in the basis of Schubert cycles are exactly the dimensions of irreducible representations of the group $GL(n-k)$.

(3.9.1). Let us review the homology theory of the Grassmannian $G(p, q) = G(p, \mathbb{C}^q)$ (see [24] for more details). Let $\alpha = (\alpha_1 \geqslant \cdots \geqslant \alpha_p \geqslant 0)$, $\alpha_i \leqslant q-p$, be a decreasing sequence of nonnegative integers. We visualize α as a Young diagram in which α_i are the lengths of rows. Because of inequalities $\alpha_i \leqslant q-p$, the diagram α lies inside the rectangle $[0, q-p] \times [0, p] \subset \mathbb{R}^2$:

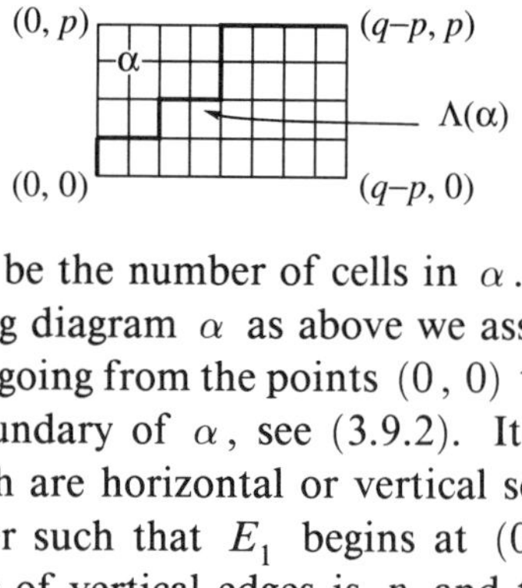

(3.9.2)

We define $|\alpha| = \sum \alpha_i$ to be the number of cells in α.

(3.9.3). To each Young diagram α as above we associate the lattice path $\Lambda(\alpha) \subset [0, q-p] \times [0, p]$ going from the points $(0, 0)$ to the point $(q-p, p)$. This path is just the boundary of α, see (3.9.2). It consists of exactly q edges $E_1, \ldots, E_q$, which are horizontal or vertical segments of the lattice. We write E_i in an order such that E_1 begins at $(0, 0)$ and E_q ends at $(q-p, p)$. The number of vertical edges is p and the path is completely determined by specifying which of the E_i are vertical. So the number of possible lattice paths in $[0, q-p] \times [0, p]$ equals $\binom{q}{p}$. It is equal to the number of all Young diagrams in $[0, q-p] \times [0, p]$, if we count also the "empty" diagram $(0, \ldots, 0)$.

(3.9.4). Let $\alpha \subset [0, q-p] \times [0, p]$ be a Young diagram and $\Lambda(\alpha) = (E_1, \ldots, E_q)$ be the corresponding lattice path. Associate to α the sequence $ht(\alpha) = (0 \leqslant ht_1(\alpha) \leqslant \cdots \leqslant ht_q(\alpha) = p)$ by setting $ht_i(\alpha)$ to be the ordinate (height) of the end of the edge E_i.

(3.9.5). Let $V_\cdot = (V_1 \subset V_2 \subset \cdots \subset V_{q-1} \subset V_q = \mathbb{C}^q)$ be a complete flag of linear subspaces in $\mathbb{C}^q$, so that $\dim(V_i) = i$. Let also $\alpha \subset [0, q-p] \times [0, p]$ be a Young diagram. We define the *Schubert variety* $S_\alpha(V_\cdot) \subset G(p, q)$ to be the locus of subspaces $L \subset \mathbb{C}^q$, $\dim(L) = p$, such that $\dim(L \cap V_i) \geqslant ht_i(\alpha)$, where $ht_i(\alpha)$ was defined in (3.9.4).

It is well known [24] that $S_\alpha(V_\cdot)$ is an irreducible variety of complex dimension $|\alpha|$. The homology class in $H_{2|\alpha|}(G(p, q), \mathbb{Z})$ represented by $S_\alpha(V_\cdot)$ is independent on $V_\cdot$. This class is denoted by σ_α and called the *Schubert cycle.*

It is known that the homology group $H_{2r}(G(p, q), \mathbb{Z})$ is freely generated by cycles σ_α, where α runs over all Young diagrams with r cells contained in the rectangle $[0, q-p] \times [0, p]$.

(3.9.6). Any Young diagram α (not necessarily contained in a given rectangle) defines the *Schur functor* Σ^α on the category of vector spaces [38]. By definition, for a vector space V the space $\Sigma^\alpha(V)$ is the space of irreducible representations of the group $GL(V)$ with the highest weight α. It can be defined, e.g., as the image of the Young symmetrizer h_α in the tensor space $V^{\otimes|\alpha|}$. In particular, for $\alpha = (m, 0, \dots, 0)$ (a horizontal strip of length m) the functor Σ^α is the symmetric power S^m. For a vertical strip $\alpha = (1^m) = (1, \dots, 1, 0, \dots, 0)$ (m ones) the functor Σ^α is the exterior power $\wedge^m$.

(3.9.7). For a Young diagram α we denote by α^* the dual (or transposed) Young diagram defined by $\alpha_i^* = \mathrm{Card}\{j \colon \alpha_j \geqslant i\}$. The rows on α^* correspond to columns of α and vice versa.

Now we can formulate the main result of this section.

(3.9.8). THEOREM. *Let $X \subset G(k-1, n-1)$ be a special $(k-1)$-dimensional Veronese variety and $\Delta \in H_{2k-2}(G(k-1, n-1), \mathbb{Z})$ be the homology class of X. Then the decomposition of Δ in the basis of Schubert cycles has the form*

$$\Delta = \sum_{|\alpha|=k-1} m_\alpha \cdot \sigma_\alpha, \qquad \textit{where} \quad m_\alpha = \dim(\Sigma^{\alpha^*}(\mathbb{C}^{n-k})).$$

(3.9.9). To prove Theorem 3.9.8, we first take X to be the variety of $(k-2)$-dimensional chords of a Veronese curve $C \subset P^{n-2}$, which is a particular case of Veronese varieties, see § 3.8. Then we degenerate C into a union of lines.

More precisely, let $e_1, \dots, e_{n-1} \in P^{n-2}$ be the points corresponding to standard basis vectors of $\mathbb{C}^{n-1}$. Consider the reducible curve $D = D_1 \cup \dots \cup D_{n-2}$, where D_i is the line $\langle e_i, e_{i+1}\rangle$. Since D can be obtained as a limit position of Veronese curves, we find that Δ is equal to the homology class of the chordal variety $Ch_{k-2}(D)$, see (3.8.1) for the notation.

(3.9.10). The variety $Ch_{k-2}(D)$ is reducible and splits into $\binom{n-2}{k-1}$ components $X_{i_1, \dots, i_{k-1}}$ that correspond to sequences $1 \leqslant i_1 < \dots < i_{k-1} \leqslant n-2$. The component $X_{i_1, \dots, i_{k-1}}$ is the locus of chordal subspaces $\langle x_{i_1}, \dots, x_{i_{k-1}}\rangle$, where x_{i_ν} lies on the line $\langle e_{i_\nu}, e_{i_\nu+1}\rangle \subset D$. Therefore our homology class

Δ is the sum of homology classes $[X_{i_1,\dots,i_{k-1}}]$ of all the components of $Ch_{k-2}(D)$.

(3.9.11). Let us introduce a different, more suitable for our purposes, combinatorial labeling of components of $Ch_{k-2}(D)$.

Denote by $W(n-k, k-1)$ the set of all (not necessarily decreasing) sequences $\lambda = (\lambda_1, \dots, \lambda_{n-k}) \in \mathbb{Z}_+^{n-k}$ of nonnegative integers such that $\sum \lambda_i = k-1$. We call elements of $W(n-k, k-1)$ *weights* (in the sense of representation theory) of degree $k-1$ in $n-k$ variables.

To any sequence $1 \leqslant i_1 < \cdots < i_{k-1} \leqslant n-2$ we associate a weight $\lambda(i_1, \dots, i_{k-1}) = (\lambda_1, \dots, \lambda_{n-k}) \in W(n-k, k-1)$ as follows. Write all elements of the set $\{1, \dots, n-2\} - \{i_1, \dots, i_{k-1}\}$ in increasing order, $j_1, \dots, j_{n-k-1}$. Set also $j_{n-k} = n-1$. Now define

$$\lambda(i_1, \dots, i_{k-1}) = (\lambda_1, \dots, \lambda_{n-k}), \quad \text{where } \lambda_\nu = j_\nu - j_{\nu-1} - 1. \quad (3.9.12)$$

Numbers λ_ν are just the lengths of arithmetic progressions with increment one into which the sequence $(i_1, \dots, i_{k-1})$ splits.

The correspondence $(i_1, \dots, i_{k-1}) \mapsto \lambda(i_1, \dots, i_{k-1})$ establishes a bijection between the set of all $(k-1)$-element subsets in $\{1, \dots, n-2\}$ and the set $W(n-k, k-1)$. This bijection is the labeling we need.

If $\lambda \in W(n-k, k-1)$ is labeled as $\lambda(i_1, \dots, i_{k-1})$, we denote the component $X(\lambda) \subset Ch_{k-2}(D)$ by $X(i_1, \dots, i_{k-1})$.

Now Theorem 3.9.8 will be a consequence of the following fact.

(3.9.13). THEOREM. *Let α, $|\alpha| = k-1$, be a Young diagram in the rectangle $[0, n-k-1] \times [0, k-1]$ and let $\lambda \in W(n-k, k-1)$ be any weight. Then the homology class of the component $X(\lambda) \subset G(k-1, n-1)$ has the form*

$$[X(\lambda)] = \sum_{|\alpha|=k-1} K_{\lambda,\alpha^*} \cdot \sigma_\alpha,$$

where K_{λ,α^*} is the multiplicity of weight λ in the irreducible representation $\Sigma^{\alpha^*}(\mathbb{C}^{n-k})$ (the Kostka number).

We concentrate on the proof of Theorem 3.9.13.

(3.9.14). PROPOSITION. *Let $\lambda = (\lambda_1, \dots, \lambda_{n-k}) \in W(n-k, k-1)$ be a weight. The component $X(\lambda)$ is isomorphic to the product of projective spaces $\prod P^{\lambda_j}$. It is embedded into the Grassmannian as the image of the direct sum map*

$$\oplus\colon \prod P^{\lambda_j} = \prod G(\lambda_j, \lambda_j + 1) \hookrightarrow G\Big(\sum \lambda_j, \sum(\lambda_j + 1)\Big) = G(k-1, n-1).$$

PROOF. Let $1 \leqslant i_1 < \cdots < i_{k-1} \leqslant n-2$ be the sequence of integers to which λ is associated, see (3.9.10). The component $X(\lambda)$ consists of chords which join points of lines $\langle e_{i_\nu}, e_{i_\nu+1}\rangle$. Let us split the sequence $(i_1, \dots, i_{k-1})$ into segments that are arithmetic progressions with increment 1. Then λ_ν are precisely lengths of these segments. Now let $i, i+1, \ldots, i+\lambda_\nu$ be any such segment. The $(\lambda_\nu - 1)$-dimensional chords of the subcurve $\langle e_i, e_{i+1}\rangle \cup \langle e_{i+1}, e_{i+2}\rangle \cup \cdots \cup \langle e_{i+\lambda_\nu}, e_{i+\lambda_\nu+1}\rangle$ are just arbitrary hyperplanes in the projective subspace $\langle e_i, e_{i+1}, \ldots, e_{i+\lambda_\nu+1}\rangle$. Any $(k-2)$-dimensional chord from our component $X_{i_1, \ldots, i_{k-1}} = X(\lambda)$ is therefore the projective span of hyperplanes in the independent projective subspaces $P^{\lambda_j} = P(\mathbb{C}^{\lambda_\nu+1})$, as required

(3.9.15). PROPOSITION. *The coefficient at the Schubert cycle σ_α in the decomposition of the class $[X(\lambda)]$ equals the multiplicity of the irreducible representation $\Sigma^\alpha\mathbb{C}^{n-k}$ in the tensor product $\wedge^{\lambda_1}(\mathbb{C}^{n-k}) \otimes \cdots \otimes \wedge^{\lambda_{n-k}}(\mathbb{C}^{n-k})$ of exterior powers.*

The proof is based on the following (known) fact.

(3.9.16). LEMMA. *For three Young diagrams α, β, γ such that $|\gamma| = |\alpha| + |\beta|$ let $c^\gamma_{\alpha\beta}$ be the multiplicity of Σ^γ in $\Sigma^\alpha \otimes \Sigma^\beta$ (the Littlewood-Richardson number). Then:*

a) *The image of the cycle $\sigma_\alpha \otimes \sigma_\beta$ under the direct sum map*
$$\varphi\colon G(p_1, V_1) \times G(p_2, V_2) \to G(p_1 + p_2, V_1 \oplus V_2)$$
equals $\sum_\gamma c^\gamma_{\alpha\beta}\sigma_\gamma$.

b) *If A and B are finite-dimensional vector spaces then for any Young diagram γ we have the isomorphism of $GL(A) \times GL(B)$-modules*
$$\Sigma^\gamma(A \oplus B) \cong \bigoplus_{|\alpha|+|\beta|=|\gamma|} c^\gamma_{\alpha\beta}\Sigma^\alpha(A) \otimes \Sigma^\beta(B).$$

PROOF OF (3.9.16). In part a) it suffices to consider the "stable" case when V_i have infinite dimension. We shall assume that it is so.

Denote by $H^\cdot(G(p, \infty), \mathbb{Z})$ the cohomology ring of $G(p, \infty)$ and by $\mathrm{Rep}(GL(p))$ the Grothendieck ring of polynomial representations of $GL(p)$. Let also $\Lambda_p = \mathbb{Z}[x_1, \ldots, x_p]^{S_p}$ denote the ring of symmetric polynomials in p variables $x_1, \ldots, x_p$. There are isomorphisms of rings
$$\Lambda_p \cong H^\cdot(G(p, \infty), \mathbb{Z}) \cong \mathrm{Rep}(GL(p))$$
that take elementary symmetric function $e_j \in \Lambda_p$ into the jth Chern class of the tautological bundle on $G(p, \infty)$ and into the representation $\wedge^j(\mathbb{C}^p) \in \mathrm{Rep}(GL(p))$.

For any Young diagram α denote by $s_\alpha(x_1, \ldots, x_p) \in \Lambda_p$ the *Schur polynomial* (see [38]). It corresponds to the following elements of the above two rings:

- The cocycle $\sigma^\alpha \in H^{2|\alpha|}(G(p, \infty), \mathbb{Z})$ dual to the Schubert cycle σ_α (i.e., $(\sigma^\alpha, \sigma_\beta) = \delta_{\alpha\beta}$).
- The irreducible representation $\Sigma^\alpha\mathbb{C}^p$ whose character is s_α.

Consider the tensor product $\Lambda_{p_1} \otimes \Lambda_{p_2}$. It can be regarded as a ring of polynomials $f(x_1, \dots, x_{p_1}, y_1, \dots, y_{p_2})$ symmetric with respect to x_i and with respect to y_i. Therefore we have an embedding

$$\delta : \Lambda_{p_1+p_2} \to \Lambda_{p_1} \otimes \Lambda_{p_2}$$

(it is a part of Hopf algebra structure on the limit $\Lambda = \lim \Lambda_p$, see [38]).

The homology space of $G(p, \infty)$ is dual to Λ_p. The map φ_* induced on homology by the direct sum map φ from part a) of the lemma is known to be dual to δ.

Similarly, if we assume in part b) of the lemma that $\dim(A) = p_1$, $\dim(B) = p_2$ then the restriction map

$$\mathrm{Rep}(GL(p_1 + p_2)) \to \mathrm{Rep}(GL(p_1)) \otimes \mathrm{Rep}(GL(p_2))$$

is identified with δ.

Therefore in part a) we must find matrix elements of the dual map $\delta^* : \Lambda^* \otimes \Lambda^* \to \Lambda^*$ in the basis dual to that of s_α. In part b) we have to find matrix elements of δ in the basis of s_α. So both parts follow from the equality

$$s_\gamma(x_1, \dots, x_{p_1}, y_1, \dots, y_{p_2}) = \sum_{|\alpha|+|\beta|=|\gamma|} c^\gamma_{\alpha\beta} s_\alpha(x_1, \dots, x_{p_1}) s_\beta(y_1, \dots, y_{p_2}),$$

which is a reformulation of [38, Chapter I, formula 5.9].

(3.9.17). PROOF OF PROPOSITION 3.9.15. Note that the fundamental class of the projective space P^{λ_i} considered as $G(\lambda_i, \mathbb{C}^{\lambda_i+1})$ is the Schubert cycle corresponding to the Young diagram (1^{λ_i}), i.e., to the vertical strip of length λ_i. The Schur functor corresponding to this diagram is the exterior power. Now the result follows from Lemma 3.9.16.

(3.9.18). PROOF OF THEOREM 3.9.13. By Proposition 3.9.15, it suffices to show that the weight multiplicity K_{λ, α^*} equals the multiplicity of Σ^α in $\wedge^{\lambda_1}(\mathbb{C}^{n-k}) \otimes \cdots \otimes \wedge^{\lambda_{n-k}}(\mathbb{C}^{n-k})$. By Young duality this is equivalent to saying that for any α the number $K_{\lambda,\alpha}$ equals the multiplicity of Σ^α in the product of symmetric powers $S^{\lambda_1}(\mathbb{C}^{n-k}) \otimes \cdots \otimes S^{\lambda_{n-k}}(\mathbb{C}^{n-k})$.

To see that this latter statement holds, decompose $\mathbb{C}^{n-k}$ into a sum of 1-dimensional subspaces $L_1 \oplus \cdots \oplus L_{n-k}$. Then the decomposition of $\Sigma^\alpha(\mathbb{C}^{n-k})$ as a $(GL(L_1) \times \cdots \times GL(L_{n-k}))$-module is just the weight decomposition. On the other hand, applying repeatedly Lemma 3.9.16 b) we find that $K_{\lambda,\alpha}$, i.e., the multiplicity of $S^{\lambda_1}(L_1) \otimes \cdots \otimes S^{\lambda_{n-k}}(L_{n-k})$ in $\Sigma^\alpha(L_1 \oplus \cdots \oplus L_{n-k})$, equals the multiplicity of Σ^α in $S^{\lambda_1} \otimes \cdots \otimes S^{\lambda_{n-k}}$.

Theorems 3.9.13 and 3.9.8 are completely proved.

(3.9.19). REMARK. A different expression for the coefficients m_α in Theorem 3.9.8 can be obtained from Klyachko's formula (Proposition 1.1.8) for the homology class of the entire Lie complex $Z \subset G(k, n)$. Our Veronese variety is just the visible contour of Z, i.e., the intersection $Z \cap G(k-1, n-1)_p$, see (3.1.1). The intersection map

$$H_r(G(k, n), \mathbb{Z}) \to H_{r-n+k}(G(k-1, n-1)_p, \mathbb{Z})$$

is easy to describe. It takes a Schubert cell $\sigma_{\beta_1, \dots, \beta_k}$ to $\sigma_{\beta_2, \dots, \beta_k}$ if $\beta_1 = n-k$ and to 0 otherwise. This leads to the formula

$$m_\alpha = \sum_{i=0}^{k} (-1)^i \binom{n}{i} \dim \left(\Sigma^{n-k, \alpha_1, \dots, \alpha_{k-1}}(\mathbb{C}^{k-i})\right).$$

According to Theorem 3.9.8 this expression equals $\dim \Sigma^{\alpha^*}(\mathbb{C}^{n-k})$; however we do not know a straightforward proof of this fact.

(3.9.20). REMARK. Let $\alpha = (\alpha_1, \dots, \alpha_{k-1})$, $|\alpha| = k-1$, be a Young diagram in the rectangle $[0, n-k] \times [0, k-1]$. Denote by $\overline{\alpha} = (n-k-\alpha_{k-1}, \dots, n-k-\alpha_1)$ the diagram complementary to α in this rectangle. It is known [24] that for any Young diagram β with $|\beta| = (k-1)(n-k-1)$ the intersection index $\sigma_\alpha \cdot \sigma_\beta$ equals 1 if $\beta = \overline{\alpha}$ and to 0 otherwise. Hence the coefficient m_α in the expansion of the cycle represented by the Veronese variety $S \subset G(k-1, n-1)$ over Schubert cycles equals $S \cdot \sigma_{\overline{\alpha}}$.

Let us realize $\sigma_{\overline{\alpha}}$ as the class of the Schubert variety $S_{\overline{\alpha}}(V_.)$ for a generic flag $V_.$. Take the Veronese variety S to be the chordal variety of the Veronese curve. We obtain the following restatement of Theorem 3.9.8:

The number of chordal $(k-1)$-dimensional subspaces of a Veronese curve in P^{n-2} satisfying any given Schubert condition equals the dimension of some irreducible representation of $GL(n-k)$! The dimension of any representation can be realized in this way.

It would be interesting to find a conceptual explanation of this fact, e.g., define a $GL(n-k)$-action on the vector space freely generated by points from $S \cap S_{\overline{\alpha}}(V_.)$. Let us also point out a series of papers of A. N. Kirillov and N. Yu. Reshetikhin (see [33, 34] and references therein) on new combinatorial formulas for weight multiplicities $K_{\lambda, \alpha}$. Their construction is based on an interpretation of any $K_{\lambda, \alpha}$ as the number of solutions of some special system of algebraic equations (the equations of Bethe-Ansatz). This interpretation seems to be related to the one given above.

(3.9.21). EXAMPLE. It is well known that the number of nodes of a plane rational curve of degree d equals $(d-1)(d-2)/2$. We can obtain this as a particular case of Theorem 3.9.8. Let $C \subset P^d$ be a Veronese curve, $L \subset P^d$ a projective subspace of dimension $d-3$, and $\pi \colon P^d - L \to P^2$ the projection with center L. Nodes of the plane curve $\pi(C)$ correspond to 1-dimensional chords of C intersecting L. Let $X = Ch_1(C) \subset G(2, d+1)$ be the surface of chordal lines of C. By Theorem 3.9.8, its homology class has the form

$$[X] = \dim(S^2\mathbb{C}^{d-1}) \cdot \sigma_{1,1} + \dim(\wedge^2\mathbb{C}^{d-1}) \cdot \sigma_{2,0}.$$

The coefficient at $\sigma_{2,0}$ equals the intersection index of S with the Schubert cycle $\sigma_{d-1, \dots, d-1, d-3}$. The corresponding Schubert variety is the locus of all lines intersecting a given $(d-3)$-dimensional subspace in P^d, for example, L. So we find the number of nodes of $\pi(C)$ to be $\dim(\wedge^2\mathbb{C}^{d-1}) = (d-1)(d-2)/2$.

(3.9.22). EXAMPLE. The number of 4-secant lines of a spatial rational curve $X \subset P^3$ of degree d can be found using arguments similar to those in the above example. This number equals

$$\dim(\Sigma^{2,2}\mathbb{C}^{d-3}) = (d-2)(d-3)^2(d-4)/12. \tag{3.9.23}$$

The right-hand side of (3.9.23) is a well-known formula for the number of quadrisecants, see, e.g., [24, Chapter 2, §5].

(3.9.24). EXAMPLE. The number of trisecant lines of a rational curve of degree d in P^4 equals

$$\dim(\wedge^3\mathbb{C}^{n-2}) = (n-2)(n-3)(n-4)/6. \tag{3.9.25}$$

Chapter 4. The Chow quotient of $G(2, n)$ and the Grothendieck-Knudsen moduli space $\overline{M_{0,n}}$

In this chapter we study in detail the Chow quotient $G(2, n)//H$ of the Grassmannian $G(2, n)$ of lines in P^{n-1}. We establish the isomorphism of this Chow quotient with the moduli space $\overline{M_{0,n}}$ of stable n-punctured curves of genus 0 introduced by A. Grothendieck [12] and later by F. Knudsen [37]. In particular, $G(2, n)//H$ is a smooth variety and the complement to the open stratum is a divisor with normal crossing. The relation of the space $\overline{M_{0,n}}$ to the Grassmannian permits us to represent this space as an iterated blow-up of the projective space P^{n-3}.

(4.1). The space $G(2, n)//H$ and stable curves.

(4.1.1). According to Theorem 2.2.4, we have an isomorphism

$$G(2, n)//H = (P^1)^n//GL(2).$$

In other words, our Chow quotient compactifies the space

$$M_{0,n} = ((P^1)^n - \bigcup\{x_i = x_j\})/GL(2)$$

of projective equivalence classes of n-tuples of distinct points on P^1. The space $M_{0,n}$ can be viewed as the moduli space of systems $(C, x_1, \dots, x_n)$, where C is a smooth curve of genus 0 and x_i are distinct points on C.

(4.1.2). There is a well-known compactification of $M_{0,n}$ by means of so-called stable n-pointed curves of genus 0 introduced by Grothendieck and Knudsen [12, 37]. Let us recall the definitions.

(4.1.3). DEFINITION. A stable n-pointed curve of genus 0 is a connected (but possibly reducible) curve C over k together with n smooth distinct points $x_1, \dots, x_n \in C$, satisfying the following conditions:

(1) C has only ordinary double points and every irreducible component of C is isomorphic to the projective line P^1.

(2) The arithmetic genus of C is equal to 0.
(3) On each component of C there are at least three points which are either marked or double.

Points of C which are either marked or double will be called special.

Condition (2) is equivalent to the condition that the graph formed by components of C is a tree. We prefer the following "dual" point of view on this tree.

(4.1.4). Definition. Let $(C, x_1, \ldots, x_n)$ be a stable n-pointed curve of genus 0. Its tree $\mathcal{T}(C, x_1, \ldots, x_n)$ has the following vertices:

(1) Endpoints (1-valent vertices) $A_1, \ldots, A_n$ corresponding to points $x_0, \ldots, x_n$.
(2) Vertices corresponding to all components of C.

Two vertices of type (2) are joined by an edge if the corresponding components intersect. An endpoint A_i is joined by a new edge to the vertex of type (2) corresponding to the unique component containing the point x_i.

Definition 4.1.4 is illustrated in (4.1.5).

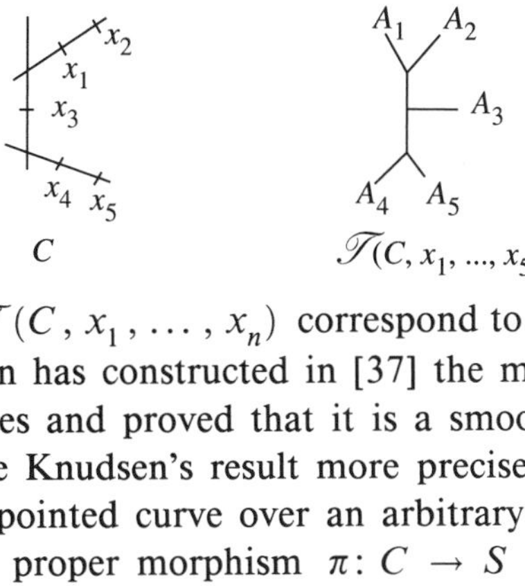

x_2 x_1 x_3 x_4 x_5 A_1 A_2 A_3 A_4 A_5

C $\qquad$ $\mathcal{T}(C, x_1, \ldots, x_5)$ $\qquad$ (4.1.5)

Therefore, edges of $\mathcal{T}(C, x_1, \ldots, x_n)$ correspond to special points of C.

(4.1.6). F. Knudsen has constructed in [37] the moduli space $\overline{M_{0,n}}$ of stable n-pointed curves and proved that it is a smooth compact algebraic variety. To formulate Knudsen's result more precisely, let us introduce a notion of a stable n-pointed curve over an arbitrary base scheme S. By definition, it is a flat proper morphism $\pi: C \to S$ together with n distinguished sections $s_1, \ldots, s_n: S \to C$ such that for any geometric point $s \in S$ the fiber $C_s = \pi^{-1}(s)$ is a reduced (i.e., without nilpotents) algebraic curve and $(C_s, s_1(s), \ldots, s_n(s))$ is a stable n-pointed curve of genus 0. An isomorphism between two such objects $(\pi: C \to S, s_1, \ldots, s_n)$ and $(\pi': C' \to S, s_1', \ldots, s_n')$ over the same base S is just an isomorphism $f: C \to C'$ commuting with projections and taking s_i to s_i'.

(4.1.7). Theorem [37]. *There exists a smooth projective complex algebraic variety $\overline{M_{0,n}}$ such that for any scheme S over $\mathbb{C}$ the set of isomorphism classes of stable n-pointed curves of genus 0 over S is naturally identified with* $\mathrm{Hom}(S, \overline{M_{0,n}})$.

An open subset $M_{0,n} \subset \overline{M_{0,n}}$ is formed by n-pointed curves $(C, x_1, \ldots, x_n)$ such that C is smooth, i.e., $C \cong P^1$.

Now we can formulate the complete description of the Chow quotient of $G(2, n)$.

(4.1.8). THEOREM. *The Chow quotients* $G(2, n)//H$ *and* $(P^1)^n//GL(2)$ *are isomorphic to the moduli space* $\overline{M_{0,n}}$.

To prove Theorem 4.1.8 note that Theorem 3.3.14 together with Example 3.4.9 implies the following description of $G(2, n)//H$.

(4.1.9). COROLLARY. *Take* n *points* $p_1, \dots, p_n$ *in* P^{n-2} *in general position. Let* $V_0(p_1, \dots, p_n)$ *be the space of all Veronese curves in* P^{n-2} *through* p_i. *Denote by* $V(p_1, \dots, p_n)$ *the closure of* $V_0(p_1, \dots, p_n)$ *in the Chow variety and by* $W(p_1, \dots, p_n)$ *the closure in the Hilbert scheme. Then* $V(p_1, \dots, p_n) \cong W(p_1, \dots, p_n) \cong G(2, n)//H$.

By [29] (Theorem 0.1) $V(p_1, \dots, p_n)$ and $W(p_1, \dots, p_n)$ are isomorphic to $\overline{M_{0,n}}$. More precisely, any subscheme from $W(p_1, \dots, p_n)$ is in fact reduced and, regarded together with p_i, is a stable n-pointed curve of genus 0. Theorem 4.1.8 is proved.

(4.1.10). REMARK. It seems to be difficult to prove directly that the Chow quotient $(P^1)^n//GL(2)$ coincides with $\overline{M_{0,n}}$. However, the Grassmannian picture (i.e., the Gelfand-MacPherson isomorphism, see Chapter 2) leads to stable curves very naturally: these curves are just visible contours of generalized Lie complexes.

(4.1.11). Now we can translate general constructions of Chapter 1 to the language of stable curves.

The combinatorial invariant of a stable n-pointed curve $(C, x_1, \dots, x_n) \in \overline{M_{0,n}}$ is its tree $\mathcal{T}(C)$ (Definition 4.1.4). For each tree $\mathcal{T}$ bounding the endpoints $1, \dots, n$ we define the stratum $M(\mathcal{T}) \subset \overline{M_{0,n}}$ consisting of stable curves C with $\mathcal{T}(C) = \mathcal{T}$. In particular, to a 1-vertex tree corresponds the open stratum $M_{0,n} \subset \overline{M_{0,n}}$.

The combinatorial invariant of a generalized Lie complex $Z \subset G(k, n)//H$ is the corresponding matroid decomposition of the hypersimplex $\Delta(k, n)$ (Proposition 1.2.4). The Chow strata in $G(k, n)//H$ were defined (Definition 1.2.16) as the loci of Z for which the corresponding matroid decomposition is fixed.

It was proved in § 1.3 that matroid decompositions of the hypersimplex $\Delta(2, n)$ correspond exactly to trees bounding n endpoints $A_1, \dots, A_n$, see Theorem 1.3.6. Thus we obtain the following corollary.

(4.1.12). COROLLARY. *All matroid decompositions of the hypersimplex* $\Delta(2, n)$ *are realizable, i.e., come from nonempty Chow strata in* $G(2, n)//H$. *These Chow strata have the form* $M(\mathcal{T}) \subset \overline{M_{0,n}}$. *The stratum* $M(\mathcal{T})$ *is isomorpic to the product* $\prod M_{0,e(v)}$, *where* e *runs over* v *of* $\mathcal{T}$ *and* $e(v)$ *is the number of edges containing* v.

(4.1.13). Forgetting the ith point on any stable n-pointed curve $(C, x_1, \ldots, x_n) \in \overline{M_{0,n}}$ gives a new $(n-1)$-pointed curve. This curve might be unstable, i.e., the condition d) of Definition 4.1.3 might be violated. (Clearly, this happens in the case when the component of C containing x_i contains only two other double or marked points.) Blowing down this component defines a stable curve $\pi_i(C)$ pointed with images of x_j, $j \neq i$, see [37]. It was shown in [37] that π_i defines a morphism $\overline{M_{0,n}} \to \overline{M_{0,n-1}}$ which identifies $\overline{M_{0,n}}$ with the universal family of curves over $\overline{M_{0,n-1}}$.

It was shown in [29] that π_i corresponds to the geometric projection of (degenerate) Veronese curves in $V(p_1, \ldots, p_n)$ (Corollary 4.1.9) from the point p_i. In terms of generalized Lie complexes (=points of $G(2, n)//H$) the projection π_i is described as follows.

(4.1.14). Proposition. *Let $Z \subset G(2, n)$ be a generalized Lie complex. Let $G(2, n-1)^i$ be the space of lines in P^{n-1} that in fact lie in the $(n-2)$-dimensional projective subspace spanned by basis vectors e_j, $j \neq i$. Then $Z \cap (G, 2, n-1)^i$ is a generalized Lie complex in $G(2, n-1)_i$. The operation of intersection with $G(2, n-1)^i$ corresponds, under the identification of Theorem* 4.1.8, *to the projection $\pi_i \colon \overline{M_{0,n}} \to \overline{M_{0,n-1}}$.*

(4.2). The birational maps $\sigma_i \colon \overline{M_{0,n}} \to P^{n-3}$.

(4.2.1). The Grothendieck-Knudsen space $\overline{M_{0,n}}$ can be seen as a "high-brow" compactification of the space $M_{0,n}$ of projective equivalence classes of n-tuples of distinct points on P^1, see (4.1.1). On the other hand, every three distinct points on P^1 can be brought to the points $0, 1, \infty$ by a unique projective transformation. Doing this with the first three points of any n-tuple, we find that

$$M_{0,n} \cong \{(x_4, \ldots, x_n) \in \mathbb{C}^{n-3} \colon x_i \neq 0, 1 \text{ for all } i, \ x_i \neq x_j \text{ for all } i \neq j\}.$$

So $M_{0,n}$ is an open subset in $\mathbb{C}^{n-3}$. This suggests a "naive" compactification of $M_{0,n}$ which is just the projective space P^{n-3} compactifying $\mathbb{C}^{n-3}$. One expects then that $\overline{M_{0,n}}$, being the finer compactification, maps to P^{n-3} by a regular birational map.

(4.2.2). According to [29], a regular map $\overline{M_{0,n}} \to P^{n-3}$ can be constructed as follows. Realize $\overline{M_{0,n}}$ as the space $V(p_1, \ldots, p_n)$ of limit position of Veronese curves in P^{n-2} containing given generic points $p_1, \ldots, p_n$. For any curve $C \in V(p_1, \ldots, p_n)$, all p_i are smooth points of C. Fix some i and consider the projective space P_i^{n-3} of all lines in P^{n-2} through p_i. By associating to any curve $C \in V(p_1, \ldots, p_n)$ its embedded tangent line $T_{p_i}C$ one gets a regular map

$$\sigma_i \colon \overline{M_{0,n}} \to P_i^{n-3}.$$

It was demonstrated in [29] that P_i^{n-3} is exactly the "naive" compactification of $M_{0,n}$ mentioned in (4.2.1). Its dependence on i is easy to explain: we need to specify which point of an n-tuple is set to be ∞. So the construction of (4.2.1) corresponds to $i = 3$.

Here we are going to study the maps σ_i in more detail.

(4.2.3). Let L_i, $i = 1, \ldots, n$, be the line bundle on $\overline{M_{0,n}}$ whose fiber at a pointed curve $(C, x_1, \ldots, x_n)$ is $T^*_{x_i}C$, the cotangent space to C at x_i. Clearly $L_i \cong \sigma_i^*(\mathscr{O}_{P_i^{n-3}}(1))$. The following fact was proved in [29].

(4.2.4). Proposition. *For any $i \in \{1, \ldots, n\}$ the space $H^0(\overline{M_{0,n}}, L_i)$ has dimension $n-2$. The corresponding morphism γ_{L_i} is everywhere regular, birational, and, moreover, one-to one outside the subvariety in P_i^{n-3} formed by lines that lie in a hyperplane spanned by p_i. In the Veronese picture the space $P(H^0(\overline{M_{0,n}}, L_i)^*)$ is identified with P_i^{n-3} and γ_{L_i} is identified with σ_i.*

(4.2.5). Let us give a description of maps σ_i in the language of generalized Lie complexes. So we start with the standard coordinatized projective space $P^{n-1} = P(\mathbb{C}^n)$. For any point $x \in P^{n-1}$ denote by $P^{n-2}(x) \subset G(2, n)$ the space of lines in P^{n-1} meeting x. Let $e_i \in P^{n-1}$ be the ith basis vector.

(4.2.6). Proposition. *Any Lie complex (and hence any generalized Lie complex) in $G(2, n)$ contains each projective space $P^{n-2}(e_i)$.*

Proof. Let Z be a Lie complex. The intersection $Z \cap P^{n-2}_{e_i}$ contains the closure of a generic torus orbit in $P^{n-2}(e_i)$. Since this generic orbit is dense in $P^{n-2}(e_i)$, the assertion follows.

(4.2.7). Note that the dimension of a (generalized) Lie complex Z exceeds the dimension of $P^{n-2}(e_i)$ just by 1. Hence at a generic point l of $P^{n-2}(e_i)$ the tangent space T_lZ represents a line in the normal space $N_l(Z/G(2, n)) = T_lG(2, n)/T_lZ$. In the construction of the Veronese curve corresponding to Z (as the visible contour, see § 3.1) we considered the set of all lines in Z meeting a given point u or, in other words, the intersection $Z \cap P^{n-2}(u)$. More precisely, we specialized to $u = e = (1, \ldots, 1)$.

(4.2.8). Proposition. *The space P_i^{n-3} is naturally identified with the projectivization of the normal space to $P^{n-2}(e_i)$ in $G(2, n)$ at the point $p_i = \langle e_i, u\rangle$. The map σ_i is identified to the map taking any generalized Lie complex Z to the line in the above normal space given by the subspace T_lZ.*

Proof. The subvarieties $P^{n-2}(e_i)$ and $P^{n-2}(u)$ in $G(2, n)$ are of middle dimension and intersect transversely in the point $p_i = \langle e_i, u\rangle$. Therefore the normal space in question is naturally identified with the tangent space at p_i to

$P^{n-2}(u)$. Since σ_i is defined by considering the tangent line to $Z \cap P^{n-2}(u)$ at p_i, the assertion follows.

(4.2.9). The advantage of the description in (4.2.8) is that it clearly states the dependence on the choice of a point u. It also shows how to obtain a more invariant description of σ_i. To do this, one should consider all the 1-dimensional subspaces in the normal spaces to $P^{n-2}(e_i)$ in $G(2, n)$ at all the generic points or, in other words, the corresponding subbundle in the normal bundle. Let us describe these bundles.

(4.2.10). PROPOSITION. *Let $x \in P^{n-1}$ be any point. Then the normal bundle of the subvariety $P^{n-2}(x) \subset G(2, n)$ is naturally isomorphic to the twisted tangent bundle $T_{P^{n-2}(x)} \otimes \mathcal{O}_{P^{n-2}(x)}(-1)$.*

PROOF. Let $l \in P_x^{n-2}$ be any line in P^{n-1} containing x and let $N = N_{l/P^{n-1}}$ be the normal bundle of l. The tangent space $T_l G(2, n)$ is identified (by Kodaira-Spencer) with the space $H^0(L, N)$, i.e., with the space of normal vector fields on l. The subspace $T_l P_x^{n-2} \subset T_l G(2, n)$ consists of those fields v which vanish at x, i.e., $v(x) = 0$. Hence we have a linear map

$$(N_{P^{n-2}(x)/G(2,n)})_l \mapsto (N_{l/P^{n-1}})_x, \qquad v \mapsto v(x).$$

It is immediate to check that this map is in fact an isomorphism. Let now E be the vector bundle on $P^{n-2}(x)$ whose fiber over a line l is the normal space $(N_{l/P^{n-1}})_x$. We have proved that $N_{P^{n-2}(x)/G(2,n)}$ is isomorphic to E. Let us regard P_x^{n-2} as the projectivization of the vector space $W = T_x P^{n-1}$. Then we have the following description of the bundle E on $P(W)$: the fiber of E at a 1-dimensional linear subspace $\Lambda \subset W$ is W/Λ. This is the standard description of $T_{P(W)} \otimes \mathcal{O}_{P(W)}(-1)$, the so-called Euler sequence, see [45].

(4.2.11). Proposition 4.2.10 implies that the projectivization of the normal bundle $N_{P^{n-2}(x)/G(2,n)}$ is the same as that of the tangent bundle $T_{P^{n-2}(x)}$. So any (generalized) Lie complex defines a 1-dimensional subbundle in the tangent bundle of $P^{n-2}(e_i)$ (this subbundle can be defined only over generic points; over some special points of $P^{n-2}(e_i)$ it may have singularities, i.e., become a non-locally-free coherent sheaf). In other words, we have a field of directions in $P^{n-2}(e_i)$. Let us denote this field by $\Sigma_i(Z)$ in $P^{n-2}(e_i)$. Let H_i be the coordinate hyperplane in P^{n-1} opposite to e_i. The projection from e_i identifies H_i with $P^{n-2}(e_i)$ so we can consider the direction field $\Sigma_i(Z)$ as being defined on H_i. It can be regarded as the choice-free materialization of $\sigma_i(C)$, where C is the stable curve corresponding to Z.

(4.2.12). Let us describe the direction field $\Sigma_i(Z)$ geometrically. Note that since Z is H-invariant, the visible contour Z_p does not change, up to isomorphism, for points p lying in one torus orbit. We shall look how Z_p splits when p goes to a point on H_i not lying on other H_j.

(4.2.13). PROPOSITION. *The field of directions* $\Sigma_i(Z)$ *is well defined at any point of* H_i *not lying on coordinate hyperplanes. For such a point* x *the visible cone* $Z_x = Z \cap P^{n-2}(x)$ *splits into a stable curve (family of lines) in* $P^{n-2}(x)$ *with all lines lying in* H_i *and a plane pencil of lines containing* $\langle x, e_i \rangle$. *The direction* $\Sigma_i(Z)(x)$ *is given by the unique line of this pencil lying in* H_i.

PROOF. Let C be the visible contour of Z at the point $e = (1, \dots, 1)$. Let K be the union of all lines from C. This is a cone with vertex e containing all lines $\langle e, e_i \rangle$. All points of $\langle e, e_i \rangle$ except e are smooth points of K. Let L be the embedded tangent (2-) plane to K along $\langle e, e_i \rangle$. The L lies in $P^{n-2}(e_i)$ and is, by definition, equal to $\sigma_i(C)$. To prove the proposition, it suffices to consider the case when $x \in H_i$ is the barycenter of the coordinate simplex, i.e., all homogeneous coordinates of x are equal to 1 except the ith coordinate, which is equal to 0. Consider the transformation $\gamma(t) = (1, \dots, 1, t, 1, \dots, 1) \in (\mathbb{C}^*)^n$ (t is on the ith place). Then we have $x = \lim_{t \to 0} \gamma(t) \cdot e$. Hence the visible cone at x of the (torus-invariant) complex Z equals $\lim_{t \to 0} \gamma(t) \cdot K$. But the latter limit is the union of the 2-plane L (which is preserved under all $\gamma(t)$) and some other part that lies inside H_i. It remains to show that the intersection $L \cap H_i$ is precisely the line in H_i whose direction is the value at x of the direction field $\Sigma_i(Z)$. This verification is left to the reader.

(4.3). Representation of the space $\overline{M_{0,n}} = G(2, n)//H$ as a blow-up. In the previous sections we have constructed regular birational morphisms σ_i from $\overline{M_{0,n}} = G(2, n)//H$ to projective spaces. When such a morphism is found, it is always desirable to decompose it into simpler ones. Standard examples of "simplest" regular birational morphisms are provided by blow-ups.

(4.3.1). Recall [24, 25] that the *blow-up* (or monoidal transformation, or sigma-process) $\mathrm{Bl}_Y X$ is defined for any smooth closed subvariety Y (which is called the *center*) in a smooth variety X. This is a new smooth variety equipped with a canonical morphism p to X. The morphism p is one-to-one outside Y and for any $y \in Y$ the preimage $p^{-1}(y)$ is canonically identified with the projectivization of the normal space $T_y X / T_y Y$. If $Z \subset X$ is another submanifold not contained in Y, then the *strict preimage* (or proper transform) of Z is the closure $\tilde{Z}$ of $p^{-1}(Z - Y)$ in $\mathrm{Bl}_Y X$. The subvariety $\tilde{Y} \subset \mathrm{Bl}_Y X$ can be, in its turn, blown up, thus giving an iterated blow-up which is abusively denoted $\mathrm{Bl}_Z \mathrm{Bl}_Y X$. This blow-up does not, in general, coincide with $\mathrm{Bl}_Y \mathrm{Bl}_Z(X)$ (it does, when Y and Z are disjoint). Similar constructions can be performed for several subvarieties $Y_1, \dots, Y_r$.

(4.3.2). Our aim in this section is to decompose the morphism σ_i into a sequence of monoidal transformations and therefore to give a "constructive" definition of $\overline{M_{0,n}} = G(2, n)//H$ as an iterated blow-up of a projective space. Proposition 4.2.10 suggests that in order to obtain $\overline{M_{0,n}}$ we should blow up $n-1$ generic points $q_1, \ldots, q_{n-1}$ in P^{n-3}, and all projective subspaces spanned by them. However, an iterated blow-up depends on the ordering of the centers, so the question is delicate.

(4.3.3). THEOREM. *Choose $n-1$ generic points $q_1, \ldots, q_{n-1}$ in P^{n-3}. The variety $\overline{M_{0,n}}$ can be obtained from P^{n-3} by a series of blow-ups of all the projective spaces spanned by q_i. The order of these blow-ups can be taken as follows:*

(1) *Points $q_1, \ldots, q_{n-2}$ and all the projective subspaces spanned by them in order of the increasing dimension;*
(2) *Point q_{n-1}, all lines $\langle q_1, q_{n-1}\rangle, \ldots, \langle q_{n-3}, q_{n-1}\rangle$ and subspaces spanned by them in order of the increasing dimension;*
(3) *Line $\langle q_{n-2}, q_{n-1}\rangle$, planes $\langle q_i, q_{n-2}, q_{n-1}\rangle$, $i \neq n-3$, and all subspaces spanned by them in order of the increasing dimension; etc.*

(4.3.4). REMARK. A representation of $\overline{M_{0,n}}$ as an iterated blow-up of the Cartesian power $(P^1)^{n-3}$ was given by S. Keel [32]. Still another representation of $\overline{M_{0,n}}$ as a blow-up of P^{n-3}, different from the one given here, can be deduced from a more general construction of W. Fulton and R. MacPherson [15]. In the Fulton-MacPherson construction all the centers of the blow-ups have codimension 2.

(4.3.5). The rest of this section will be devoted to the proof of Theorem 4.3.3.

In Proposition 1.4.12 we have constructed certain regular birational morphisms f_I of the general Chow quotient of the Grassmannian $G(k, n)$ to the "secondary variety" of the product of two simplices $\Delta^{k-1} \times \Delta^{n-k-1}$. We want now to analyze these morphisms for the Chow quotient of $G(2, n)$ which is $\overline{M_{0,n}}$ in order to use them as a halfway approximation to a required sequence of blow-ups.

(4.3.6). Recall that for every two-element subset $I = \{i, j\} \subset \{1, \ldots, n\}$ the coordinate subspace $\mathbb{C}^I = \mathbb{C}e_i \oplus \mathbb{C}e_j \subset \mathbb{C}^n$ is a fixed point for the torus action on $G(2, n)$ and hence our torus H acts in the tangent space $T_I = T_{\mathbb{C}^I} G(2, n)$ and on its projectivization. To each H-orbit in $G(2, n)$ whose closure contains $\mathbb{C}^I$ is therefore associated an H-orbit in the projective space $P(T_I)$. The map f_I from $G(2, n)//H = \overline{M_{0,n}}$ to $P(T_I)//H$ is induced by this correspondence.

(4.3.7). As we explained in § 0.2, the Chow quotient of a projective space by a torus H is a toric variety corresponding to the secondary polytope of the point configuration given by the characters of H defining the action. In our case the space T_I is identified with the space of 2 by $n-2$ matrices $\|a_{ij}\|$, $i \in I$, $j \in \overline{I}$, and the action of a torus element $(t_1, \ldots, t_n)$ on such a matrix gives a new matrix $\|t_i^{-1} t_j a_{ij}\|$, $i \in I$, $j \in \overline{I}$. The H-characters are therefore identified with vectors $e_j - e_i$, $i \in I$, $j \in \bar{I}$, in $\mathbb{Z}^n$. These vectors are vertices of the simplicial prism $\Delta^1 \times \Delta^{n-3}$, which we shall also denote $\Delta^I \times \Delta^{\bar{I}}$ to emphasize the dependence on I.

For the case of simplicial prisms $\Delta^1 \times \Delta^k$ triangulations have a complete and simple description.

(4.3.8). Note that the symmetric group S_{k+1} acts on $\Delta^1 \times \Delta^k$ by permuting the vertices of the second factor and hence acts on the triangulations of $\Delta^1 \times \Delta^k$.

Let us describe the *standard* triangulation of $\Delta^1 \times \Delta^k$ used in combinatorial topology [16]. It depends on the numberings of the vertices of factors. To fix these numberings denote the vertices of our prism by pairs (a, b), where $0 \leqslant a \leqslant 1$, $0 \leqslant b \leqslant k$. The triangulation T_{st} consists of the simplices Δ_i, $0 \leqslant i \leqslant k$, where Δ_i is the convex hull of $(0, j)$, $j \leqslant k$, and $(1, j)$, $j \geqslant k$. The characteristic function of this triangulation (i.e., the corresponding vertex of the secondary polytope, see §0.2) equals $\varphi_{st}(i, j) = j + 1$.

(4.3.9). PROPOSITION. *There exist exactly* $(k+1)!$ *regular triangulations of the prism* $\Delta^1 \times \Delta^k$. *They can all be obtained from the standard one by the action of* S_{k+1}.

In fact, all the triangulations of the prism are regular, but we do not need this.

PROOF. Let Σ be the secondary polytope of $\Delta^1 \times \Delta^k$. Its vertices are functions $\varphi_T(i, j)$, $i = 0, 1$; $j = 0, \dots, k$, where T runs over all the regular triangulations. Let us use the original interpretation of the secondary polytope as the Newton polytope of the principal determinant [22, 23]. In our situation this means the following.

Consider a $2 \times (k+1)$ matrix

$$A = \begin{pmatrix} a_{00} & a_{01} & \cdots & a_{0k} \\ a_{10} & a_{11} & \cdots & a_{1k} \end{pmatrix}$$

with indeterminate entries. Consider the polynomial $E(A) = (\prod_{p,j} a_{pj}) \cdot \prod_{0 \leqslant i < j \leqslant k} D_{ij}(A)$, where $D_{ij}(A) = a_{0i}a_{1j} - a_{0j}a_{1i}$ is the minor of A on the ith and jth columns. Then, as shown in [23], Σ is the Newton polytope of E, i.e., the convex hull in $\mathrm{Mat}(2 \times (k+1), \mathbb{R})$ of integral points $\omega = \|\omega_{pj}\| \in \mathrm{Mat}(2 \times (k+1), \mathbb{Z}_+)$ such that the monomial $\prod_{p,j} a_{pj}^{\omega_{pj}}$ enters $E(A)$ with nonzero coefficient. On the other hand, $E(A)$ can be found explicitly by means of the Vandermonde determinant:

$$E(A) = \left(\prod_{p,j} a_{pj}\right) \cdot \det \begin{pmatrix} a_{00}^k & a_{01}^k & \cdots & a_{0k}^k \\ a_{00}^{k-1} a_{10} & a_{01}^{k-1} a_{11} & \cdots & a_{0k}^{k-1} a_{1k} \\ a_{00}^{k-2} a_{10}^2 & a_{01}^{k-2} a_{11}^2 & \cdots & a_{0k}^{k-2} a_{1k}^2 \\ \vdots & \vdots & \vdots & \vdots \\ a_{10}^k & a_{11}^k & \cdots & a_{1k}^k \end{pmatrix}$$

The exponent vector of any monomial in this polynomial is obtained from the vector φ_{st} described in (4.3.8), by a permutation of columns. Proposition 4.3.9 is proved.

(4.3.10). COROLLARY. *The secondary polytope of* $\Delta^1 \times \Delta^k$ *is linearly isomorphic to the convex hull of the* S_{k+1}*-orbit of the point* $(1, 2, \ldots, k+1) \in \mathbb{Z}^{k+1}$.

This polytope is known as the k-dimensional *permutohedron* and denoted $\mathscr{P}_k$. It is a particular case of so-called general hypersimplices associated to homogeneous spaces G/P by I. M. Gelfand and V. V. Serganova [21].

(4.3.11). The toric variety corresponding to $\mathscr{P}_k$ is called the k-dimensional *permutohedral space* and denoted Π^k. The following is one description of this space which generalizes to arbitrary G/P-hypersimplices.

(4.3.12). PROPOSITION [21]. *The permutohedral space* Π^k *is isomorphic to the closure of a generic orbit of torus* $(\mathbb{C}^*)^{k+1}$ *in the space of complete flags of linear subspaces of* $\mathbb{C}^{k+1}$.

We will be interested in a slightly different point of view on Π^k, realizing it as an explicit blow-up of a projective space P^k.

(4.3.13). PROPOSITION. *The permutohedral space* Π^k *can be obtained from the projective space* P^k *by the following sequence of blow-ups. First blow up* $k+1$ *generic points (the projectivizations of basis vectors), then blow up the strict preimages of all coordinate lines joining them, then the strict preimages of coordinate planes, etc.*

PROOF. Let F_i be the space of $(1, 2, \ldots, i)$-flags in $\mathbb{C}^{k+1}$. Let $X = X_k$ be the closure of a generic orbit of $(\mathbb{C}^*)^k$ in F_k and X_i the projection of X to F_i. Then X_1 is the projective space $F_1 = P^k$. It is straightforward to see that each projection $X_i \to X_{i-1}$ realizes X_i as the blow-up of strict preimages of all $(i-1)$-dimensional projective subspaces spanned by the basis vectors of $\mathbb{C}^{k+1}$.

(4.3.14). REMARKS. a) Note that the orbit closure $\Pi^k = X \subset F_k$ can be mapped as well to the projective space of hyperplanes in $\mathbb{C}^{k+1}$. Considering the decomposition of this projection through spaces of $(i, i+1, \ldots, k)$-flags we find that X is represented as the blow-up of the dual projective space $P^{k\vee}$ similar to that of Proposition 4.3.13. The corresponding birational map from P^k to $P^{k\vee}$ is the standard Cremona inversion [24]. Thus the permutohedral space provides an explicit decomposition of the Cremona inversion into sigma-processes and their inverses.

b) In the correspondence between convex polytopes and toric varieties, blowing up the closure of an orbit corresponds to chiseling of the face corresponding to this orbit, see [49]. Proposition 4.3.13 amounts to the following construction of permutohedron from the simplex. First cut out all vertices, then all edges, etc.

(4.3.15). Let us relate the regular birational morphisms $\sigma_i \colon \overline{M_{0,n}} \to P_i^{n-3}$ and $f_{ij} \colon \overline{M_{0,n}} \to \Pi_{ij}^{n-3}$.

(4.3.16). PROPOSITION. *There exist regular birational morphisms* $\tau_{ij} \colon \Pi_{ij}^{n-3} \to P_i^{n-3}$ *such that the composite morphisms*

$$\overline{M_{0,n}} \xrightarrow[f_{ij}]{} \Pi_{ij}^{n-3} \xrightarrow[\tau_{ij}]{} P_i^{n-3}$$

coincide with σ_i.

PROOF. The choice of coordinates identifies the tangent space to $G(2, n)$ at the fixed point $\langle e_i, e_j \rangle$ with the open Schubert cell in $G(2, n)$ consisting of all lines not intersecting the span of points e_m, $m \neq i, j$. By considerations of §1.4 for a generalized Lie complex Z the point $\sigma_i(Z)$ can be read off from the normal spaces to Z at a generic point of $P^{n-2}(e_i)$. Such a generic point can be contracted, by the action of the torus H, to the point $\langle e_i, e_j \rangle$. Therefore, our normal space can be recovered from the part of Z that can be contracted to this fixed point, i.e., from $f_{ij}(Z)$.

(4.3.17). PROPOSITION. *The space* $\overline{M_{0,n}}$ *coincides with the closure of the open stratum* $M_{0,n}$ *in the inverse limit of* Π_{ij}^{n-3} *and* P_i^{n-3}.

PROOF. First let us show that the natural map of $\overline{M_{0,n}}$ into the said inverse limit is injective. This means that if two generalized Lie complexes Z, Z' induce the same algebraic cycles in the projectivizations of all the tangent spaces $T_{\langle e_i, e_j \rangle} G(2, n)$, then they coincide. This is obvious since any H-orbit has a fixed point in its closure.

Hence we have a regular morphism (denote it ψ) of $\overline{M_{0,n}}$ to the inverse limit in question, and this morphism is bijective on $\mathbb{C}$-points. To show that ψ is in fact an isomorphism of algebraic varieties, it suffices to show that the differential of ψ does not annihilate nonzero tangent vectors to $\overline{M_{0,n}}$. This is done similarly to the proof of Theorem 3.3.14.

(4.3.18). PROPOSITION. *The map*

$$f_{ij} \times \pi_i \colon \overline{M_{0,n}} \to \Pi_{ij}^{n-3} \times \overline{M_{0,n-1}}$$

is an embedding of algebraic varieties.

PROOF. We shall check only the injectivity on $\mathbb{C}$-points, leaving the injectivity on tangent vectors to the reader. Let $\mathcal{T}$ be a tree bounding the endpoints $1, \dots, n$ and $M(\mathcal{T})$ be the corresponding stratum of $\overline{M_{0,n}}$ or, what is the same, the corresponding Chow stratum in $G(2, n)//H$. Let Z be a generalized Lie complex from $M(\mathcal{T})$ and $C = Z_u$ be the corresponding stable n-pointed curve of genus 0.

The value of $f_{ij}(Z)$ depends only of the components of Z containing the fixed point $\langle e_i, e_j\rangle$. The components correspond to vertices of $\mathcal{T}$ lying on the path $[ij]$ with the shortest edge path joining the ith and the jth end-points. Denote these vertices, in natural order from i to j, by $v_1, \dots, v_r$. Let s_ν be the number of edges meeting v_ν. To the chain of vertices v_ν corresponds a chain of irreducible components $C_1, \dots, C_r$ of C and on each C_ν we have s_ν marked points. The projective configurations of these groups of points are precisely what is taken into account by the map f_{ij} on the curve from $M(\mathcal{T})$. Now our assertion means that the isomorphism class of the stable n-pointed curve C can be recovered from two groups of data:

a) The isomorphism class of the stable $(n-1)$-pointed curve $\pi_i(C)$;
b) The isomorphism class of the stable curve

$$C' = C_1 \cup \cdots \cup C_r \subset C$$

pointed by x_i, x_j and all the marked and double points of C lying on C'.

This is obvious and Proposition 4.3.18 is proved.

(4.3.19). Let us now connect the spaces $\overline{M_{0,n-1}}$ and Π_{ij}^{n-3}. We view the latter space as the blow-up of P_i^{n-3} at all vertices, edges, etc., of the coordinate simplex formed by points q_m, $m \neq j$. Projecting these points from q_j gives a circuit in the space P_{ij}^{n-4} of lines in P_i^{n-3} meeting q_j. The space P_{ij}^{n-4} is in the same relation to $\overline{M_{0,n-1}} = \pi_i(\overline{M_{0,n}})$ as P_i^{n-3} was to $\overline{M_{0,n}}$. In particular, we have the regular birational morphism $\sigma_{j/i}$ from $\overline{M_{0,n-1}}$ to P_{ij}^{n-4}. On the other hand, consider the blow-up $\mathrm{Bl}_{p_i}\Pi_{ij}^{n-3}$. It also possesses a projection to P_{ij}^{n-4}. Proposition 4.3.18 implies the following corollary:

(4.3.20). Corollary. *The space $\overline{M_{0,n}}$ coincides with the closure of $M_{0,n}$ in the fiber product of $\overline{M_{0,n-1}}$ and $\mathrm{Bl}_{p_j}\Pi_{ij}^{n-3}$ over P_{ij}^{n-4}.*

This corollary can be reformulated as follows. Suppose we know the way of constructing $\overline{M_{0,n-1}}$ as an iterated blow-up of the projective space P_{ij}^{n-4} whose centers are proper transforms of smooth subvarieties $Y_1, \dots, Y_r$. Then we have in $\mathrm{Bl}_{p_j}P_j^{n-3}$ the varieties $\tilde{Y}_\nu$ that are blow-ups of cones over Y_ν with apex p_j. The corollary means that if we perform the sequence of blow-ups of $\mathrm{Bl}_{p_j}\Pi_{ij}^{n-3}$ with centers in proper transforms of $\tilde{Y}_\nu$ then we obtain $\overline{M_{0,n}}$.

In other words, the problem of recovering $\overline{M_{0,n}}$ from the partial blow-up $\mathrm{Bl}_{p_i}\Pi_{ij}^{n-3}$ is equivalent to the problem of recovering $\overline{M_{0,n-1}}$ from the projective space. This gives an inductive proof of Theorem 4.3.3.

REFERENCES

1. M. F. Atiyah, *The geometry and physics of knots*, Cambridge Univ. Press, Cambridge, 1991.
2. H. Baker, *Principles of geometry*, vols. 3, 4, Cambridge Univ. Press, Cambridge, 1925, republished Ungar, New York, 1963.
3. D. Barlet, *Espace analytique reduit des cycles analytiques comlpexes compacts*, Lecture Notes in Math., vol. 482, Springer-Verlag, Berlin, Heidelberg, 1975, pp. 1–158.
4. ———, *Note on the "Joint theorem"*, preprint (1991).
5. D. Bayer and I. Morrison, *Gröbner bases and geometric invariant theory*. I, J. Symbolic Comput. **6** (1988), 209–217.
6. D. Bayer and M. Stillman, *A theorem on refining division orders by the reverse lexicographic order*, Duke Math. J. **55** (1987), 321–328.
7. A. A. Beilinson, R. D. MacPherson, and V. V. Schechtman, *Notes on motivic cohomology*, Duke Math. J. **54** (1977), 679–710.
8. L. J. Billera and B. Sturmfels, *Fiber polytopes*, Ann. of Math. (2) **135** (1992), 527–549.
9. L. M. Brown, *The map of* S_m *by means of its* n*-ic primals*, J. London Math. Soc. **5** (1930), 168–176.
10. A. Bialynicki-Birula and A. J. Sommese, *A conjecture about compact quotients by tori*, Adv. Studies in Pure Math. **8** (1987), 59–68.
11. A. B. Coble, *Algebraic geometry and theta-functions*, Colloquium Publ., vol. 10, Amer. Math. Soc., Providence, RI, 1928, Third edition, 1969.
12. P. Deligne, *Resumé des premièrs exposés de A. Grothendieck*, SGA 7, Expose I, Lecture Notes in Math., vol. 288, Springer-Verlag, Heidelberg, Berlin, 1972, pp. 1–24.
13. ———, *Theorie de Hodge* II, Inst. Hautes Etudes Sci. Publ. Math. **40** (1971), 5–58.
14. I. Dolgachev and D. Ortland, *Point sets in projective spaces and theta functions*, Astérisque **165** (1988).
15. W. Fulton and R. D. MacPherson, *The compactification of configuration spaces*, preprint (1991).
16. P. Gabriel and M. Zisman, *Calculus of fractions and homotopy theory*, Springer-Verlag, New York, 1967.
17. A. M. Gabrielov, I. M. Gelfand, and M. V. Losik, *Combinatorial calculation of characteristic classes*. I, Funktsional. Anal. i Prilozhen. **9** (1975), no. 2, 12–28; English transl., Functional Anal. Appl. **9** (1975), 103–115, reprinted in: *Collected papers of I. M. Gelfand*, vol. 3, Springer-Verlag, Berlin, Heidelberg, 1989, pp. 407–419.
18. I. M. Gelfand, *General theory of hypergeometric functions*, Dokl. Akad. Nauk SSSR **288** (1986), no. 1, 14–18; English transl., Soviet Math. Dokl. **33** (1986), 573–577, reprinted in: *Collected papers of I. M. Gelfand*, vol. 3, Springer-Verlag, Berlin, Heidelberg, 1989, pp. 877–881.
19. I. M. Gelfand, R. M. Goresky, R. W. MacPherson, and V. V. Serganova, *Combinatorial geometries, convex polyhedra and Schubert cells*, Adv. in Math **63** (1978), 301–316.
20. I. M. Gelfand and R. W. MacPherson, *Geometry in Grassmannians and a generalization of the dilogarithm*, Adv. in Math. **44** (1982), 279–312.
21. I. M. Gelfand and V. V. Serganova, *Combinatorial geometries and torus strata on compact homogeneous spaces*, Uspekhi Mat. Nauk **42** (1987), 107–134; English transl. in Russian Math. Surveys **42** (1987), reprinted in: *Collected papers of I. M. Gelfand*, vol. 3, Springer-Verlag, Berlin, Heidelberg, 1989, pp. 926–958.
22. I. M. Gelfand, A. V. Zelevinsky, and M. M. Kapranov, *Discriminants of polynomials in several variables and triangulations of Newton polytopes*, Algebra i Analiz **2** (1990); English transl. in Leningrad Math. J. **2** (1991).

23. ______, *Newton polytopes of principal A-determinants*, Dokl. Akad. Nauk. SSSR **308** (1989), no. 1, 20–23; English transl., Soviet Math. Dokl. **40** (1990), 278–281.

24. Ph. Griffiths and J. Harris, *Principles of algebraic geometry*, Wiley-Interscience, New York, 1978.

25. R. Hartshorne, *Algebraic geometry*, Graduate Texts in Math., vol. 52, Springer-Verlag, New York, Heidelberg, Berlin, 1977.

26. Y. Hu, *Geometry and topology of quotient varieties*, Thesis, MIT, 1991.

27. C. M. Jessop, *A treatise on the line complex*, Cambridge Univ. Press, Cambridge, 1903, republished, Chelsea, New York, 1969.

28. F. Junker, *Über symmetrische Funktionen von mehreren Reihen von Veränderlichen*, Math. Ann. **43** (1893), 225–270.

29. M. M. Kapranov, *Veronese curves and Grothendieck-Knudsen moduli space* $\overline{M_{0,n}}$, J. Algebraic Geom. **2** (1993) (to appear).

30. M. M. Kapranov, B. Sturmfels, and A. V. Zelevinsky, *Chow polytopes and general resultants*, Duke Math. J. (to appear).

31. ______, *Quotients of toric varieties*, Math. Ann. **290** (1991), 643–655.

32. S. Keel, *Intersection theory of moduli spaces of stable n-pointed curves of genus zero*, Trans. Amer. Math. Soc. **330** (1992), 545–574.

33. A. N. Kirillov and N. Yu. Reshetikhin, *Multiplicity of weights in irreducible representations of a general linear superalgebra*, Funktsional. Anal. i Prilozhen. **22** (1988); English transl., Functional Anal. Appl. **22** (1988), 328–330.

34. ______, *The Bethe Ansatz and the combinatorics of Young tableaux*, Zap. Nauchn. Sem. Leningrad. Otdel. Mat. Inst. Steklov. (LOMI) **155** (1986), 65–115; English transl., J. Soviet. Math. **41** (1988), 925–955.

35. F. C. Kirwan, *Partial desingularizations of quotients of nonsingular varieties and their Betti numbers*, Ann. of Math. (2) **122** (1985), 41–85.

36. A. A. Klyachko, *Orbits of the maximal torus on the flag space*, Funktsional. Anal. i Prilozhen. **19** (1985), no. 1, 77–78; English transl. in Functional Anal. Appl. **19** (1985), no. 1.

37. F. F. Knudsen, *The projectivity of moduli spaces of stable curves*. II: *The stacks* $M_{g,n}$, Math. Scand. **52** (1983), 161–199.

38. I. Macdonald, *Symmetric functions and Hall polynomials*, Clarendon Press, Oxford, 1979.

39. P. A. MacMahon, *Memoir on symmetric functions of the roots of systems of equations*, Phil. Trans. **181** (1890), 481–536, reprinted in: *Collected Papers*, vol. 2, MIT Press, Cambridge, MA, 1986, pp. 32–84.

40. R. D. MacPherson, *The combinatorial formula of Gabrielov, Gelfand and Losik for the first Pontrjagin class*, Sem. Bourbaki no. 497 (1976–77), Lecture Notes in Math., vol. 677, Springer-Verlag, Berlin, Heidelberg, New York, 1978, pp. 105–124.

41. K. Matsumoto, T. Sasaki, and M. Yoshida, *The monodromy of the period map of a 4-parameter family of* $K3$ *surfaces and the hypergeometric function of type* $G(3, 6)$, Internat. J. Math. **3** (1992), 1–164.

42. D. Mumford and J. Fogarty, *Geometric invariant theory*, Springer-Verlag, New York, 1982.

43. M. Nagata, *On the normality of the Chow variety of positive 0-cycles in an algebraic variety*, Mem. College. Sci. Univ. Kyoto **29** (1955), 165–176.

44. A. Neeman, *Zero-cycles in* P^n, Adv. in Math. **89** (1991), 217–227.

45. K. Okonek, M. Schneider, and H. Spindler, *Vector bundles on complex projective spaces*, Progr. Math., vol. 3, Birkhäuser, Boston, 1980.

46. T. G. Room, *The geometry of determinantal loci*, Cambridge Univ. Press, Cambridge, 1938.

47. E. Sernesi, *Topics on families of projective schemes*, Queen's Papers in Pure and Appl. Math. **73** (1986).

48. B. Sturmfels, *Gröbner bases of toric varieties*, Tôhoku Math. J. **43** (1991), 249–261.

49. T. Oda, *Convex bodies and algebraic geometry*, Springer-Verlag, Berlin, Heidelberg, New York, 1988.

50. B. L. van der Waerden, *Einführung in die algebraische Geometrie*, Springer-Verlag, Berlin, 1955.

DEPARTMENT OF MATHEMATICS, NORTHWESTERN UNIVERSITY, EVANSTON, ILLINOIS 60208
E-mail address: kapranov@math.nwu.edu

ADVANCES IN SOVIET MATHEMATICS
Volume 16, Part 2, 1993

Reconstructing Monoidal Categories

DAVID KAZHDAN AND HANS WENZL

Introduction

Let $\mathfrak{g}$ be a simple complex Lie algebra, $\mathscr{C}_{\mathfrak{g}}$ the tensor category of finite dimensional $\mathfrak{g}$-representations of $\mathfrak{g}$. In his remarkable paper [D1] (see also [D2] for a simplified version) V. Drinfeld has shown that any formal deformation $\mathfrak{F}$ of $\mathscr{C}_{\mathfrak{g}}$ as a braided category is equivalent to the category $\mathscr{C}_{\mathfrak{g},q}$ of finite dimensional representations of the corresponding quantum group $U_q(\mathfrak{g})$. S. Shnider (see [S]) has shown that any deformation of $\mathscr{C}_{\mathfrak{g}}$ as a monoidal category is also equivalent to $\mathscr{C}_{\mathfrak{g},q}$. The proofs are based on a deformation technique and on the computation of the cohomology groups describing the deformations. It is natural to ask if it is possible to characterize the categories $C_{\mathfrak{g},q}$ intrinsically. One of the results of this paper is the following

THEOREM A_∞. *Let $\mathscr{C}$ be a rigid monoidal semisimple $\mathbb{C}$-category such that the Grothendieck semiring $[\mathscr{C}]$ of $\mathscr{C}$ is isomorphic to the semiring $R_{N,\infty}$ of representations of the Lie algebra SL_N. Then $\mathscr{C}$ is equivalent to a twist (see the end of § 1) of $\mathscr{C}_{SL_N,q}$ for some $q \in \mathbb{C}^*$, and either $q = 1$ or q is not a root of unity. Moreover q is uniquely determined up to $q \to q^{-1}$.*

To describe the extension of Theorem A_∞ to the case when q is a primitive root of unity of order ℓ, $\ell > 1$, we need some notations. For any $q \in \mathbb{C}^*$ we denote by $\mathscr{R}_{N,q}$ the Hopf algebra of functions on the quantum group $SL_{N,q}$ (see [FRT, 1.3]) and denote by $\widetilde{\mathscr{C}}_{N,q}$ the category of finite dimensional $\mathscr{R}_{N,q}$-comodules. If $q = 1$ or q is not a root of unity, then $\widetilde{\mathscr{C}}_{N,q}$ is a semisimple braided category and we write $\mathscr{C}_{N,q} = \widetilde{\mathscr{C}}_{N,q}$. If q is a primitive root of unity of order $\ell > N$, then there exists a braided "subquotient" category $\mathscr{C}_{N,q}$ of $\widetilde{\mathscr{C}}_{N,q}$ (see [A, 3.7]) such that the isomorphism classes of simple objects of $\mathscr{C}_{N,q}$ are in one-to-one correspondence with the

1991 *Mathematics Subject Classification*. Primary 18D10; Secondary 17B37.
The first author was partially supported by NSF Grant 33-576-7956-2.
The second author was partially supported by NSF Grant DMS91-08009.

highest weights λ of SL_N satisfying the condition $\langle\lambda, \rho\rangle < \ell$. Let $R_{N,\ell}$ be the Grothendieck semiring of $\mathscr{C}_{N,q}$. It is easy to see that $R_{N,\ell}$ does not depend on a choice of a primitive root q of order ℓ.

Theorem A_ℓ. *Let $\mathscr{C}$ be a rigid monoidal semisimple $\mathbb{C}$-category such that $[\mathscr{C}] = R_{N,\ell}$. Then $\mathscr{C}$ is equivalent to a "twist" of $\mathscr{C}_{N,q}$ for some primitive root q of order ℓ of 1 and $q \in \mathbb{C}$ is uniquely determined up to $q \to q^{-1}$.*

Remark. Theorems A_∞ and A_ℓ in the case $N = 2$ are proven in [K].

At the end of the introduction we discuss the structure of the proof. The body of the proof is in § 4, while in the first three sections we prove some auxiliary results. Let $X \in \mathscr{C}$ be the object corresponding to the standard representation of SL_N. Then $X^{\otimes 2}$ is a direct sum of two objects, $X^{\otimes 2} = X_a \oplus X_s$. Let $a \in \mathrm{End}(X^{\otimes 2})$ be the projection onto X_a. We define elements $a_1, a_2 \in \mathrm{End}(X^{\otimes 3})$ by the formula $a_1 = a \otimes \mathrm{id}_X$, $a_2 = \mathrm{id}_X \otimes a$. It is easy to see that there exists $\gamma \in \mathbb{C}$ such that $a_1 a_2 a_1 - \gamma a_1 = a_2 a_1 a_2 - \gamma a_2$. We show in Proposition 4.2 (using Proposition 1.1) that the rigidity of $\mathscr{C}$ implies that $\gamma \neq 0$. Therefore we can find a number $q \in \mathbb{C}$ such that $q + q^{-1} = \gamma^{-1} - 2$. We define $T \in \mathrm{End}(X^{\otimes 2})$ as the linear combination $T = q\,\mathrm{id} - (1+q)a$. For any $n, i \in \mathbb{Z}$, $n > i$, we define $T_i^{(n)} = \mathrm{id}_{X^{\otimes i-1}} \otimes T \otimes \mathrm{id}_{X^{\otimes n-i-1}} \in \mathrm{End}(X^{\otimes n})$. It is easy to see that endomorphisms $T_i^{(n)}$, $1 \leq i \leq n$, satisfy the Hecke relations and therefore we have defined a morphism of the Hecke algebra $H_n(q)$ to $\mathrm{End}(X^{\otimes n})$. Using the trace on the algebra $\mathrm{End}(X^{\otimes n})$ and the results of § 3 we show that this morphism is surjective. Any object in $\mathscr{C}$ is a direct summand of $X^{\otimes n}$ and therefore it is not surprising that we can reconstruct the category $\mathscr{C}$ from the information on $\mathrm{End}(X^{\otimes n})$. For this we develop some general reconstruction technique in § 2.

Acknowledgment. Hans Wenzl would like to thank the Mathematics Department of Harvard University for its hospitality and support.

§ 1. Monoidal categories

A monoidal category $\mathscr{C}$ is a triple $\mathscr{C} = (\tilde{\mathscr{C}}, \otimes, a)$, where $\tilde{\mathscr{C}}$ is a category, $\otimes : \tilde{\mathscr{C}} \times \tilde{\mathscr{C}} \to \tilde{\mathscr{C}}$ is a functor, and a is the natural transformation between the functors $\otimes(\otimes \times \mathrm{id})$ and $\otimes(\mathrm{id} \times \otimes)$ from $\tilde{\mathscr{C}} \times \tilde{\mathscr{C}} \times \tilde{\mathscr{C}}$ to $\tilde{\mathscr{C}}$, $a = \{a_{X,Y,Z} : (X \otimes Y) \otimes Z \longrightarrow X \otimes (Y \otimes Z),\ X, Y, Z \text{ in } \tilde{\mathscr{C}}\}$, such that all $a_{X,Y,Z}$ are isomorphisms, the pentagon axiom is satisfied, and there exists an object U in $\tilde{\mathscr{C}}$ and an isomorphism $u : U \otimes U \xrightarrow{\sim} U$ such that the functors $X \to X \otimes U$ and $X \to U \otimes X$ are autoequivalences of $\tilde{\mathscr{C}}$ (see [DM]). We call such a pair (U, u) the identity object of $\mathscr{C}$. It is clear that the identity object is determined uniquely up to a unique isomorphism.

Let $\mathscr{C} = (\tilde{\mathscr{C}}, \otimes, a)$ be a monoidal category, and (U, u) the identity object in $\mathscr{C}$.

LEMMA 1.1. a) *For any object* X *in* $\widetilde{\mathscr{C}}$ *there exist unique isomorphisms* $r_X : X \xrightarrow{\sim} X \otimes U$, $l_X : X \to U \otimes X$ *such that*

$$a_{X,U,U} \circ (r_X \otimes \mathrm{id}_U) = \mathrm{id}_X \otimes u, \quad \mathrm{id}_U \otimes l_X = a_{U,U,X} \circ (\mathrm{id}_X \otimes u).$$

b) $r_U = u$.

c) *For any morphism* $\alpha : Y \to X$ *in* $\widetilde{\mathscr{C}}$ *the diagrams*

$$\begin{array}{ccc} Y & \xrightarrow{\alpha} & X \\ \downarrow{\scriptstyle r_Y} & & \downarrow{\scriptstyle r_X} \\ Y \otimes U & \xrightarrow{\alpha\otimes 1} & X \otimes U \end{array} \qquad \begin{array}{ccc} Y & \xrightarrow{\alpha} & X \\ \downarrow{\scriptstyle l_Y} & & \downarrow{\scriptstyle l_X} \\ U \otimes Y & \xrightarrow{\alpha\otimes 1} & U \otimes X \end{array}$$

are commutative.

PROOF. Easy.

We say that a monoidal category $\mathscr{C} = (\widetilde{\mathscr{C}}, \otimes, a)$ is strict if for any X, Y, Z in $\widetilde{\mathscr{C}}$ we have $(X\otimes Y)\otimes Z = X\otimes(Y\otimes Z)$, $a_{X,Y,Z} = \mathrm{id}$, and there exists an identity object $(\mathbb{1}, u)$ in $\mathscr{C}$ such that $\mathbb{1} \otimes X = X \otimes \mathbb{1} = X$ for all X in $\widetilde{\mathscr{C}}$ and $u: \mathbb{1} \to \mathbb{1}$ is the identity morphism. Mac Lane's theorem says that any monoidal category is equivalent to a strict one (see [M, Chapter 7]). To simplify formulas we will often assume that our monoidal categories are strict. Let $\mathscr{C}$ be a strict monoidal category. We denote by $\overline{\mathscr{C}}$ the set of isomorphism classes of objects in $\widetilde{\mathscr{C}}$. We say that an object X in $\widetilde{\mathscr{C}}$ is rigid if there exists an object Y in $\widetilde{\mathscr{C}}$ and a pair of morphisms

$$i_X : \mathbb{1} \to X \otimes Y, \qquad e_X : Y \otimes X \to \mathbb{1},$$

such that the compositions

$$X = \mathbb{1} \otimes X \xrightarrow{i_X\otimes \mathrm{id}} X \otimes Y \otimes X \xrightarrow{\mathrm{id}\otimes e_X} X,$$

$$Y = Y \otimes \mathbb{1} \xrightarrow{\mathrm{id}\otimes i_X} Y \otimes X \otimes Y \xrightarrow{e_X\otimes \mathrm{id}} Y$$

are equal to id_X and id_Y correspondingly. In this case we say that Y is dual to X. It is easy to see that such a triple (Y, i_X, e_X), if it exists, is unique up to a unique isomorphism. We say that the category $\mathscr{C}$ is rigid if all its objects are rigid and the induced map from $\overline{\mathscr{C}}$ to $\overline{\mathscr{C}}$ is a bijection. In this case there exists an equivalence $* : \widetilde{\mathscr{C}} \to \widetilde{\mathscr{C}}^\circ$ of categories such that X^* is dual to X for all X in $\widetilde{\mathscr{C}}$ and for any $\alpha \in \mathrm{Hom}_{\widetilde{\mathscr{C}}}(X, Y)$ the morphism $\alpha^* \in \mathrm{Hom}_{\widetilde{\mathscr{C}}}(Y^*, X^*)$ is the composition

$$Y^* = Y^* \otimes \mathbb{1} \xrightarrow{\mathrm{id}\otimes i_X} Y^* \otimes X \otimes X^* \xrightarrow{\mathrm{id}\otimes\alpha\otimes\mathrm{id}} Y^* \otimes Y \otimes X^* \xrightarrow{e_Y\otimes\mathrm{id}} \mathbb{1} \otimes X^* = X^*$$

and such an equivalence is unique up to a unique isomorphism.

Moreover, the functor $*$ has a natural monoidal structure. That is, for any X, Y in $\widetilde{\mathscr{C}}$ the morphisms $i_{X\otimes Y}$, $e_{X\otimes Y}$ defined as the compositions

$$i_{X\otimes Y} : \mathbb{1} \xrightarrow{i_X} X \otimes X^* = X \otimes \mathbb{1} \otimes X^* \xrightarrow{\mathrm{id}\otimes i_Y\otimes\mathrm{id}} (X \otimes Y) \otimes (Y^* \otimes X^*),$$

$$e_{X\otimes Y} : (Y^* \otimes X^*) \otimes (X \otimes Y) \xrightarrow{\mathrm{id}\otimes e_X\otimes\mathrm{id}} Y^* \otimes \mathbb{1} \otimes Y = Y^* \otimes Y \xrightarrow{e_Y} \mathbb{1}$$

define isomorphisms of $(X \otimes Y)^*$ with $Y^* \otimes X^*$. We will freely use this identification and will, for example, identify $(X^{\otimes n})^*$ with $(X^*)^{\otimes n}$ for all X in $\widetilde{\mathscr{C}}$, $n \in \mathbb{Z}^+$.

It is easy to see that for any X, Y, Z in $\widetilde{\mathscr{C}}$ the map $\mathrm{Hom}_{\widetilde{\mathscr{C}}}(X \otimes Y, Z) \to \mathrm{Hom}_{\widetilde{\mathscr{C}}}(Z, Z \otimes Y^*)$ given by $\alpha \to (\mathrm{id}_X \otimes i_Y) \circ (\alpha \otimes \mathrm{id}_{Y^*})$ is an isomorphism for all X, Y, Z in $\widetilde{\mathscr{C}}$. In particular, we have an isomorphism $\mathrm{Hom}_{\widetilde{\mathscr{C}}}(X \otimes Y, \mathbb{1}) \xrightarrow{\sim} \mathrm{Hom}_{\widetilde{\mathscr{C}}}(X, Y^*)$. From now on we will always assume that the category $\widetilde{\mathscr{C}}$ is abelian and that $\otimes$ is an additive functor. Then $\otimes$ is an exact functor and the object $\mathbb{1}$ is semisimple (see [DM, 1.17]). We assume in addition that the object $\mathbb{1}$ is simple. In this case the ring $\mathrm{End}_{\widetilde{\mathscr{C}}}(\mathbb{1})$ is a field (see [DM, 1.17]) and we assume that it is an algebraically closed field, say F, of characteristic zero.

The isomorphism $\mathbb{1} \otimes X = X$ induces a structure of an F-module on each group $\mathrm{Hom}_{\widetilde{\mathscr{C}}}(X, Y)$, X, Y in $\widetilde{\mathscr{C}}$. We define

$$h(X, Y) = \dim_F \mathrm{Hom}_{\widetilde{\mathscr{C}}}(X, Y).$$

We assume in addition that $h(X, Y) < \infty$ for all X, Y in $\widetilde{\mathscr{C}}$. Then for all simple objects X in $\widetilde{\mathscr{C}}$ we have $\mathrm{End}_{\widetilde{\mathscr{C}}}(X) = F$. A strict monoidal category $\mathscr{C} = (\widetilde{\mathscr{C}}, \otimes, \mathbb{1})$ satisfying these conditions will be called a monoidal F-category.

Let $\mathscr{C}$ be a semisimple monoidal F-category, and X a simple object in $\mathscr{C}$. Since X is irreducible, we have $h(\mathbb{1}, X \otimes X^*) = h(X, X) = 1$. Since $\mathscr{C}$ is semisimple, we have $h(X \otimes X^*, \mathbb{1}) = h(\mathbb{1}, X \otimes X^*) = 1$. Let $\widetilde{\Gamma}_X \in \mathrm{Hom}_{\mathscr{C}}(X \otimes X^*, \mathbb{1})$ be the unique morphism such that $\widetilde{\Gamma}_X \circ i_X = \mathrm{id}$. Using the isomorphism $\mathrm{Hom}_{\widetilde{\mathscr{C}}}(X \otimes X^*, \mathbb{1}) \xrightarrow{\sim} \mathrm{Hom}_{\widetilde{\mathscr{C}}}(X, X^{**})$ we define $\Gamma_X \in \mathrm{Hom}_{\widetilde{\mathscr{C}}}(X, X^{**})$ to be the image of $\widetilde{\Gamma}_X$. Since X, X^{**} are simple and $\Gamma_X \neq 0$ we see that Γ_X is an isomorphism. It is clear from the definition that the composition $\mathbb{1} \xrightarrow{i_X} X \otimes X^* \xrightarrow{\Gamma_X \otimes \mathrm{id}} X^{**} \otimes X^* \xrightarrow{e_{X^*}} \mathbb{1}$ is equal to the identity morphism. Let Y be a simple object in $\widetilde{\mathscr{C}}$ and $n \in \mathbb{Z}^+$ a number such that $h(Y, X^{\otimes n}) = 1$. Then we define the number $\eta_{Y|X^{\otimes n}} \in F^*$ in the following way. Fix a nonzero morphism $j : Y \to X^{\otimes n}$ and let $p : X^{\otimes n} \to X^{\otimes n}$ be the unique projection on $j(Y)$ and $\widetilde{p} \overset{\mathrm{df}}{=} j^{-1} \circ p : X^{\otimes n} \to Y$. We define $\eta_{Y|X^{\otimes n}}$ as the composition

$$\mathbb{1} \xrightarrow{i_{X^{\otimes n}}} X^{\otimes n} \otimes (X^{\otimes n})^* \xrightarrow{p \otimes \mathrm{id}} X^{\otimes n} \otimes (X^{\otimes n})^*$$
$$\xrightarrow{\Gamma_X^{\otimes n} \otimes \mathrm{id}} (X^{\otimes n})^{**} \otimes (X^{\otimes n})^* \xrightarrow{e_{(X^{\otimes n})^*}} \mathbb{1}.$$

It is clear that the definition does not depend on the choice of j.

Proposition 1.1. *We have* $\eta_{Y|X^{\otimes n}} \neq 0$.

Proof. Since Γ_X is an isomorphism, we have $h(Y, (X^{\otimes n})^{**}) = 1$ and therefore the morphisms $j^{**} \circ \Gamma_Y$, $\Gamma_{X^{\otimes n}} \circ j \in \operatorname{Hom}_{\widetilde{\mathscr{C}}}(Y, (X^{\otimes n})^{**})$ are nonzero and proportional. Therefore there exists $\gamma \in F^*$ such that $\gamma j^{**} \circ \Gamma_Y = \Gamma_X^{\otimes n} \circ j$. It is easy to see that the diagram

$$\begin{array}{ccccccccc}
\mathbb{1} & \xrightarrow{i_{X^{\otimes n}}} & X^{\otimes n} \otimes (X^{\otimes n})^* & \xrightarrow{p \otimes \mathrm{id}} & X^{\otimes n} \otimes (X^{\otimes n})^* & \xrightarrow{\Gamma_{X^{\otimes n}} \otimes \mathrm{id}} & (X^{**})^{\otimes n} \otimes (X^{\otimes n})^* & \xrightarrow{e_{(X^{\otimes n})^*}} & \mathbb{1} \\
\uparrow{\scriptstyle \mathrm{id}} & & & & \uparrow{\scriptstyle j \otimes \mathrm{id}} & & \uparrow{\scriptstyle j^{**} \otimes \mathrm{id}} & & \\
\mathbb{1} & \xrightarrow{i_Y} & Y \otimes Y^* & \xrightarrow{\mathrm{id} \otimes \widetilde{p}^*} & Y \otimes (X^{\otimes n})^* & \xrightarrow{\Gamma_Y} & Y^{**} \otimes (X^{\otimes n})^* & &
\end{array}$$

is commutative. Therefore $\eta_{Y|X^{\otimes n}} = \gamma$. Proposition 1.1 is proved.

Corollary. *If $Z \overset{\mathrm{df}}{=} \ker p \subset X^{\otimes n}$ is irreducible, then $\eta_{Y|X^{\otimes n}} \neq 1$.*

Proof. In this case we can define an element $\eta_{Z|X^{\otimes n}} \in F^*$ and it is clear that $\eta_{Y|X^{\otimes n}} + \eta_{Z|X^{\otimes n}} = 1$. Corollary is proved.

Let $\mathscr{C} = (\widetilde{\mathscr{C}}, \otimes, \mathrm{id})$ be a strict monoidal category. A braiding s on $\mathscr{C}$ is a natural transformation between the functors $\otimes$ and $\otimes \circ P$, where $P\colon \widetilde{\mathscr{C}} \times \widetilde{\mathscr{C}} \to \widetilde{\mathscr{C}} \times \widetilde{\mathscr{C}}$ is the permutation $(X, Y) \to (Y, X)$ (so that $s = (s_{X,Y})$, where $s_{X,Y} \in \operatorname{Hom}(X \otimes Y, Y \otimes X)$ for X, Y in $\widetilde{\mathscr{C}}$) such that

a) all $s_{X,Y}$ are isomorphisms;

b) for any X, Y, Z in $\widetilde{\mathscr{C}}$ we have

$$s_{X \otimes Y, Z} = (s_{X,Z} \otimes \mathrm{id}_Y) \circ (\mathrm{id}_X \otimes s_{Y,Z}),$$
$$s_{X, Y \otimes Z} = (\mathrm{id}_X \otimes s_{X,Z}) \circ (s_{X,Y} \otimes \mathrm{id}_Z);$$

c) $s_{x,\mathbb{1}} = s_{\mathbb{1},X} = \mathrm{id}_X$ for all X in $\widetilde{\mathscr{C}}$.

Let $\mathscr{C}$ be a semisimple monoidal category, $\Lambda_{\mathscr{C}}$ the set of equivalence classes of simple objects in $\mathscr{C}$. We denote by $[\mathscr{C}]$ the free abelian semigroup generated by elements e_i, $i \in \Lambda_{\mathscr{C}}$, and for any $i, j, k \in \Lambda_{\mathscr{C}}$ we choose representatives X_i, X_j, X_k in $\widetilde{\mathscr{C}}$ and define

$$C_{ij}^k = \dim_F \operatorname{Hom}(X_k, X_i \otimes X_j).$$

We introduce a semiring structure on $[\mathscr{C}]$ in such a way that $e_i e_j = \sum_{k \in \Lambda_{\mathscr{C}}} C_{ij}^k e_k$.

It is clear that the semiring $[\mathscr{C}]$ is associative and that $[\mathscr{C}]$ is commutative if $\mathscr{C}$ admits a braiding.

Let $[\mathscr{C}]^+$ be the free abelian group with generators e_i, $i \in \Lambda_{\mathscr{C}}$. We define the ring structure on $[\mathscr{C}]^+$ in such a way that $e_i e_j = \sum_{k \in I_{\mathscr{C}}} C_{ij}^k e_k$.

Examples. 1) Let $\mathscr{R} = (\mathscr{R}, m, \Delta, 1, \varepsilon)$ be a Hopf algebra over F, $\widetilde{\mathscr{C}} = \mathscr{C}_{\mathscr{R}}$ be the category of finite dimensional $\mathscr{R}$-comodules and the morphism $\otimes : \widetilde{\mathscr{C}} \times \widetilde{\mathscr{C}} \to \widetilde{\mathscr{C}}$ is the usual tensor product. We denote by $\mathbb{1}$ the $\mathscr{R}$-comodule

$(F, \rho: F \to F \otimes_F \mathscr{R} = \mathscr{R})$, where $\rho(t) = t$. Then $\mathscr{C} = (\widetilde{\mathscr{C}}, \otimes, \mathrm{id}, \mathbb{1})$ is a strict rigid monoidal category.

2) Fix a number $N \in \mathbb{Z}^+$. Let $\mathscr{R}_N$ be the ring of regular functions on the algebraic F-group SL_N. The ring $\mathscr{R}_N$ has a natural structure of a Hopf algebra and we define $\mathscr{C}_N \overset{\mathrm{df}}{=} \mathscr{C}_{\mathscr{R}_N}$, $\Lambda_N \overset{\mathrm{df}}{=} \Lambda_{\mathscr{C}_N}$. As is well known, Λ_N is the set of heighest weights of SL_N. In other words, $\Lambda_N = \{(m_1, \dots, m_{N-1}) \in \mathbb{Z}^{N-1} \mid m_1 \geq m_2 \geq \cdots \geq m_{N-1} \geq 0\}$. We define $R_{N,\infty} \overset{\mathrm{df}}{=} [\mathscr{C}_N]$, $R^+_{N,\infty} \overset{\mathrm{df}}{=} [\mathscr{C}_N]^+$. As is well known, $R^+_{N,\infty}$ is isomorphic to the ring $\mathbb{Z}[e_1, \dots, e_{N-1}]$, where $e_i = (\underbrace{1, \dots, 1}_{i \text{ times}}, 0, \dots, 0)$.

3) For any $q \in F^*$ we denote by $\mathscr{R}_{N,q}$ the Hopf algebra of functions on the quantum group $SL_{N,q}$ (see [FRT, § 1]).

We say that q is a primitive root of 1 of order ∞ if either $q = 1$ or q is not a root of unity. In this case we define $\mathscr{C}_{N,q} \overset{\mathrm{df}}{=} \mathscr{C}_{\mathscr{R}_{N,q}}$. As is well known, the category $\mathscr{C}_{N,q}$ is semisimple and $[\mathscr{C}_{N,q}] = R_{n,\infty}$ (see, for example, [L]).

4) If q is a primitive root of 1 of order ℓ, $1 < \ell < \infty$, the category $\mathscr{C}_{\mathscr{R}_{N,q}}$ is not semisimple. As in the case of a generic q we can define objects X_λ in $\mathscr{C}_{\mathscr{R}_{N,q}}$ for any $\lambda = (m_1, \dots, m_{N-1}) \in \Lambda_N$ (see [L]), but the X_λ's are not necessarily simple objects.

We define $\Lambda_{N,\ell} = \{(m_1, \dots, m_{N-1}) \in \Lambda_n \mid m_1 \leq \ell - N\}$. For any $\lambda \in \Lambda_{N,\ell}$ the object X_λ is simple. Let $\widetilde{\mathscr{C}}_{N,q}$ be the full subcategory of $\mathscr{C}_{\mathscr{R}_{N,q}}$ of direct sums of X_λ, $\lambda \in \Lambda_{N,\ell}$. As explained in [A, 3.7], there exists a canonical structure of a rigid monoidal category of $\widetilde{\mathscr{C}}_{N,q}$. We denote by $\mathscr{C}_{N,q}$ the corresponding monoidal category. It is easy to see that the semiring $[\mathscr{C}_{N,q}]$ does not depend on the choice of the primitive root q of order ℓ.

We can describe the semiring $[\mathscr{C}_{N,q}]$ as follows (see [A]). Let $I_\ell \subset R^+_{N,\infty}$ be the ideal generated by e_λ, $\lambda = (\ell - N + 1, m_2, \dots, m_{N-1}) \in \Lambda_N$ and $R^+_{N,\ell} = R^+_{N,\infty}/I_\ell$. Then the subsemigroup $R_{N,\ell}$ in $R^+_{N,\ell}$, generated by the images of X_λ, $\lambda \in \Lambda_{N,\ell}$, is closed under multiplication and the semiring $[\mathscr{C}_{N,q}]$ is isomorphic to $R_{N,\ell}$.

Given $N \in \mathbb{Z}^+$, $\ell \in \mathbb{Z}^+ \cup \infty$, $\ell > N$, we define $\lambda_1 \overset{\mathrm{df}}{=} (1, 0, \dots, 0) \in \Lambda_{N,\ell}$, and consider an element $e^n_{\lambda_1} \in R_{N,\ell}$. Since e_λ, $\lambda \in \Lambda_{N,\ell}$, form a basis of the free abelian semigroup $R_{N,\ell}$, we can find nonnegative integers $d^n_{N,\ell}(\lambda)$ such that $e^n_{\lambda_1} = \sum_{\lambda \in \Lambda_{N,\ell}} d^n_{N,\ell}(\lambda) e_\lambda$.

It will be useful for us to have an inductive formula for $d^n_{N,\ell}(\lambda)$. For this we denote by $\widetilde{\Lambda}_N$ the set of all nonincreasing sequences of N integers $\widetilde{\lambda} = (\widetilde{m}_1 \geq \widetilde{m}_2 \geq \cdots \geq \widetilde{m}_N \geq 0)$ and define the projection $p : \widetilde{\Lambda}_N \to \Lambda_N$ by $(\widetilde{m}_1, \dots, \widetilde{m}_N) \to (m_1, \dots, m_{N-1})$, where $m_i \overset{\mathrm{df}}{=} \widetilde{m}_1 - \widetilde{m}_N$. For any $\widetilde{\lambda} \in \widetilde{\Lambda}_N$ let $|\widetilde{\lambda}| = \sum_{i=1}^N \widetilde{m}_i$. We define $\widetilde{\Lambda}_{N,\ell} = \{\widetilde{\lambda} \in \widetilde{\Lambda}_N \mid \widetilde{m}_1 - \widetilde{m}_N \leq \ell - N\}$.

Given $\widetilde{\lambda} = (m_1, \dots, m_N)$, $\widetilde{\mu} = (s_1, \dots, s_N) \in \widetilde{\Lambda}_N$, $|\widetilde{\lambda}| = |\widetilde{\mu}| + 1$, we say $\widetilde{\mu} < \widetilde{\lambda}$ if $m_i \geq s_i$ for all i, $1 \leq i \leq N$.

LEMMA 1.2. a) *For any* $\widetilde{\lambda} \in \widetilde{\Lambda}_{N,\ell}$, $|\widetilde{\lambda}| = n$, *we have*

$$d^n_{N,\ell}(p(\widetilde{\lambda})) = \sum_{\substack{\widetilde{\mu} \in \widetilde{\Lambda}_{N,\ell} \\ |\widetilde{\mu}| = |\widetilde{\lambda}| - 1 \\ \widetilde{\mu} < \widetilde{\lambda}}} d^{n-1}_{N,\ell}(p(\widetilde{\mu})).$$

b) *If* $|\lambda| \not\equiv n (\mathrm{mod}\ N)$ *then* $d^n_{N,\ell}(\lambda) = 0$.

PROOF. It is easy to see that for any $\widetilde{\mu} \in \widetilde{\Lambda}_{N,\ell}$, $|\widetilde{\mu}| = n - 1$, we have

$$e_{\lambda_1} \cdot e_{p(\widetilde{\mu})} = \sum_{\substack{\widetilde{\lambda} \in \widetilde{\Lambda}_{N,\ell} \\ |\widetilde{\lambda}| = n \\ \widetilde{\lambda} > \widetilde{\mu}}} e_{p(\widetilde{\lambda})}.$$

The lemma follows from this equality by induction in n.

We say that an abelian semisimple rigid monoidal category $\mathscr{C} = (\widetilde{\mathscr{C}}, \otimes, a)$ is of $SL_{N,\ell}$-type (or an $SL_{N,\ell}$ category) if the semirings of semirings $R_{N,\ell}$ and $[\mathscr{C}]$ are isomorphic. If $\mathscr{C}$ is a category of $SL_{N,\ell}$-type, we will always fix an isomorphism $\varphi\colon R_{N,\ell} \xrightarrow{\sim} [\mathscr{C}]$ (which is unique up to an automorphism of the semiring $R_{N,\ell}$) and for any partition $\lambda \in \Lambda_{N,\ell}$ choose a representative X_λ in the isomorphism class $\varphi(\Gamma_\lambda)$.

For $\lambda \in \Lambda_N - \Lambda_{N,\ell}$ we define X_λ to be the zero object in $\widetilde{\mathscr{C}}$. For any Nth root of unity τ we define the "twisted" functorial isomorphisms $a^\tau_{T,Y,Z}\colon (T \otimes Y) \otimes Z \xrightarrow{\sim} T \otimes (Y \otimes Z)$, T, Y, Z in $\widetilde{\mathscr{C}}$, by the formula

$$a^\tau_{X_\lambda, X_\mu, X_\nu} = \tau^{\omega(|\lambda|, |\mu|)\nu} a_{X_\lambda, X_\mu, X_\mu}, \qquad \lambda, \mu, \nu \in \Lambda_{N,\ell},$$

where $\omega : \mathbb{Z} \times \mathbb{Z} \to \mathbb{Z}$ is the function given by

$$\omega(a, b) = \left[\frac{a+b}{N}\right] - \left[\frac{a}{N}\right] - \left[\frac{b}{N}\right]$$

and $[x]$ is the integral part of x, $x \in \mathbb{Q}$.

LEMMA 1.3. *The morphisms* $a^\tau_{X,Y,Z}$ *satisfy the pentagon axiom.*

PROOF. Clear.

The goal of this paper is to prove the following result.

THEOREM A_ℓ. *Let* $\mathscr{C}$ *be an abelian semisimple rigid monoidal category of* $SL_{N,\ell}$*-type. Then there exists an* N*th root of unity* τ *such that* $\mathscr{C}$ *is equivalent to the category* $\mathscr{C}^\tau_{N,q}$ *for some primitive root* q *of unity of order* ℓ.

§ 2. Monoidal algebras

Let F be a field and let $\mathscr{C}$ be a semisimple strict abelian monoidal F-category. In particular, $F = \mathrm{End}_{\mathscr{C}}(\mathbb{1})$, where $\mathbb{1}$ is a unit object in $\mathscr{C}$. Let X be an object in $\mathscr{C}$. We define a family $A_n^m = A_n^m(X) = \mathrm{Hom}_{\mathscr{C}}(X^{\otimes m}, X^{\otimes n})$ of F-vector spaces. The composition of morphisms in $\mathscr{C}$ defines the linear mappings

$$\circ = \circ(m, n, p) \colon A_p^n \otimes A_n^m \to A_p^m,$$
$$(\alpha \otimes \beta) \longmapsto \alpha \circ \beta.$$

By the definition of a monoidal structure of $\mathscr{C}$ we also have morphisms

$$\widehat{\otimes} : (\alpha \otimes \beta) \in A_n^m \otimes A_q^p \longmapsto \alpha \widehat{\otimes} \beta \in A_{n+q}^{m+p}.$$

It is clear that this family of linear F-spaces and maps satisfy the following conditions (a)–(f).

(a) There exists an isomorphism $F \cong A_0^0$, $a \longmapsto a_0$ such that $(ab)_0 = a_0 b_0$ and for any $\alpha \in A_n^m$ we have $1_0 \widehat{\otimes} \alpha = \alpha \widehat{\otimes} 1_0 = \alpha$.

(b) Let $A_n = A_n^n$. Then the linear map $\circ : A_n \otimes A_n \to A_n$ defines on A_n the structure of an associative semisimple F-algebra with unit which we denote $1_n \in A_n$.

(c) For a finite subset S of $\mathbb{Z}^+$ define the linear space $A_S = \bigoplus_{m, n \in S} A_n^m$ and the linear map $\circ_S : A_S \otimes A_S \to A_S$, where

$$(\circ_S)\,|_{A_p^n \otimes A_n^m} = \circ(m, n, p),$$
$$(\circ_S)\,|_{A_n^m \otimes A_q^p} = 0 \quad \text{if } m \neq q.$$

Then $\circ_S$ defines the structure of an associative semisimple F-algebra on A_S with unit $1_S = \sum_{n \in S} 1_n$.

(d) For any $\alpha_i \in A_{n_i}^{m_i}$, $1 \leq i \leq 3$, we have

$$(\alpha_1 \widehat{\otimes} \alpha_2) \widehat{\otimes} \alpha_3 = \alpha_1 \widehat{\otimes} (\alpha_2 \widehat{\otimes} \alpha_3).$$

(e) For any $\alpha \in A_n^m$, $\beta \in A_p^n$, $\gamma \in A_r^q$, $\delta \in A_s^r$ we have

$$(\beta \circ \alpha) \widehat{\otimes} (\delta \circ \gamma) = (\beta \widehat{\otimes} \delta) \circ (\alpha \widehat{\otimes} \gamma).$$

(f) Given $\vec{\alpha} = (\alpha_1, \ldots, \alpha_u)$, $\alpha_i \in A_n^{m_i}$, $\vec{\beta} = (\beta_1, \beta_2, \ldots, \beta_v)$, $\beta_j \in A_n^{l_j}$, we say that $\vec{\alpha} > \vec{\beta}$ if for any $\gamma \in A_p^n$ such that $\gamma \circ \alpha_i = 0$ for all i, $1 \leq i \leq u$, we have $\gamma \circ \beta_j = 0$ for all j, $1 \leq j \leq v$. Assume that $\vec{\alpha} > \vec{\beta}$. Then for any $\delta \in A_t^s$ we have $\vec{\alpha} \widehat{\otimes} \delta > \vec{\beta} \widehat{\otimes} \delta$ and $\delta \widehat{\otimes} \vec{\alpha} > \delta \widehat{\otimes} \vec{\beta}$.

A family $\mathscr{A}$ of F-vector spaces $(A_n^m)_{m, n \in \mathbb{N}}$ and F-linear morphisms

$$\circ = \circ(m, n, p) : A_p^n \otimes A_n^m \to A_p^m,$$
$$\widehat{\otimes}_{n,q}^{m,p} : (\alpha \otimes \beta) \in A_n^m \otimes A_q^p \longmapsto \alpha \widehat{\otimes} \beta \in A_{n+q}^{m+p}$$

satisfying the conditions (a)–(f) is called a *monoidal F-algebra*. It is easy to see that the restriction to the vector spaces A_n^n, $n \in \mathbb{N}$, gives us a monoidal subalgebra, which we call the *diagonal* monoidal subalgebra.

We have seen that to a semisimple strict monoidal F-category $\mathscr{C}$ and to an object X in $\mathscr{C}$ we can associate a monoidal F-algebra $\mathscr{A} = \mathscr{A}_{\mathscr{C},X}$ such that $A_n^m = \operatorname{Hom}_{\mathscr{C}}(X^{\otimes m}, X^{\otimes n})$. It is easy to see that given two categories $\mathscr{C}$, $\mathscr{C}'$ and two objects X, X' in $\mathscr{C}$, $\mathscr{C}'$, any monoidal equivalence $\mathscr{F}$ between $\mathscr{C}$ and $\mathscr{C}'$ such that $\mathscr{F}(X) = X'$ induces an isomorphism between $\mathscr{A}_{\mathscr{C},X}$ and $\mathscr{A}_{\mathscr{C}',X'}$. Since by the Mac Lane theorem any monoidal category is equivalent to a strict monoidal category, we can associate a monoidal algebra $\mathscr{A}_{\mathscr{C},X}$ to any pair consisting of a semisimple monoidal F-category $\mathscr{C}$ and an object X in $\mathscr{C}$.

Conversely, let $\mathscr{A} = (A_n^m, \circ, \widehat{\otimes})$ be a monoidal algebra. Consider the set $\overline{\overline{\mathscr{C}}}$ of data $Y = (\vec{m}, n, \vec{\alpha})$, where $\vec{m} = (m_1, \dots, m_u)$, $\vec{\alpha} = (\alpha_1, \dots, \alpha_u)$, $m_i, n \in \mathbb{Z}^+$, $\alpha_i \in A_n^{m_i}$, $1 \le i \le u$ for some integer $u \ge 0$. If $u = 0$ we write $Y = (n)$. For any $Y = (\vec{m}, n, \vec{\alpha})$, $Z = (\vec{p}, q, \vec{\beta})$ in $\overline{\overline{\mathscr{C}}}$ we define $\operatorname{Hom}_{\mathscr{C}}(Y, Z) = L_1(Y, Z)/L_0(Y, Z)$, where $L_0(Y, Z)$, $L_1(Y, Z)$ are subspaces of A_q^n defined by

$$L_0(Y, Z) = \{\delta \in A_q^n, \vec{\beta} > \delta\},$$
$$L_1(Y, Z) = \{\delta \in A_q^n, \ \vec{\beta} > \delta \circ \vec{\alpha}\}.$$

Given $Y = (\vec{m}, n, \vec{\alpha})$, $Z = (\vec{p}, q, \vec{\beta})$, and $T = (\vec{s}, t, \vec{\gamma})$ in $\widetilde{\mathscr{C}}$ we define the map $\operatorname{Hom}_{\widetilde{\mathscr{C}}}(Y, Z) \times \operatorname{Hom}_{\widetilde{\mathscr{C}}}(Z, T) \to \operatorname{Hom}_{\widetilde{\mathscr{C}}}(Y, T)$ by $(\delta_1, \delta_2) \longmapsto \delta_2 \circ \delta_1$. It is clear that we have defined a category $\widetilde{\mathscr{C}}$ with $\overline{\overline{\mathscr{C}}}$ as the set of objects. We define $\otimes : \overline{\overline{\mathscr{C}}} \times \overline{\overline{\mathscr{C}}} \to \overline{\overline{\mathscr{C}}}$ to be the map such that

$$(\vec{m}, n, \vec{\alpha}) \otimes (\vec{p}, q, \vec{\beta}) = (\vec{s}, n + q, \vec{\gamma}),$$

where $\vec{s} \overset{\text{df}}{=} (\vec{m}q, \vec{p}n)$, $\vec{\gamma} \overset{\text{df}}{=} (\vec{\alpha}\widehat{\otimes}1_q, 1_n\widehat{\otimes}\vec{\beta})$; here $\vec{m}q = (m_1q, \dots, m_nq)$ and similarly for $\vec{n}p$. For any objects

$$X_i = (\vec{m}_i, n_i, \vec{\alpha}_i), \quad Y_i = (\vec{p}_i, q_i, \vec{\beta}_i), \qquad i = 1, 2,$$

and morphisms $\delta_i \in \operatorname{Hom}_{\widetilde{\mathscr{C}}}(X_i, Y_i)$ define the element

$$\delta_1 \otimes \delta_2 \in \operatorname{Hom}_{\widetilde{\mathscr{C}}}(X_1 \otimes X_2, Y_1 \otimes Y_2)$$

in the following way. Choose representatives $\widehat{\delta}_i$ of δ_i in $L_1(X_i, Y_i) \subset A_{q_i}^{n_i}$ for $i = 1, 2$. Then, as follows from the condition (f), we have

$$\widehat{\delta}_1 \widehat{\otimes} \widehat{\delta}_2 \in L_1(X_1 \otimes X_2, Y_1 \otimes Y_2).$$

We denote by $\delta_1 \otimes \delta_2$ the image of $\widehat{\delta}_1 \widehat{\otimes} \widehat{\delta}_2$ in $\operatorname{Hom}_{\widetilde{\mathscr{C}}}(X_1 \otimes X_2, Y_1 \otimes Y_2)$. It is easy to see that $\delta_1 \otimes \delta_2$ does not depend on a choice of $\widehat{\delta}_1$, $\widehat{\delta}_2$, that we have constructed a functor $\otimes : \widetilde{\mathscr{C}} \times \widetilde{\mathscr{C}} \longrightarrow \widetilde{\mathscr{C}}$, and that this functor defines the structure of a strict monoidal category on $\widetilde{\mathscr{C}}$ such that $\mathbb{1} = (0)$.

Let $\mathscr{C}_{\mathscr{A}}$ be a unique additive category that contains $\widetilde{\mathscr{C}}$ as a full subcategory and such that any object of $\mathscr{C}$ is a direct sum of objects from $\widetilde{\mathscr{C}}$. Then there exists a unique structure of a strict monoidal category on $\mathscr{C}$ compatible with the one on $\widetilde{\mathscr{C}}$.

Let $\mathscr{C}$ be a strict semisimple abelian monoidal F-category, X an object in $\mathscr{C}$, $\mathscr{A} = \mathscr{A}_{\mathscr{C},X}$, $\mathscr{D} \stackrel{\text{df}}{=} \mathscr{C}_{\mathscr{A}}$. We define a monoidal functor $\mathscr{F}$ from $\mathscr{D}$ to $\mathscr{C}$ as follows. For any object $Y = (\vec{m}, n, \vec{\alpha})$ in $\mathscr{D}$ we define $\widetilde{\mathscr{F}}(Y) \stackrel{\text{df}}{=} X^{\otimes n}/I_Y$, where $I_Y \subset X^{\otimes n}$ is the subobject generated by $\operatorname{Im}(\alpha_i) \subset X^{\otimes n}$, $1 \leq i \leq n$. Given a morphism $\delta \in \operatorname{Hom}(Y, Z)$, $Y = (\vec{m}, n, \vec{\alpha})$, $Z = (\vec{p}, q, \vec{\beta})$ in $\mathscr{D}$, we define the morphism $\mathscr{F}(\delta) \in \operatorname{Hom}_{\mathscr{C}}(\widetilde{\mathscr{F}}(Y), \widetilde{\mathscr{F}}(Z))$ as follows. Let $\widehat{\delta} \in L_1(Y, Z) \subset A_q^n = \operatorname{Hom}_{\mathscr{C}}(X^{\otimes n} X^{\otimes q})$ be a representative of δ. Since $\widehat{\delta}(I_Y) \subset I_Z$, $\widehat{\delta}$ defines a morphism $\mathscr{F}(\delta) : \widetilde{\mathscr{F}}(Y) \to \widetilde{\mathscr{F}}(Z)$. It is clear that $\mathscr{F}(\delta)$ does not depend on the choice of $\widehat{\delta}$ and that we have constructed a functor $\mathscr{F}$ from $\mathscr{D}$ to $\mathscr{C}$. To define a monoidal structure on $\mathscr{F}$ we have to define for any Y, Z in $\mathscr{C}$ an isomorphism $C_{Y,Z} : \mathscr{F}(Y) \otimes \mathscr{F}(Z) \to \mathscr{F}(Y \otimes Z)$. But our definition of the tensor product in $\mathscr{D} = \mathscr{C}_{\mathscr{A}}$ was chosen in such a way that a natural morphism $X^{\otimes n} \otimes X^{\otimes q} \to X^{\otimes(n+q)}$ extends to an isomorphism

$$X^{\otimes n}/I_Y \otimes X^{\otimes q}/I_Z \xrightarrow{\sim} X^{\otimes(n+q)}/I_{Y \otimes Z},$$

which we take for $C_{Y,Z}$. It is easy to see that for any Y, Z, and T in $\mathscr{D}$ we have

$$C_{Y \otimes Z, T} \circ (C_{Y,Z} \otimes \operatorname{id}_T) = C_{Y, Z \otimes T} \circ (\operatorname{id}_Y \otimes C_{Z,T}).$$

Therefore we have constructed a monoidal functor $\mathscr{F} = (\widetilde{\mathscr{F}}, C)$ between monoidal categories $\mathscr{D}$ and $\mathscr{C}$.

Proposition 2.1. *If any simple object in $\mathscr{C}$ is a subobject of $X^{\otimes n}$ for some $n \in \mathbb{Z}^+$ then $\mathscr{F}$ is an equivalence between the monoidal categories $\mathscr{D}$ and $\mathscr{C}$.*

Proof. Since any object in $\mathscr{C}$ is a quotient of $X^{\otimes n}$ for some $n \in \mathbb{Z}^+$ it is clear that $\widetilde{\mathscr{F}}$ defines a surjection from the set of isomorphism classes $[\mathscr{D}]$ of objects in $\mathscr{D}$ to $[\mathscr{C}]$. By [M, 4.4.1] it is sufficient to show that for any $Y = (\vec{m}, n, \vec{\alpha})$, $Z = (\vec{p}, q, \vec{\beta})$ in $\mathscr{D}$ the map $\widetilde{\mathscr{F}}_{Y,Z}$: $\operatorname{Hom}_{\mathscr{D}}(Y, Z) \to \operatorname{Hom}_{\mathscr{C}}(\mathscr{F}(Y), \mathscr{F}(Z))$ is an isomorphism of F-vector spaces. To show this we describe the inverse map from $\operatorname{Hom}_{\mathscr{C}}(\mathscr{F}(Y), \mathscr{F}(Z))$ to $\operatorname{Hom}_{\mathscr{D}}(Y, Z)$. By definition, we have $\mathscr{F}(Y) = X^{\otimes n}/I_Y$, $\mathscr{F}(Z) = X^{\otimes p}/I_Z$. Since $\mathscr{C}$ is semisimple, any morphism $\theta \in \operatorname{Hom}_{\mathscr{C}}(\mathscr{F}(Y), \mathscr{F}(Z))$ can be lifted to a morphism $\widehat{\theta} \in \operatorname{Hom}_{\mathscr{C}}(X^{\otimes n}, X^{\otimes p}) = A_p^n$ such that $\widehat{\theta}(I_Y) \subset I_Z$. It is easy to see that $\widehat{\theta}$ belongs to $L_1(Y, Z) \subset A_p^n$, the image $\overline{\theta}$ of $\widehat{\theta}$ in $\mathrm{Hom}_{\mathscr{D}}(Y, Z)$ does not depend on a choice of $\widehat{\theta}$, and $\theta \to \overline{\theta}$ is the map inverse to $\widetilde{\mathscr{F}}_{Y,Z}$.

Proposition 2.1 is proved.

We say that a monoidal algebra $\mathscr{A} = (A_n^m, \circ, \widehat{\otimes})$ is *braided* if for any $m, n \in \mathbb{Z}^+$ we are given an invertible element $S_{m,n} \in A_{m+n}^{m+n}$ such that

a) for any $a \in A_n^m$, $b \in A_q^p$ we have

$$S_{n,q} \circ (a \widehat{\otimes} b) = (b \widehat{\otimes} a) S_{m,p};$$

b) for any $m, n, p \in \mathbb{Z}^+$ we have

$$(1_n \widehat{\otimes} S_{m,p}) \circ (S_{m,n} \widehat{\otimes} 1_p) = S_{m,n+p},$$
$$(S_{m,n} \widehat{\otimes} 1_n) \circ (1_m \widehat{\otimes} S_{n,p}) = S_{m+n,p};$$

c) $S_{0,m} = S_{m,0} = 1_m$.

Lemma 2.1. a) *If $(\mathscr{C}, \otimes, \mathbb{1}, s)$ is a strict braided category, $\mathscr{A} = \mathscr{A}_{\mathscr{C}}$ is the corresponding monoidal algebra, and $S_{m,n} = s_{X^{\otimes m}, X^{\otimes n}}$, then $(\mathscr{A}, S_{m,n})$ is a braided monoidal algebra.*

b) *If $(\mathscr{A}, S_{m,n})$ is a braided monoidal algebra, then the monoidal category $\mathscr{C}_{\mathscr{A}}$ has a unique braiding s such that $s_{X^{\otimes m}, X^{\otimes n}} = S_{m,n} \in \mathscr{A}_{m+n}^{m+n} =$ $\mathrm{End}_{\mathscr{C}_{\mathscr{A}}}(m+n)$.*

c) *If $\mathscr{C}$ is semisimple and $\mathscr{F} : \mathscr{D} \to \mathscr{C}$ is the monoidal equivalence as in Proposition 2.1, then $\mathscr{F}$ is an equivalence of braided categories.*

Proof. Easy.

Remark. It is clear that a braiding $S_{m,n}$ on a monoidal algebra $\mathscr{A}$ is completely determined by $S_{1,1} \in A_2^2$.

We say that a monoidal algebra $\mathscr{A}$ is diagonal if $A_n^m = \{0\}$ if $m \neq n$. For any monoidal algebra $\mathscr{A} = (A_n^m, \circ, \widehat{\otimes})$ we denote by $\Delta\mathscr{A}$ the diagonal monoidal subalgebra $(A_n^n, \circ, \widehat{\otimes})$ of $\mathscr{A}$. We say that a monoidal algebra $\mathscr{A}$ with a braiding $S_{m,n}$ on $\Delta\mathscr{A}$ is an algebra of type N if the following three conditions are satisfied:

a) $A_n^m = \{0\}$ if $m \not\equiv n \pmod N$;

b) $\dim A_N^0 = \dim A_0^N = 1$.

We fix a nonzero element $\nu \in A_N^0$ and define $\pi \in A_0^N$ to be the element such that $\pi \circ \nu = 1_0$. We define $\Pi \stackrel{\mathrm{df}}{=} \nu \circ \pi \in A_N^N$.

c) $S_{N,1}(\nu \widehat{\otimes} 1_1) = (1_1 \widehat{\otimes} \nu)$.

Proposition 2.2. a) *Let $(\mathscr{A}, S_{m,n})$ be a monoidal algebra of type N. Then $S_{m,n}$ defines a braiding on $\mathscr{A}$.*

b) *Let $(\mathscr{A}, S_{m,n})$, $(\mathscr{A}', S'_{m,n})$ be monoidal algebras of type N and there exists an isomorphism $\psi : \Delta\mathscr{A} \xrightarrow{\sim} \Delta\mathscr{A}'$ of the braided algebras such that $\psi(\Pi) = \Pi'$. Then $\mathscr{A}$ and $\mathscr{A}'$ are isomorphic.*

PROOF. We start with the following result. For any $n, p \in \mathbb{Z}^+$ define $\nu_{n,p} \overset{\mathrm{df}}{=} 1_n \widehat{\otimes} \nu \widehat{\otimes} 1_p \in A^{n+p}_{n+p+N}$, $\pi_{n,p} \overset{\mathrm{df}}{=} 1_n \widehat{\otimes} \pi \widehat{\otimes} 1_p \in A^{n+p+N}_{n+p}$.

LEMMA 2.2. a) *For all* $n \in \mathbb{Z}^+$ *we have*

$$(S_{N,n} \widehat{\otimes} 1_p) \circ (\nu \widehat{\otimes} 1_{n+p}) = \nu_{n,p}.$$

b) *In general, for any* $a \in A^{\ell}_{n+p}$ *we have*

$$(S_{N,n} \widehat{\otimes} 1_p) \circ (\nu \widehat{\otimes} a) = \nu_{n,p} \circ a.$$

c) *For any* $a \in A^{n+p}_{\ell}$ *we have*

$$(\pi \widehat{\otimes} a) \circ S_{n,N} = \pi_{n,p} \circ a.$$

PROOF OF LEMMA. a) We prove a) by induction in n. If $n = 1$, the claim follows from condition c) on S. Since $S_{N,n} = (1_{n-1} \otimes S_{N,1}) \circ (S_{N,n-1} \otimes 1_1)$ we have

$$\begin{aligned} S_{N,n}(\nu \widehat{\otimes} 1_n) &= (1_{n-1} \widehat{\otimes} S_{N,1}) \circ (S_{N,n-1} \widehat{\otimes} 1_1) \circ (\nu \widehat{\otimes} 1_{n-1} \widehat{\otimes} 1_1) \\ &= (1_{n-1} \widehat{\otimes} S_{N,1}) \circ (1_{n-1} \widehat{\otimes} (\nu \widehat{\otimes} 1_1)) = 1_n \widehat{\otimes} \nu. \end{aligned}$$

b) Since $\nu \widehat{\otimes} a = (\nu \widehat{\otimes} 1_n) \circ a$ we have

$$S_{N,n} \circ (\nu \widehat{\otimes} a) = A_{N,n} \circ (\nu \widehat{\otimes} 1_n) \circ a = \nu_{n,p} \circ a.$$

c) The proof of c) is similar to the proof of b).

Lemma 2.2 is proved.

To prove part a) of Proposition 2.2 we must show that for any $a \in A^m_n$, $b \in A^p_q$ we have $S_{n,q} \circ (a \otimes b) = (b \widehat{\otimes} a) \circ S_{m,p}$. If $a, b \neq 0$ we have $m = n + \alpha N$, $p = q + \beta N$, $\alpha, \beta \in \mathbb{Z}$. We assume for simplicity that $\alpha = 1$, $\beta = 0$. So $m = n + N$, $p = q$. Other cases are analogous and could be proved by induction. Let $a' = \nu \otimes a \in A^m_n$. Since $S_{m,p}$ is a braiding on $\Delta\mathscr{A}$ we know that

$$S_{m,p} \circ (a' \widehat{\otimes} b) = (b \widehat{\otimes} a') \circ S_{m,p}.$$

By definition, $(b \widehat{\otimes} a') \circ S_{m,p} = \nu_{p,n} \circ (b \widehat{\otimes} a) S_{m,p}$. On the other hand,

$$\begin{aligned} S_{m,p} \circ (a' \widehat{\otimes} b) &= S_{m,p} \circ (\nu \widehat{\otimes} (a \widehat{\otimes} b)) = (S_{N,p} \widehat{\otimes} 1_n) \circ (1_N \widehat{\otimes} S_{n,p}) \circ (\nu \widehat{\otimes} \circ (a \widehat{\otimes} b)) \\ &= (S_{n,p} \widehat{\otimes} 1_n) \circ (\nu \widehat{\otimes} (S_{n,p} \circ (a \widehat{\otimes} b))) = \nu_{p,n} \circ (S_{n,p} \circ (a \widehat{\otimes} b)). \end{aligned}$$

Since the map $\nu_{p,n} \circ \colon A^{p+n}_{p+n+N} \longrightarrow A^{p+n+N}_{p+n+N}$ is an imbedding, we see that $(b \widehat{\otimes} a) \circ S_{m,p} = S_{n,p} \circ (a \widehat{\otimes} b)$. Part a) is proved.

b) We start the proof of b) with the following result. Given $m, n, p \in \mathbb{Z}^+$ such that $m, n \leq p$ and $p = m + \alpha N = n + \beta N$, $\alpha, \beta \in \mathbb{Z}^+$, we define a map $A^m_n \xrightarrow{\mathscr{H}^{m,n}_p} A^p_p$ as follows:

$$a \longmapsto (\nu^{\otimes \beta} \widehat{\otimes} 1_n) \circ a \circ (\pi^{\otimes \alpha} \widehat{\otimes} 1_m)$$

and define a subspace $\Sigma_p^{m,n}$ in A_p^p,

$$\Sigma_p^{m,n} = \{a \in A_p^p \mid (\pi^{\otimes\beta}\widehat{\otimes}1_n) \circ a = a \circ (\pi^{\otimes\alpha}\widehat{\otimes}1_m) = a\}.$$

LEMMA 2.3. a) *$\mathscr{H}_p^{m,n}$ defines an isomorphism between the linear spaces A_n^m and $\Sigma_p^{m,n}$.*

b) *For any $t \in \mathbb{Z}^+$, $t \le p$, $t \equiv p \ (\mathrm{mod} N)$ and any $a \in A_n^m$, $b \in A_t^n$ we have*

$$\mathscr{H}_p^{m,t}(b \circ a) = \mathscr{H}_p^{n,t}(b) \circ \mathscr{H}_p^{m,n}(a).$$

PROOF. a) Since $\Pi \circ \nu = \nu$ and $\pi \circ \Pi = \pi$ we see that $\mathrm{Im}\,\mathscr{H}_p^{m,n} \subset \Sigma_p^{m,n}$. Define a linear map $\theta_{m,n}^p \colon \Sigma_p^{m,n} \to A_n^m$ as the composition $a \longrightarrow (\pi^{\otimes\beta}\widehat{\otimes}1_n) \circ a \circ (\nu^{\otimes\alpha}\widehat{\otimes}1_m)$. It is clear that the map $\theta_{m,n}^p$ is the inverse of $\mathscr{H}_p^{m,n}$.

b) Follows from the equality $\pi \circ \nu = 1_0$.

Lemma 2.3 is proved.

Let $\mathscr{A}$ be a monoidal algebra of type N. For any $a \in A_n^m$ we define $a' = \pi\widehat{\otimes}a \in A_n^{m+N}$, $a'' = \nu\widehat{\otimes}a \in A_{n+N}^m$.

LEMMA 2.4. *For any $a \in A_n^m$, $b \in A_q^p$ we have*

$$\begin{aligned}
a'\widehat{\otimes}b &= (a\widehat{\otimes}b)', \\
a''\widehat{\otimes}b &= (a\widehat{\otimes}b)'', \\
a\widehat{\otimes}b' &= (a\widehat{\otimes}b)' \circ (S_{N,m}\widehat{\otimes}1_p), \\
a''\widehat{\otimes}b &= (S_{N,n} \otimes 1_q) \circ (a\widehat{\otimes}b)''.
\end{aligned}$$

PROOF. By definition,

$$\begin{aligned}
a'\widehat{\otimes}b &= (\pi\widehat{\otimes}a)\widehat{\otimes}b = \pi\widehat{\otimes}(a\widehat{\otimes}b) = (a\widehat{\otimes}b)', \\
a''\widehat{\otimes}b &= (\nu\widehat{\otimes}a)\widehat{\otimes}b = \nu\widehat{\otimes}(a\widehat{\otimes}b) = (a\widehat{\otimes}b)'', \\
a\widehat{\otimes}b' &= a\widehat{\otimes}(\pi\widehat{\otimes}b) = (a\widehat{\otimes}\pi)\widehat{\otimes b} = ((\pi\widehat{\otimes}a)\widehat{\otimes}b) \circ (S_{N,m}\widehat{\otimes}1_p) \\
&= (a\widehat{\otimes}b)' \circ (S_{N,m}\widehat{\otimes}1_p), \\
a\widehat{\otimes}b'' &= (a\widehat{\otimes}\nu) \otimes b'' = (S_{N,m} \otimes 1_q) \circ \nu\widehat{\otimes}a\widehat{\otimes}b \\
&= (S_{N,n} \otimes 1_q) \circ (a\widehat{\otimes}b)''.
\end{aligned}$$

Lemma 2.4 is proved.

Now we can complete the proof of Proposition 2.2. Let $\mathscr{A}$, $\mathscr{A}'$ be monoidal algebras of type N and $\psi \colon \Delta\mathscr{A} \xrightarrow{\sim} \Delta\mathscr{A}'$ be an isomorphism such that $\psi(\Pi) = \Pi'$. For any $m, n \in \mathbb{Z}^+$, $m \equiv n$ (mod N) we choose $p > m, n$ such that $p \equiv m$ mod N. Then $\psi \circ \mathscr{H}_p^{m,n}(A_n^m) \subset \Sigma_p^{\prime m,n}$ so that there exists a unique linear map $\psi_n^m : A_n^m \to A_n^{\prime m}$ such that $\psi \circ \mathscr{H}_p^{m,n} = \mathscr{H}_p^{m,n} \circ \psi_n^m$.

Lemmas 2.3 and 2.4 imply that $\{(\psi_n^m)\}$, $m, n \in \mathbb{Z}$, defines an isomorphism of monoidal algebras $\mathscr{A}$ and $\mathscr{A}'$. Proposition 2.2 is proved.

§ 3. Hecke algebras

The Hecke algebra H_n of type A_{n-1} is an F-algebra with generators T_i, $1 \le i < n$, and relations

$$\begin{aligned} &T_iT_j = T_jT_i && \text{if } |i-j| > 1, \\ &T_iT_{i+1}T_i = T_{i+1}T_iT_{i+1} && \text{if } 1 \le i < n-1, \\ &(T_i - q)(T_i + 1) = 0 && \text{for } 1 \le i < n. \end{aligned}$$

For any element σ of the symmetric group S_n we can define an element $T_\sigma \in H_n$ in the following way. Choose any presentation of σ as a product of elementary transpositions $\sigma_i = (i, i+1) \in S_n$, $\sigma = \sigma_{i_1} \cdots \sigma_{i_\ell}$, that has the shortest length $\ell = \ell(\sigma)$ and define $T_\sigma \overset{\text{df}}{=} T_{i_1} \cdots T_{i_\ell}$. As is well known (see [Bou, Prop. 4.1.5]) the element T_σ does not depend on a choice of a shortest decomposition $\sigma = \sigma_{i_1} \cdots \sigma_{i_\ell}$.

We denote by $\psi_n', \psi_n'': H_n \to H_{n+1}$ the algebra homomorphisms defined by

$$\psi_n'(T_i) = T_i, \quad \psi_n''(T_i) = T_{i+1}, \qquad 1 \le i < n.$$

Consider the following elements in $H_n(q)$:

$$\begin{aligned} \alpha_n &= \sum_{\sigma \in S_n} (-q)^{-\ell(\sigma)} T_\sigma, \\ \beta_n &= (1 - q^{-1}T_{\sigma_{n-1}} + q^{-2}T_{\sigma_{n-1}\sigma_{n-2}} - \cdots + (-q^{-1})^{n-1}T_{\sigma_{n-1}\sigma_{n-2}\cdots\sigma_1}), \\ \gamma_n &= (1 - q^{-1}T_{\sigma_1} + q^{-2}T_{\sigma_2\sigma_1} - \cdots + (-q^{-1})^{n-1}T_{\sigma_{n-1}\sigma_{n-2}\cdots\sigma_1}). \end{aligned}$$

Lemma 3.1. a) *$\alpha_n = \alpha_{n-1}'\beta_{n+1} = \gamma_n\alpha_{n-1}''$, where $\alpha_{n-1}' = \psi_{n-1}'(\alpha_{n-1})$, $\alpha_n'' = \psi_{n-1}''(\alpha_{n-1})$.*

b) *$T_\sigma\alpha_n = \alpha_nT_\sigma = (-1)^{\ell(\sigma)}\alpha_n$ for all $\sigma \in S_n$.*

c) *If $q^n = 1$, then $\alpha_n^2 = 0$.*

Proof. a) Easy.

b) It is enough to consider the case $\sigma = \sigma_i$, $1 \le i \le n-1$, and the result follows from part a) by induction.

c) Follows from b).

Using ψ_n' we will consider H_n as a subalgebra of H_{n+1}. Let

$$C: H_{n-1} \oplus (H_{n-1} \otimes_{H_{n-2}} H_{n-1}) \to H_n$$

be the linear map given by

$$C(h + h' \otimes h'') = h + h'T_{n-1}h'', \qquad h, h', h'' \in H_{n-1}.$$

The map C is well defined since T_{n-1} commutes with H_{n-2}. It is easy to see that C is surjective, and it is injective by the dimension count.

For any $\mu \in F$ we define the linear map $\varphi_\mu^{n-1,n} : H_n \to H_{n-1}$ by the formula

$$\varphi_\mu^{n-1,n}(h + h'_{n-1}h'') = h + \mu h'h'', \qquad h, h', h'' \in H_{n-1}.$$

Lemma 3.2. *We have* $\varphi_\mu^{n-1,n}(\alpha_n) = (1 + \mu - q^{-(n-1)}\mu[n]_q)\alpha_{n-1}$, *where* $[n]_q = \frac{q^n-1}{q-1}$ *if* $q \neq 1$ *and* $[n]_1 = n$.

Proof. Lemma 3.1 a) implies

$$\varphi_\mu^{n-1,n}(\alpha_n) = \alpha_{n-1}(1 - q^{-1}\mu + q^{-2}\mu T_{\sigma_{n-2}} - \cdots + (-q^{-1})^{n-1}\mu T_{\sigma_{n-2}} \cdots T_{\sigma_1}).$$

The statement of Lemma 3.2 follows immediately from Lemma 3.1 b).

We define a linear functional $\varphi_\mu^n : H_n \to F$ as the composition

$$\varphi_\mu^n \stackrel{\text{df}}{=} \varphi_\mu^2 \circ \cdots \circ \varphi_\mu^{n-1,m} : H_n \to H_1 = F.$$

It is well known (see [FYHLMO]) that for any $\mu \in \mathbb{C}$ we have $\varphi_\mu^n(h_1h_2) = \varphi_\mu^n(h_2h_1)$ for all $h_1, h_2 \in H_n$. Therefore the subspace

$$I_\mu^n \stackrel{\text{df}}{=} \{h \in H_n | \varphi_\mu^n(hh') = 0 \quad \text{for all } h' \in H_n\}$$

is a two-sided ideal in H_n.

Proposition 3.1. *Assume that* q *is a primitive root of unity of order* ℓ, $2 \leq \ell \leq \infty$.

a) *If* $\ell > n$, *then* $H_n(q)$ *is semisimple and is isomorphic to the group algebra* $F[S_n]$. *Moreover,*

a′) *if* $\mu \neq \frac{q^n}{[m]_q}$ *for all* m, $1 < |m| \leq n$, *then* $I_\mu^n = \{0\}$ *and*

$$H_n(q) = \bigoplus_{\substack{\lambda \in \widetilde{\Lambda}_\infty \\ |\lambda| = n}} M_{d(\lambda)}(F),$$

where d_λ *is the dimension of the representation of* S_n *corresponding to the partition* λ *of* n;

a″) *if* $\mu = \frac{q^m}{[m]_q}$ *for some* m, $1 \leq |m| < n$, *then*

$$H_n(q)/I_\mu^n = \bigoplus_{\substack{\lambda \in \widetilde{\Lambda}_m \\ |\lambda| = n}} M_{d_{m,\infty}^n(\lambda)},$$

where $d_{m,\infty}^n(\lambda)$ *is defined at the end of* § 1.

b) *If* $\ell \leq n$ *and* $\mu = \frac{q^m}{[m]_q}$ *for some* m, $1 < m < \ell - 1$, *then* $H_n(q)/I_\mu^n$ *is semisimple and it is isomorphic to*

$$\bigoplus_{\substack{\lambda \in \widetilde{\Lambda}_{m,\ell} \\ |\lambda| = n}} M_{d_{m,\ell}^n(\lambda)}(F).$$

PROOF. We will consider only the more difficult case b). For any $\lambda \in \widetilde{\Lambda}_{m,\ell}$, $|\lambda| = n$, we denote by $\pi_\lambda : H_n(q) \to \operatorname{End} N_\lambda$ the irreducible representation as in [W, Corollary 2.5] and by $w_\lambda^{(n)} \in F$ the number

$$w_\lambda^{(n)} = \frac{q^{-c(\lambda, m)}}{[m]_q^n} \prod_{1 \le r < s \le m} \frac{[\lambda_r - \lambda_s + s - r]_q}{[s - r]_q},$$

where $c(\lambda, m)$ is chosen in such a way that $w_\lambda^{(n)}$ does not change under the substitution $q \to q^{-1}$, and define $w_\lambda^{(n)} = 0$ if $\lambda \in \widetilde{\Lambda} - \widetilde{\Lambda}_{m,\ell}$, $|\lambda| = n$. It is known that $c(\lambda, m)$ is an integer (see, for example, the equivalent formula [W, (3.8)]). We denote by $\varphi^n : H_n(q) \to F$ the functional given by

$$\varphi^n(h) = \sum_{\substack{\lambda \in \widetilde{\Lambda}_{m,\ell} \\ |\lambda| = n}} w_\lambda^{(n)} \cdot \operatorname{Tr}(\pi_\lambda(h)).$$

LEMMA 3.3. *We have* $\varphi^n = \varphi_\mu^n$.

PROOF OF THE LEMMA. As follows from Lemma 3.3 (b) in [W] it is sufficient to check that the sequence φ^n, $n \in \mathbb{Z}^+$, of functionals satisfies the conditions of this Lemma 3.3 (b) in [W]. But this follows easily from Lemmas 3.4 and 3.5 (a),(c) in [W]. Lemma 3.3 is proven.

Proposition 3.1 follows now from the inequality $w_\lambda \neq 0$ for $\lambda \in \widetilde{\Lambda}_{m,\ell}$, $|\lambda| = n$.

§ 4. $SL_{N,\ell}$-categories

By a $\mathbb{Z}/N\mathbb{Z}$-category we mean a pair $(\mathscr{C}, \varphi_{\mathscr{C}})$, where $\mathscr{C}$ is a semisimple monoidal rigid F-category such that any simple object in it is invertible, and $\varphi_{\mathscr{C}}$ is an isomorphism $\varphi_{\mathscr{C}} : \mathbb{Z}/N\mathbb{Z} \xrightarrow{\sim} \Gamma_{\mathscr{C}}$, where $\Gamma_{\mathscr{C}}$ is the group of the isomorphism classes of invertible objects in $\mathscr{C}$ with the composition induced by the tensor product in $\mathscr{C}$. We say that two $\mathbb{Z}/N\mathbb{Z}$-categories $\mathscr{C}'$, $\mathscr{C}''$ are equivalent if there exists a monoidal equivalence between $\mathscr{C}'$ and $\mathscr{C}''$ compatible with $\varphi_{\mathscr{C}'}$, $\varphi_{\mathscr{C}''}$. To any $\mathbb{Z}/N\mathbb{Z}$-category $(\mathscr{C}, \varphi_{\mathscr{C}})$ we associate a number $\delta(\mathscr{C}, \varphi_{\mathscr{C}}) \in F^*$ in the following way. Choose an irreducible object X in the isomorphism class $1 \in \mathbb{Z}/N\mathbb{Z}$ and consider the object

$$X^{(N)} \stackrel{\text{df}}{=} \underbrace{((X \otimes X) \otimes \cdots \otimes X) \otimes X}_{N \text{ times}}.$$

By definition, $X^{(N)}$ is isomorphic to an identity object $\mathbb{1}$ in $\mathscr{C}$. We fix an isomorphism $\varphi : \mathbb{1} \to X^{(N)}$ and consider the automorphism $\delta(\mathscr{C}) \in \operatorname{Aut}_{\mathscr{C}}(X) = F^*$ given as the composition

$$X \to X \otimes \mathbb{1} \xrightarrow{\text{id} \otimes \varphi} X \otimes X^{(N)} \xrightarrow{a} X^{(N)} \otimes X \xrightarrow{(\varphi \otimes 1)^{-1}} \mathbb{1} \otimes X \to X,$$

where a is defined using the associativity constraints.

LEMMA 4.1. a) *The number* $\delta(\mathscr{C}, \varphi_{\mathscr{C}})$ *depends only on the equivalence class of the* $\mathbb{Z}/N\mathbb{Z}$*-category* $(\mathscr{C}, \varphi_{\mathscr{C}})$.

b) $\delta(\mathscr{C}, \varphi_{\mathscr{C}})^N = 1$.

c) *Two* $\mathbb{Z}/N\mathbb{Z}$*-categories* $(\mathscr{C}', \varphi_{\mathscr{C}'})$, $(\mathscr{C}'', \varphi_{\mathscr{C}''})$ *such that* $\delta(\mathscr{C}', \varphi_{\mathscr{C}'}) = \delta(\mathscr{C}'', \varphi_{\mathscr{C}''})$ *are equivalent.*

PROOF. To any 3-cocycle $\alpha \in Z^3(\mathbb{Z}/N\mathbb{Z}, F^*)$ we associate a $\mathbb{Z}/N\mathbb{Z}$-category $\mathscr{C}_\alpha$ with irreducible objects X_a, $a \in \mathbb{Z}/N\mathbb{Z}$, the tensor product $X_a \otimes X_b = X_{a+b}$, and associativity constraints $a_{X_a, X_b, X_c} : (X_a \otimes X_b) \otimes X_c \to X_a \otimes (X_b \otimes X_c)$ defined as compositions

$$(X_a \otimes X_b) \otimes X_c = X_{a+b} \otimes X_c = X_{a+b+c} \xrightarrow{\alpha(a,b,c)} \\ \to X_{a+b+c} = X_a \otimes X_{b+c} = X_a \otimes (X_b \otimes X_c).$$

We define $\varphi_{\mathscr{C}_\alpha}(X_a) = a$ and write $\mathscr{C}_\alpha$ instead of $(\mathscr{C}_\alpha, \varphi_{\mathscr{C}_\alpha})$.

It is easy to verify that $\mathscr{C}_\alpha$ are indeed monoidal categories (that is, that the pentagon axiom is satisfied) and that $\mathbb{Z}/N\mathbb{Z}$-categories $\mathscr{C}_\alpha$ and $\mathscr{C}_\beta$ are equivalent if and only if the cocycles α and β define the same class in $H^3(\mathbb{Z}/N\mathbb{Z}, F^*)$. Moreover it is known (see, for example, [K, p. 285]) that any $\mathbb{Z}/N\mathbb{Z}$-category is equivalent to $\mathscr{C}_\alpha$ for some 3-cocycle α. It is easy to check that $\delta(\mathscr{C}_\alpha) = \prod_{j=0}^{N-1} \alpha(1, j, 1)$. Since the standard chain $\sum_{j=0}^{N-1}(1, j, 1)$ is a cycle and its image in $H_3(\mathbb{Z}/N\mathbb{Z}, F^*)$ is a generator of the cyclic group $H_3(\mathbb{Z}/N\mathbb{Z}, F^*)$, we see that the map $\alpha \to \delta(\mathscr{C}_\alpha)$ defines an isomorphism $H_3(\mathbb{Z}/N\mathbb{Z}, F^*) \xrightarrow{\sim} \mu_N \subset F^*$, where $\mu_N = \{z \in F^* | z^N = 1\}$. Lemma 4.1 is proved.

Let $\mathscr{C}$ be an $SL_{N,\ell}$-category. As follows from the Mac Lane theorem (see [M, 7.2.1]), we can find a strict monoidal category which is equivalent to $\mathscr{C}$ as a monoidal category. To simplify formulas we will assume that $\mathscr{C}$ is itself a strict category.

We will write $X = X_{\{1\}}$ and for any $\lambda \in \Lambda_{N,\ell}$ we define the linear space $L_\lambda \stackrel{\text{df}}{=} \operatorname{Hom}_{\widetilde{\mathscr{C}}}(X_\lambda, X^{\otimes|\lambda|})$. Since $\widetilde{\mathscr{C}}$ is semisimple, we have a natural isomorphism

$$X^{\otimes n} = \bigoplus_{\substack{\lambda \in \Lambda_{N,\ell} \\ |\lambda| = n}} L_\lambda \otimes X_\lambda.$$

To simplify the notations, we will identify L_λ with F in the case when $\dim L_\lambda = 1$ and consider X_λ as a submodule of $X^{\otimes|\lambda|}$. (We could actually define such X_λ as the corresponding submodules of $X^{\otimes|\lambda|}$.)

REMARK. It is clear that any SL_1-category is equivalent to a monoidal category with $\mathbb{1}$ as the only simple object. Therefore we will always assume that $N > 1$.

PROPOSITION 4.1. *Theorem* A_ℓ *is true for* $\ell = N + 1$.

PROOF. Let $\mathscr{C}$ be an $SL_{N,N+1}$ category and $X = X_{\{1\}}$. Since in the ring $R_{N,N+1}$ we have an equality $r_{\{1\}}^N = 1$ we see that $X^{\otimes N}$ is isomorphic to $\mathbb{1}$. This immediately implies that $(\mathscr{C}, \varphi_{\mathscr{C}})$ is a $\mathbb{Z}/N\mathbb{Z}$-category, where $\varphi_{\mathscr{C}}(r_{\{1\}}) = 1 \in \mathbb{Z}/N\mathbb{Z}$. Since the map $\Lambda_{N,N+1} \to \mathbb{Z}/N\mathbb{Z}$, $\lambda \mapsto |\lambda| \pmod N$ is a bijection, and $\alpha(a, b, c) \stackrel{\mathrm{df}}{=} \eta^{\omega(a,b)c}$ is a 3-cocycle on $\mathbb{Z}/N\mathbb{Z}$, which generates the cyclic group $H^3(\mathbb{Z}/N\mathbb{Z}, F^*)$, Proposition 4.1 follows from Lemma 4.1.

From now on we assume that $\ell > N+1$ and, as before, $N > 1$. In this case $r_{\{2\}}$, $r_{\{1^2\}}$, and $r_{\{2,1\}}$ are distinct nonzero elements in $R_{N,\ell}$. Therefore $X^{\otimes 2}$ is isomorphic to $X_{\{2\}} \otimes X_{\{1^2\}}$ and $X^{\otimes 3}$ is isomorphic to

$$\begin{aligned} X_{\{2,1\}} \otimes V \oplus X_{\{3\}} \oplus X_{\{1^3\}} \quad & \text{if } \ell > N+2 \text{ and } N > 2, \\ X_{\{2,1\}} \otimes V + X_{\{1^3\}} \quad & \text{if } \ell = N+2 \text{ and } N > 2, \\ X_{\{2,1\}} \otimes V + X_{\{3\}} \quad & \text{if } \ell > N+2 \text{ and } N = 2, \\ X_{\{2,1\}} \otimes V \quad & \text{if } \ell > N+2 \text{ and } N = 2, \end{aligned}$$

where $V \stackrel{\mathrm{df}}{=} \mathrm{Hom}_{\mathscr{C}}(X_{\{2,1\}}, X^{\otimes 3})$ is a two-dimensional complex vector space.

Let $a \in \mathrm{End}\, X^{\otimes 2}$ be the projection on $X_{\{1^2\}}$, $a_1 \stackrel{\mathrm{df}}{=} a \otimes \mathrm{id}_X \in \mathrm{End}\, X^{\otimes 3}$, and $a_2 \stackrel{\mathrm{df}}{=} \mathrm{id}_X \otimes a \in \mathrm{End}(X^{\otimes 3})$.

PROPOSITION 4.2. *There exists* $\gamma_{\mathscr{C}} \in F^*$ *such that*

$$a_1 a_2 a_1 - \gamma_{\mathscr{C}} a_1 = a_2 a_1 a_2 - \gamma_{\mathscr{C}} a_2.$$

PROOF. We will consider only the case $\ell \geq N+2$, $N \geq 2$. Other cases are similar and easier.

LEMMA 4.2. a) $a_1 \mid X_{\{1^3\}} = a_2 \mid X_{\{1^3\}} = 1$;
b) $a_1 \mid X_{\{3\}} = a_2 \mid X_{\{3\}} = 0$.

PROOF. Since $r_{\{1^2\}} \cdot r_{\{1\}} = r_{\{2,1\}} + r_{\{1^3\}}$ we see that $X_{\{1^3\}}$ is a submodule of $X_{\{1^2\}} \otimes X$. Therefore $a_1 | X_{\{1^3\}} = 1$. Other assertions of Lemma 4.2 are proved similarly.

The maps a_1, a_2 induce linear maps $f_1, f_2 \in \mathrm{End}(V)$. Since we have $\dim \mathrm{Hom}_{\mathscr{C}}(X_{\{2,1\}}, X_{\{1^2\}} \otimes X) = \dim \mathrm{Hom}_{\mathscr{C}}(X_{\{2,1\}}, X \otimes X_{\{1^2\}}) = 1$, both f_1, f_2 are rank 1 idempotents in the ring $\mathrm{End}(V)$.

LEMMA 4.3. *Let* f_1, f_2 *be two arbitrary rank* 1 *idempotents in the ring* Mat_2 *of* 2×2 *complex matrices. Then there exists* $\gamma \in F$ *such that*

$$f_1 f_2 f_1 = \gamma f_1, \quad f_2 f_1 f_2 = \gamma f_2.$$

PROOF. Easy.
Lemma 4.3 implies that

$$a_1 a_2 a_1 - \gamma a_1 = a_2 a_1 a_2 - \gamma a_2.$$

To complete the proof of Proposition 4.1 we must show that $\gamma \neq 0$. To do this we extend our definitions and for any $n, i \in \mathbb{Z}^+$, $n > i$, define an endomorphism $a_i^{(n)}$ of $X^{\otimes n}$, by the formula $a_i^{(n)} = 1_X^{\otimes(i-1)} \otimes a \otimes 1_X^{(n-i-1)}$. We will often write a_i instead of $a_i^{(n)}$. It is clear that these notations are compatible with the previous definitions $a_1, a_2 \in \operatorname{End}_{\mathscr{C}}(X^{\otimes 3})$. As follows from functoriality, $a_i a_j = a_j a_i$ if $|i-j| > 1$ and $a_i^2 = a_i$.

We assume now that $\gamma = 0$ and derive a contradiction with the assumption that $\mathscr{C}$ is a rigid category.

If $\gamma = 0$, then $f_1 f_2 f_1 = f_2 f_1 f_2$ and $a_1^{(3)} a_2^{(3)} a_1^{(3)} = a_2^{(3)} a_1^{(3)} a_2^{(3)}$. Therefore for any $n \in \mathbb{Z}^+$, $1 \le i, j \le n$, the braid relations

$$a_i^{(n)} a_j^{(n)} = a_j^{(n)} a_i^{(n)} \qquad \text{if } |i-j| > 1,$$
$$a_i^{(n)} a_j^{(n)} a_i^{(n)} = a_j^{(n)} a_i^{(n)} a_j^{(n)} \quad \text{if } |i-j| = 1$$

are satisfied. Therefore for any $\sigma \in S_n$ we can define (see the beginning of §3) $a_\sigma \in \operatorname{End}(X^{\otimes n})$. Let σ_n be the maximal length element in S_n. It is easy to see that $a_i^{(n)} a_{\sigma_n} = a_{\sigma_n} a_i^{(n)} = a_{\sigma_n}$ for all i, $1 \le i < n$, and therefore a_{σ_n} is an idempotent.

Lemma 4.4$_s$. *a_{σ_s} is the projection on $X_{\{1^s\}} \subset X^{\otimes s}$ if $s \le N$ and $a_{\sigma_s} = 0$ if $s > N$.*

Proof. We prove this lemma by induction in s. In the case $s = 2$ the statement is the definition of a. We assume that the statement is true for some s and prove it for $s+1$. Consider the idempotents in $a'_{\sigma_s} = a_{\sigma_s} \otimes \mathrm{id}_X$ and $a''_{\sigma_s} = \mathrm{id}_X \otimes a_{\sigma_s}$ in $\operatorname{End}(X^{\otimes(s+1)})$. Since $a_i^{(n)}$ satisfy the braid relations it is easy to see that there exists $b'_s, b''_s \in \operatorname{End}(X^{\otimes s})$ such that $a_{\sigma_{s+1}} = a'_{\sigma_s} b'_s = a''_{\sigma_s} b''_s$ and that we have $a_{\sigma_{s+1}} = (a'_{\sigma_s} a''_{\sigma_s})^s$.

Since $X_{\{1^s\}} \otimes X$ is isomorphic to the direct sum $X_{\{1^{s+1}\}} \oplus X_{\{21^{s-1}\}}$ and the linear space $L_{\{1^{s+1}\}}$ is at most one-dimensional, we see that $a'_{\sigma_s}|_{X_{\{1^{s+1}\}}} = a''_{\sigma_s}|_{X_{\{1^{s+1}\}}} = 1$ in the case when $s < N$ and $X_{\{1^N\}} \otimes X = X_{\{21^{N-1}\}}$. Consider the submodules $M' = X_{\{1^s\}} \otimes X$ and $M'' = X \otimes X_{\{1^s\}}$ in $X^{\otimes(s+1)}$. They both contain $X_{\{1^{s+1}\}}$. Since $M'/X_{\{1^{s+1}\}}$ and $M''/X_{\{1^{s+1}\}}$ are both isomorphic to $X_{\{21^{s-1}\}}$ and therefore are irreducible, we have either $M' \cap M'' = X_{\{1^{s+1}\}}$ or $M' = M''$. Since $a_{\sigma_{s+1}} = (a'_{\sigma_s} a''_{\sigma_s})^s$ is the projection onto $M' \cap M''$, it is sufficient to show that $M' \neq M''$. So assume that $M' = M''$. Then $a'_{\sigma_s} = a''_{\sigma_s}$. Consider the endomorphisms

$$a^i_{\sigma_s} = \mathrm{id}_{X^{\otimes i}} \otimes a_{\sigma_s} \otimes \mathrm{id}_{X^{\otimes s-i}} \in \operatorname{End}(X^{\otimes 2s}), \qquad 0 \le i \le s.$$

It is clear that the equality $a'_{\sigma_s} = a''_{\sigma_s}$ implies the equality $a^i_{\sigma_s} = a^{i+1}_{\sigma_s}$ for all

i, $0 \le i < s$. Therefore $a_s^0 = a_s^s$. But by the inductive assumption $\operatorname{Im} a_{\sigma_s} = X_{\{1^s\}} \subset X^{\otimes s}$. Therefore $\operatorname{Im}\, a_s^0 = X_{\{1^s\}} \otimes X^{\otimes s}$ and $\operatorname{Im}\, a_s^s = X^{\otimes s} \otimes X_{\{1^s\}}$ as submodules in $X^{\otimes 2s}$. But $(1 - a_1^{(2s)})a_s^0 = 0$ and $\operatorname{Im}(1 - a_1^{(2s)})a_s^s = X_{\{2\}} \otimes X^{\otimes(s-2)} \otimes X_{\{1^s\}} \neq \{0\}$. This contradiction shows that $a'_{\sigma_s} \neq a''_{\sigma_s}$. Therefore $a_{\sigma_{s+1}} = (a'_{\sigma_s} a''_{\sigma_s})^s$ is the projection onto $X_{\{1^{s+1}\}}$, and Lemma 4.4_{s+1} is proved. So Lemma 4.4_s is true for all $s > 1$.

Lemma 4.5. *The dual object X^* to X is isomorphic to $X_{\{1^{N-1}\}}$.*

Proof. Since $r_{\{1\}} r_{\{1^{N-1}\}} = 1 + r_{\{21^{N-2}\}}$ in $R_{N,\ell}$, we see that there exists a nontrivial morphism $\varphi : X_{\{1^{N-1}\}} \otimes X \to \mathbb{1}$ and therefore a nontrivial morphism

$$\widehat{\varphi}: X_{\{1^{N-1}\}} = X_{\{1^{N-1}\}} \otimes \mathbb{1} \xrightarrow{\mathrm{id} \otimes i_X} X_{\{1^{N-1}\}} \otimes X \otimes X^* \xrightarrow{\varphi \otimes \mathrm{id}} X^*.$$

Since $X_{\{1^{N-1}\}}$ and X^* are irreducible $\widehat{\varphi}$ is an isomorphism. Lemma 4.5 is proved.

Since X^* is isomorphic to a direct summand $X_{\{1^{N-1}\}}$ of $X^{\otimes(N-1)}$ there exists an imbedding $i : \mathbb{1} \to X^{\otimes N}$ and a projection $e : X^{\otimes N} \to \mathbb{1}$ such that $(\mathrm{id}_X \otimes e)(i \otimes \mathrm{id}_X) = 1_X$.

Let $X^\perp \subset X^{\otimes N}$ be the kernel of the e. It is clear that $X^\perp$ is equal to the kernel of a_{σ_N}. Let $Z \subset \operatorname{End}(X^{\otimes(N+1)})$ be the linear space of morphisms α such that $X \otimes X^\perp \subset \operatorname{Ker}\alpha$ and $\operatorname{Im}\alpha \subset X_{\{1^{N+1}\}}$. It is clear that $\dim_{\mathbb{C}} Z = 1$ and $z \overset{\mathrm{df}}{=} (i \otimes \mathrm{id}_X)(\mathrm{id}_X \otimes e)$ is a base of Z. It is clear that $z^2 = z$ and $a'_{\sigma_N} a''_{\sigma_N} \in Z$. Therefore $a'_{\sigma_N} a''_{\sigma_N} = \delta z$ for some $\delta \in F$. As follows from Lemma 4.4_{N+1}, we have $(a'_{\sigma_N} a''_{\sigma_N})^N = 0$. So $\delta = 0$ and $a'_{\sigma_N} a''_{\sigma_N} = 0$.

On the other hand, Lemma 4.4_N implies that $a'_{\sigma_N}(i \otimes \mathrm{id}_X) = i \otimes \mathrm{id}_X$ and $(\mathrm{id}_X \otimes e)a''_{\sigma_N} = (\mathrm{id}_X \otimes e)$. Therefore

$$\mathrm{id}_X = (\mathrm{id}_X \otimes e)(i \otimes \mathrm{id}_X) = (\mathrm{id}_X \otimes e)a''_{\sigma_N} a'_{\sigma_N}(i \otimes \mathrm{id}_X) = (\mathrm{id}_X \otimes e)a_{\sigma_{N+1}}(i \otimes \mathrm{id}_X).$$

But by Lemma 4.4_{N+1} we have $a_{\sigma_{N+1}} = 0$. This contradiction shows that $\gamma \neq 0$. Proposition 4.2 is proved.

To formulate a useful corollary we fix $q_{\mathscr{C}} \in \mathbb{C}^*$ such that $q_{\mathscr{C}} + q_{\mathscr{C}}^{-1} = \gamma_{\mathscr{C}}^{-1} - 2$. It is clear that $q_{\mathscr{C}}$ is uniquely defined up to $q_{\mathscr{C}} \to q_{\mathscr{C}}^{-1}$. We will write q instead of $q_{\mathscr{C}}$, when this does not create confusion. Let $T \overset{\mathrm{df}}{=} q - (1+q)a \in \operatorname{End}(X^{\otimes 2})$. For any $n, i \in \mathbb{Z}^+$, $n > i$, define

$$T_i^{(n)} = \mathrm{id}_X^{\otimes i-1} \otimes T \otimes \mathrm{id}_X^{n-i-1} \in \operatorname{End}(X^{\otimes n}).$$

Corollary 1. 1) $T_i^{(n)} T_j^{(n)} = T_j^{(n)} T_i^{(n)}$ *if* $|i-j| > 1$, $1 \le i, j \le n-1$.

2) $T_i^{(n)} T_{i+1}^{(n)} T_i^{(n)} = T_{i+1}^{(n)} T_i^{(n)} T_{i+1}^{(n)}$, $1 \le i \le n-2$.

3) $(T_i - q)(T_i + 1) = 0$, $1 \le i \le n-1$.

PROOF. Clear.

We can reformulate Corollary 1 in the following way.

COROLLARY 2. *For any* $n \in \mathbb{Z}^+$ *there exists an algebra homomorphism* $\theta_n : H_n(q) \to \operatorname{End}_{\widetilde{\mathscr{C}}} X^{\otimes n}$, $h \to \overline{h}$, *such that* $\overline{T}_i = T_i^{(n)}$, $1 \le i \le n-1$.

To simplify notations we will write H_n instead of $H_n(q)$ when it does not lead to confusion.

Let $\mathscr{C}$ be an $SL_{N,\ell}$-category (as before $N > 1$, $\ell > N+1$), $\eta_{\mathscr{C}} \overset{\text{df}}{=} \eta_{X_{\{2\}}X^{\otimes 2}}$ (see Proposition 1.1), $h \to \overline{h}$ the homomorphism $H_n(q) \to \operatorname{End}_{\mathscr{C}}(X^{\otimes n})$ as in Corollary 2 to Proposition 4.2, and $\mu_{\mathscr{C}} \overset{\text{df}}{=} q - (1+q)\eta_{\mathscr{C}}$.

REMARK. By the Corollary to Proposition 1.1 we know that $\eta_{\mathscr{C}} \neq 0, 1$. Therefore $\mu_{\mathscr{C}} \neq q, -1$.

Let $\varphi_{\mu_{\mathscr{C}}}^{n-1,n} : H_n(q) \to H_{n-1}(q)$ be the map defined as in § 3. We define a linear map $\overline{\varphi}^{n-1,n}$: $\operatorname{End}(X^{\otimes n}) \to \operatorname{End}(X^{\otimes(n-1)})$ by the formula

$$\alpha \to (\mathrm{id}_{X^{\otimes n-1}} \otimes \widetilde{\Gamma}_X) \circ (\alpha \otimes 1_{X^*}) \circ (\mathrm{id}_{X^{\otimes(n-1)}} \otimes i_X)$$

($\widetilde{\Gamma}_X$ was defined in § 1).

LEMMA 4.6. *For any* $h \in H_n$ *we have*

$$\overline{\varphi_{\mu_{\mathscr{C}}}^{n-1,n}(h)} = \overline{\varphi}^{n-1,n}(\overline{h}).$$

PROOF. It is clear from the definition that

$$\varphi_{\mu_{\mathscr{C}}}^{n-1,n}(1) = 1, \quad \varphi_{\mu_{\mathscr{C}}}^{n-1,n}(\widetilde{h}'h\widetilde{h}'') = \widetilde{h}'\varphi_{\mu_{\mathscr{C}}}^{n-1,n}(h)\widetilde{h}''$$

for $\widetilde{h}', \widetilde{h}'' \in H_{n-1}$, $h \in H_n$, $\overline{\varphi}^{n-1,n}(\mathrm{id}_{X^{\otimes n}}) = \mathrm{id}_{X^{\otimes n-1}}$, and

$$\overline{\varphi}^{n-1,n}((\widetilde{\alpha}' \otimes \mathrm{id}_X)\alpha(\widetilde{\alpha}'' \otimes \mathrm{id}_X)) = \widetilde{\alpha}'\overline{\varphi}^{n-1,n}(\alpha)\widetilde{\alpha}''$$

for $\widetilde{\alpha}', \widetilde{\alpha}'' \in \operatorname{End}(X^{\otimes(n-1)})$, $\alpha \in \operatorname{End}(X^{\otimes n})$. Therefore it is sufficient to show that $\overline{\varphi_{\mu_{\mathscr{C}}}^{n-1,n}(T_{n-1})} = \overline{\varphi}^{n-1,n}(\overline{T}_{n-1})$. By definition, $\varphi_{\mu_{\mathscr{C}}}^{n-1,n}(T_{n-1}) = \mu_{\mathscr{C}}$ and $\overline{\varphi}^{n-1,n}(\overline{T}_{n-1}) = \overline{\varphi}^{n-1,n}(T_{n-1}^{(n)}) = \overline{\varphi}^{n-1,n}(q-(1+q)a_{n-1}) = q-(1+q)\eta_{\mathscr{C}} = \mu_{\mathscr{C}}$. Lemma 4.6 is proved.

COROLLARY. *Let* $\mathscr{C}$ *be an* $SL_{N,\ell}$*-category,* $q = q_\ell$, $\alpha_n \in H_n(q)$ *are as in* § 3.

a) *If* q *is a primitive root of unity of order* $k > 1$ *and* $\overline{\alpha}_{k-1} \neq 0$ *then* $\overline{\alpha}_k \neq 0$.

b) *If* $\overline{\alpha}_n \neq 0$ *and* $\mu_{\mathscr{C}} \neq \frac{q^n}{[n]_q}$, *then* $\overline{\alpha}_{n+1} \neq 0$.

PROOF. a) Since

$$\overline{\varphi}^{k-1,k}(\overline{\alpha}_k) = \overline{\varphi_{\mu_{\mathscr{C}}}^{k-1,k}(\alpha_k)} = (1+\mu_{\mathscr{C}} - q^{-(k-1)}\mu_{\mathscr{C}}[k]_q)\overline{\alpha}_{k-1} = (1+\mu_{\mathscr{C}})\overline{\alpha}_{k-1}$$

and $\mu_{\mathscr{C}} \neq -1$, we see that $\overline{\alpha}_k \neq 0$.

b) Since $\mu_{\mathscr{C}} \neq \frac{q^n}{[n]_q}$, we have $(1+\mu_{\mathscr{C}} - q^{-n}\mu_{\mathscr{C}}[n+1]_q) \neq 0$ and the claim follows from the equality

$$\overline{\varphi}^{n;n+1}(\overline{\alpha}_{n+1}) = (1+\mu_{\mathscr{C}} - q^{-n}\mu_{\mathscr{C}}[n+1]_q)\overline{\alpha}_n.$$

The Corollary is proved.

LEMMA 4.7. *Let $\mathscr{C}$ be an $SL_{N,\ell}$-category and let $q = q_{\mathscr{C}}$ be a primitive root of unity of order k. Then for any m, $1 \leq m < \min(k, N+1)$, the endomorphism $(\overline{[m]_q!})^{-1}\overline{\alpha_m} \in \mathrm{End}_{\tilde{C}}(X^{\otimes m})$ is a projection onto $X_{\{1^m\}}$, where* $[m]_q! \overset{\mathrm{df}}{=} [1]_q[2]_q \cdots [m]_q$.

PROOF. We prove the lemma by induction on m. It is clear that it is true for $m = 1$. We assume that it is true for some m and prove it for $m+1$. Consider the subobjects $M' = X_{\{1^m\}} \otimes X$ and $M'' = X \otimes X_{\{1^m\}}$ in $X^{\otimes m+1}$.

As follows from Lemma 4.6, $(\overline{[m]_q!})^{-1}\overline{\alpha}_{m+1}$ is the projection onto $M' \cap M''$. Since $\mathscr{C}$ is an $SL_{N,\ell}$-category, we know that M' and M'' are isomorphic to $X_{\{21^{m-1}\}} \oplus X_{\{1^{m+1}\}}$ and therefore either $M' \cap M'' = X_{\{1^{m+1}\}}$ or $M' = M''$. The same arguments as in the proof of Lemma 4.4_s show that $M' \neq M''$. Therefore $M = X_{\{1^{m+1}\}}$. Lemma 4.7 is proved.

COROLLARY. a) $k > N$.
b) $\mu_{\mathscr{C}} = q^m/[m]_q$ *for some* m, $1 < m < k-1$.

PROOF. a) Assume then $k \leq N$ and consider the endomorphism $\overline{\alpha}_k \in \mathrm{End}(X^{\otimes k})$. The same arguments as in the proof of the lemma show that $\mathrm{Im}\,\overline{\alpha_K} \subset X_{\{1^k\}}$. Since $q^k = 1$ it follows from Lemma 3.1 c) and Lemma 4.6 that $\alpha_k^2 = 0$. Therefore $\overline{\alpha}_k^2 = 0$. Since $\tilde{\mathscr{C}}$ is semisimple and $X_{\{1^k\}}$ is a simple subobject of $X^{\otimes k}$, we see that $\overline{\alpha}_k = 0$. But by the Corollary to Lemma 4.6 we see that $\overline{\varphi}^{k-1,k}(\overline{\alpha}_k) \neq 0$. This contradiction shows that $k > N$.

b) Assume that $\mu_{\mathscr{C}} \neq \frac{q^m}{[m]_q}$ for all m, $1 < m < k-1$. Since we know that $\mu_{\mathscr{C}} \neq q$, or -1, we see that $\mu_{\mathscr{C}} \neq \frac{q^m}{[m]_q}$ for all $m \in \mathbb{Z}$. Therefore it follows from the Corollary to Lemma 4.6 that $\overline{\alpha_m} \neq 0$ for $m \in \mathbb{Z}$. In particular, $\alpha_k \neq 0$. But we have seen in the proof of part a) that this contradicts the semisimplicity of $\tilde{\mathscr{C}}$. The Corollary is proved.

We define a linear functional $\overline{\varphi}^n : \mathrm{End}(X^{\otimes n}) \to F$ as the composition

$$\overline{\varphi}^n \overset{\mathrm{df}}{=} \overline{\varphi}^{1,2} \circ \overline{\varphi}^{2,3} \circ \cdots \circ \overline{\varphi}^{n-1,n} \colon \mathrm{End}(X^{\otimes n}) \to \mathrm{End}(X) = F.$$

LEMMA 4.8. *For any $h \in H_n$ we have $\varphi^n_{\mu_{\mathscr{C}}}(h) = \overline{\varphi}^n(\overline{h})$.*

PROOF. Follows from Lemma 4.6.

THEOREM 4.1. *Let $\mathscr{C}$ be an $SL_{N,\ell}$-category. Then $q_{\mathscr{C}}$ is a primitive root of unity of order ℓ, $\mu_{\mathscr{C}} = \frac{q_{\mathscr{C}}^N}{[N]_{q_{\mathscr{C}}}}$, $\operatorname{Ker}\theta_n = I^n_{\mu_{\mathscr{C}}}$, and the map $h \to \overline{h}$ defines an isomorphism $H_n(q_{\mathscr{C}})/I^n_{\mu_{\mathscr{C}}} \xrightarrow{\sim} \operatorname{End}(X^{\otimes n})$, $n \in \mathbb{Z}$.*

PROOF. We first analyze the case when $q_{\mathscr{C}}$ is not a primitive root of unity of order k for any $k > 1$. If $\mu_{\mathscr{C}} \neq \frac{q_{\mathscr{C}}^m}{[m]_{q_{\mathscr{C}}}}$ for all $m \leq N$ then Proposition 3.1 a) implies that the map $H_{N+1}(q_{\mathscr{C}}) \to \operatorname{End}(X^{\otimes(N+1)})$ is an imbedding. But

$$\dim \operatorname{End}_{\widetilde{\mathscr{C}}}(X^{\otimes(N+1)}) = \sum_{\substack{\lambda\in\Lambda_{N,\mathscr{C}}\\ |\lambda|=N+1}} d^{N+1}_{N,\ell}(\lambda)^2 < \sum_{\substack{\lambda\in\Lambda\\ |\lambda|=N+1}} d(\lambda)^2 = \dim H_{N+1}(q_{\mathscr{C}}).$$

This contradiction shows that $\mu_{\mathscr{C}} = \frac{q^m}{[m]_{q_{\mathscr{C}}}}$ for some m, $1 < m \leq N$. Assume now that $m < N$. Then it follows from Lemma 4.7 that $\frac{1}{[m+1]_q!}\overline{\alpha}_{m+1}$ is the projection on a nonzero simple subject $X_{\{1^{m+1}\}}$ in $X^{\otimes(m+1)}$. It is clear that the number $\eta_{X_{\{1^{m+1}\}}, X^{\otimes(m+1)}} \in F$ is equal to $\frac{1}{[m+1]_q!}\overline{\varphi}^n(\overline{\alpha}_{m+1})$. From Proposition 1.1 we conclude that $\varphi_{\mu_{\mathscr{C}}^{m+1}}(\alpha_{m+1}) = \overline{\varphi}^{m+1}(\overline{\alpha}_{m+1}) \neq 0$. On the other hand, it is easy to see that $\alpha_{m+1} \in I^{m+1}_{\mu_{\mathscr{C}}}$ if $\mu = \frac{q^{m+1}}{[m+1]_q}$. This contradiction shows that $m = N$. As follows from Proposition 3.1 a), for any $n \in \mathbb{Z}^+$ we have the inequality $\dim H_n(q_{\mathscr{C}})/I^n_{\mu_{\mathscr{C}}} \leq \dim \operatorname{End}(X^{\otimes n})$ and we have an equality for all $n \in \mathbb{Z}^+$ if and only if $\ell = \infty$. On the other hand, it follows from Lemma 4.8 that $\ker\theta_n < I^n_{\mu_{\mathscr{C}}}$. Therefore $\operatorname{Ker}\theta_n = I^n_{\mu_{\mathscr{C}}}$, θ_n is surjective, and $\ell = \infty$.

Consider now the case when $q_{\mathscr{C}}$ is a primitive root of unity of order $k > 1$. As follows from the Corollary to Lemma 4.7 we have $\mu_{\mathscr{C}} = \frac{q^m}{[m]_q}$ for some m, $1 < m < k - 1$ and $k > N$. The same arguments as before show that $m = N$. To prove that $k = \ell$ we consider the elements $s_n = \sum_{\sigma\in S_N} T_\sigma \in H_n$ and $\overline{s}_n = \theta_n(s_n) \in \operatorname{End}(X^{\otimes n})$.

LEMMA 4.9. *For any $n \leq \min(k-1, \ell - N)$ the endomorphism $\frac{1}{[m]_q^!}\overline{s}_n$ is the projection onto $X_{\{n\}} \subset X^{\otimes n}$.*

PROOF. Similar to Lemma 4.7.

COROLLARY 1. *For any $n \leq \min(k-1, \ell-N)$ we have $s_n \notin I^n_{\mu_{\mathscr{C}}}$.*

PROOF. Follows from Proposition 1.1 and the formula for $w^{(n)}_\lambda$ in Lemma 3.3.

COROLLARY 2. *We have $k \geq \ell$.*

PROOF. Assume $k < l$. By Lemma 4.9, $(1/[k-N+1]_q!)\bar{s}_{\{k-N+1\}}$ is the projection onto $X_{\{k-N+1\}}$. By Lemma 4.8 and Lemma 3.3 we have

$$\bar{\varphi}^n(\bar{s}_{\{k-N+1\}}) = \varphi^n(s_{\{k-N+1\}}) = 0.$$

But then also $\eta_{X_{\{k-N+1\}}|X^{\otimes k-N+1}} = 0$, which contradicts Proposition 1.1.

On the other hand, it follows from Lemma 3.3 that $\operatorname{Ker}\theta_n \subset I^n_{\mu_{\mathscr{C}}}$ and therefore $\dim H_n/I^n_{\mu_{\mathscr{C}}} \le \dim \operatorname{End}_{\tilde{\mathscr{C}}}(X^{\otimes n})$ for all $n \in \mathbb{Z}^+$. As follows from Proposition 3.1 c),

$$\dim H_n/I^n_{\mu_{\mathscr{C}}} = \sum_{\substack{\lambda\in\Lambda_{N,k}\\ |\lambda|=n}} (d^n_{N,k}(\lambda))^2,$$

$$\dim \operatorname{End}_{\tilde{\mathscr{C}}}(X^{\otimes n}) = \sum_{\substack{\lambda\in\Lambda_{N,\ell}\\ |\lambda|=n}} (d^n_{N,\ell}(\lambda))^2.$$

These two equalities show that the inequality $\dim H_n/I^n_{\mu_{\mathscr{C}}} < \dim \operatorname{End}_{\tilde{\mathscr{C}}}(X^{\otimes n})$ can hold only if $k \le \ell$. Therefore $k = \ell$ and θ_n induces an isomorphism

$$H_n/I^n_{\mu_{\mathscr{C}}} \xrightarrow{\sim} \operatorname{End}_{\tilde{\mathscr{C}}}(X^{\otimes n}).$$

Theorem 4.1 is proved.

§ 5. The normalization of $SL_{N,\ell}$-categories

Let $\mathscr{C}$ be an $SL_{N,\ell}$-category. For any $m, n \in \mathbb{Z}$ we denote by $\sigma_{m,n} \in S_{m+n}$ the permutation given by

$$\sigma_{m,n}(i) = \begin{cases} i+n & \text{if } 1 \le i \le m, \\ i-m & \text{if } m+1 \le i \le m+n. \end{cases}$$

Consider the element $T_{\sigma_{m,n}} \in H_{m+n}(q)$ and define $\mathscr{S}_{m,n} = (-1)^{mn}\overline{T}_{\sigma_{m,n}} \in \operatorname{Hom}(X^{\otimes m} \otimes X^{\otimes n}, X^{\otimes n} \otimes X^{\otimes m}) = \operatorname{End}_{\tilde{\mathscr{C}}}(X^{\otimes(m+n)})$.

LEMMA 5.1. *For any* $\alpha \in \operatorname{End}_{\mathscr{C}}(X^{\otimes m})$, $\beta \in \operatorname{End}_{\mathscr{C}}(X^{\otimes n})$ *we have* $\mathscr{S}_{m,n} \circ (\alpha\otimes\beta) = (\beta\otimes\alpha)\circ\mathscr{S}_{m,n}$.

PROOF. As follows from Theorem 4.1, it is sufficient to prove this equality in the case when $\alpha = \overline{T}_{\sigma'}$, $\beta = \overline{T}_{\sigma''}$, $\sigma' \in S_m$, $\sigma'' \in S_n$. Let $\sigma_1, \sigma_2 \in S_{m+n}$ be the images of $\sigma'\times\sigma''$ and $\sigma''\times\sigma'$ under the natural imbeddings $S_m\times S_n \hookrightarrow S_{m+n}$, $S_n \times S_m \hookrightarrow S_{m+n}$. Then $\sigma_{m,n}\circ\sigma_1 = \sigma_2\circ\sigma_{m,n}$ and $l(\sigma_{m,n}\circ\sigma_1) = l(\sigma_{m,n}) + l(\sigma_1) = l(\sigma_2) + l(\sigma_{m,n})$. So we have $T_{\sigma_{m,n}}T_{\sigma_1} = T_{\sigma_2}T_{\sigma_{m,n}}$ and therefore $\mathscr{S}_{m,n}\circ\overline{T}_{\sigma_1} = \overline{T}_{\sigma_2}\circ\mathscr{S}_{m,n}$. But by definition $\overline{T}_{\sigma_1} = \overline{T}_{\sigma'}\otimes\overline{T}_{\sigma''}$, $\overline{T}_{\sigma_2} = \overline{T}_{\sigma''}\otimes\overline{T}_{\sigma'}$.

The lemma is proved.

By definition of an $SL_{N,\ell}$-category, the subobject $X_{\{1^N\}}$ of $X^{\otimes N}$ is isomorphic to $\mathbb{1}$. We choose an imbedding $\nu\colon \mathbb{1} \hookrightarrow X^{\otimes N}$ and denote by $p\colon X^{\otimes N} \to \mathbb{1}$ the unique morphism such that $p \circ \nu = 1$. We define a number $\tau = \tau_{\mathscr{C}} \in \operatorname{Hom}(X, X) = F$ as the composition

$$X = X \otimes \mathbb{1} \xrightarrow{1\otimes\nu} X \otimes X^N \xrightarrow{S_{1,N}} X^{\otimes N} \otimes X \xrightarrow{p\otimes 1} \mathbb{1} \otimes X = X.$$

It is clear that τ does not depend on a choice of an imbedding ν. For any $k \in \mathbb{Z}^+$ define $\tau_k \in \operatorname{End}(X^{\otimes k})$ as the composition

$$X^{\otimes k} = X^{\otimes k} \otimes \mathbb{1} \xrightarrow{\operatorname{id}\otimes\nu} X^k \otimes X^{\otimes N} \xrightarrow{S_{k,N}} X^{\otimes N} \otimes X^{\otimes k} \xrightarrow{p\otimes 1} \mathbb{1} \otimes X^{\otimes k} = X^{\otimes k}.$$

Lemma 5.2. *We have* $\tau_k = \tau^k \operatorname{id}_{X^{\otimes k}}$.

Proof. We will prove the lemma by induction on k. So we assume that the lemma is true for some $k \in \mathbb{Z}$ and prove it for $k+1$. Since $S_{k+1,N} = (S_{1,N} \otimes \operatorname{id}_{X^{\otimes k}})(\operatorname{id}_X \otimes S_{k,N})$ we can write τ_{k+1} in the form $\tau_{k+1} = (\tau \otimes \operatorname{id}_{X^{\otimes k}})(\operatorname{id}_{X^{\otimes k}} \otimes \tau_k)$. The lemma is proved.

Proposition 5.1. *We have* $\tau^N = 1$.

Proof. Consider the restriction $\widetilde{\tau}$ of $\tau_N : X^{\otimes N} \to X^{\otimes N}$ to the subobject $X_{\{1^N\}}$. By definition, $\widetilde{\tau}$ is the composition

$$X_{\{1^N\}} = X_{\{1^N\}} \otimes \mathbb{1} \xrightarrow{\operatorname{id}\otimes\nu} X_{\{1^N\}} \otimes X_{\{1^N\}} \xrightarrow{S_{N,N}}$$
$$\longrightarrow X_{\{1^N\}} \otimes X_{\{1^N\}} \xrightarrow{\nu^{-1}\otimes\operatorname{id}} \mathbb{1} \otimes X_{\{1^N\}} = X_{\{1^N\}},$$

where we consider ν as an isomorphism $\nu\colon \mathbb{1} \xrightarrow{\sim} X_{\{1^N\}} \subset X^{\otimes N}$. It is easy to verify that the restriction of $S_{N,N}$ to the subobject $X_{\{1^N\}} \otimes X_{\{1^N\}} \subset X^{\otimes N} \otimes X^{\otimes N}$ is equal to 1. Therefore it follows from Lemma 1.2 that $\tau_N|_{X_{\{1^N\}}} = \operatorname{id}_{X_{\{1^N\}}}$. But as follows from Lemma 5.2, $\tau_N = \tau^N \cdot \operatorname{id}_{X^{\otimes n}}$. Proposition 5.1 is proved.

We say that an $SL_{N,\ell}$-category $\mathscr{C}$ is normalized if $\tau_{\mathscr{C}} = 1$.

Corollary. *For any $S\ell_{N,\ell}$-category $\mathscr{C}$ there exists a normalized $S\ell_{N,\ell}$-category $\mathscr{C}'$ such that $\mathscr{C}$ is equivalent to $(\mathscr{C}')^{\tau_{\mathscr{C}}}$ (see the end of § 1).*

This corollary implies that it is sufficient to prove Theorem A_ℓ for normalized $SL_{N,\ell}$-categories.

It is clear that for any normalized $SL_{N,\ell}$-category $\mathscr{C}$ the corresponding monoidal algebra $\mathscr{A}_{\mathscr{C}}$ is a monoidal algebra of type N. Therefore it follows from Theorem 4.1 and Propositions 2.1, 2.2 that $\mathscr{C}$ is equivalent to the monoidal category $\mathscr{C}_{N,q}$ for some primitive root q of 1 of order ℓ. Theorem A_ℓ is proved.

REFERENCES

[A] H. Anderson, *Quantum groups, invariants of 3-manifolds and semisimple tensor categories*, Preprint Matem. Inst. Aarhus Univ., No. 15, 1992.

[Bou] N. Bourbaki, *Groupes et algebres de Lie.* Chap. IV–VI, Hermann, Paris, 1968.

[D1] V. Drinfeld, *Quantum groups*, Proc. Intern. Congress Math., Berkeley, 1988, vol. 2, Amer. Math. Soc., Providence, RI, 1989, pp. 798–820.

[D2] ——, *On the structure of the quasitriangular quasi-Hopf algebras*, Funktsional. Analiz i Prilozhen. **26** (1992), no. 1, 78–81; English transl. in Functional Anal. Appl. **26** (1992), no. 1.

[DM] P. Deligne and J. Milne, *Tannakian categories*, Lecture Notes in Math., vol. 900, Springer-Verlag, Berlin and New York.

[FRT] L. Faddeev, N. Reshetikhin, and L. Takhtajan, *Quantization of Lie groups and Lie algebras*, Algebra i Analiz **1** (1989), no. 1; English transl. in Leningrad Math. J. **1** (1990), no. 1.

[FYHLMO] P. Freyd, D. Yetter, J. Hoste, W. Lickorish, K. Millet, and A. Ocneanu, *A new polynomial invariant of knots and links*, Bull. Amer. Math. Soc. **12** (1985), 239–246.

[K] T. Kerler, *Quantum groups, quantum categories and quantum field theory*, Dissertation, ETH, Zurich, 1992.

[L] G. Lusztig, *Quantum deformations of certain simple modules over enveloping algebras*, Adv. in Math. **90** (1988), 237–249.

[M] S. Mac Lane, *Categories for working mathematicians*, Springer-Verlag, Berlin and New York, 1971.

[S] S. Shnider, *Deformation cohomology for bialgebras and quasi-bialgebras*, Deformation Theory and Quantum Groups with Applications in Mathematical Physics (M. Gerstenhaber and J. Stasheff, eds.), Contemp. Math., vol. 134, Amer. Math. Soc., Providence, RI, 1992, pp. 259–297.

[W] H. Wenzl, *Hecke algebras of type* A_n *and subfactors*, Invent. Math. **92** (1988), 349–383.

DEPARTMENT OF MATHEMATICS, HARVARD UNIVERSITY, CAMBRIDGE, MASSACHUSETTS 02138

ADVANCES IN SOVIET MATHEMATICS
Volume 16, Part 2, 1993

Vassiliev's Knot Invariants

MAXIM KONTSEVICH

To my teacher I. M. Gelfand on the occasion of his 80th birthday

V. Vassiliev [V1, V2] defined a broad class of knot invariants using a kind of infinite-dimensional Alexander duality. Some comments on the main idea of V.Vassiliev are contained in § 1. The vector space V of Vassiliev's invariants has a filtration by natural numbers. We want to mention here several features of these invariants:

(1) The space of invariants of a fixed degree is finite-dimensional and there exists an a priori upper bound on its dimension. Moreover, this space is algorithmically computable. Unfortunately, the only known method to compute this space for a fixed degree takes a super-exponential time.
(2) For any of Vassiliev's invariants there exists a polynomial-time algorithm for computing this invariant for arbitrary knots.
(3) It is not hard to prove that if all Vassiliev's invariants for two knots coincide, then their (Alexander, Conway, Jones, Kaufmann, HOMFLY, etc.) polynomial invariants coincide. In a sense, Vassiliev's invariants are stronger than any invariant coming from the solution of the Yang-Baxter equation that can be deformed to the trivial solution. It seems likely that Vassiliev's invariants can distingush any two different knots.

Here the reader will find only a short exposition of the theory. We recommend a very detailed review by D. Bar-Natan [BN] containing most of what is written here and much more.

§0. Two formulas

Let $K : S^1 \hookrightarrow \mathbb{R}^3$ be a parametrized closed curve in $\mathbb{R}^3$. We shall write two different formulas for the simplest nontrivial Vassiliev invariant of the oriented knot $K(S^1)$. They arise from the perturbative Chern-Simons theory and from the (perturbative) Knizhnik-Zamolodchikov equation.

1991 *Mathematics Subject Classification*. Primary 57M25; Secondary 81T40.

First formula. Denote by $\omega(x)$ the closed 2-form $\frac{1}{8\pi}\varepsilon_{ijk}\frac{x^i dx^j \wedge dx^k}{|x|^3}$ on $\mathbb{R}^3\backslash\{0\}$ (the standard volume element on S^2 written in homogeneous coordinates). This form appears in the Gauss formula for the linking number of two nonintersecting oriented curves $L_1, L_2 \subset \mathbb{R}^3$:

$$\#(L_1, L_2) = \int_{x\in L_1, y\in L_2} \omega(x-y).$$

For the knot $K(S^1)$, where $S^1 = [0, 1]/\{0, 1\}$, the following sum

$$\int_{0<l_1<l_2<l_3<l_4<1} \omega(K(l_1)-K(l_3)) \wedge \omega(K(l_2)-K(l_4))$$
$$+ \int_{0<l_1<l_2<l_3<1, x\in\mathbb{R}^3\backslash K(S^1)} \omega(K(l_1)-x)\wedge\omega(K(l_2)-x)\wedge\omega(K(l_3)-x) - \frac{1}{24}$$

is an invariant, i.e., does not change when we continuously vary K in the class of embeddings.

The convergence of integrals above is almost clear (at least for real analytic knots). For example, for the second integral we can define a map from the integration domain to $(S^2)^3$ by sending (l_1, l_2, l_3, x) to the triple of directions from x to l_i. The image of this map is a real analytic subset of $(S^2)^3$ of dimension 6. The integral is equal to the part of volume of $(S^2)^3$ covered by this subset, counting, of course, multiplicities and orientations.

The invariance of the sum of 4-dimensional and 6-dimensional integrals above follows from Stokes's theorem and the following properties of the form ω:

(1) $\omega(\lambda x) = \mathrm{sgn}(\lambda)\omega(x)$ for any $\lambda \in \mathbb{R}^*$,
(2) $\int_{S^2}\omega(x) = 1$,
(3) $\int_{y\in\mathbb{R}^3\backslash\{x,z\}} \omega(x-y)\wedge\omega(y-z) = 0$ as a 1-form in variables $x \neq z \in \mathbb{R}^3$.

Second formula. Introduce coordinates t, z in $\mathbb{R}^3$, $t = x_1 \in \mathbb{R}$, $z = x_2 + ix_3 \in \mathbb{C}$. Suppose that our knot is such that $t\circ K$ is a Morse function on $S^1 \subset \mathbb{R}^2$. Let us consider the knot $K(S^1)$ as a graph of multivalued functions $\mathbb{R}^1 \longrightarrow \mathbb{C}^1$. The following sum of 2-dimensional and 0-dimensional integrals

$$\frac{1}{4\pi^2}\int_{t_1<t_2}\sum_{\{z,z'\}} \frac{dz_1 - dz_1'}{z_1 - z_1'} \wedge \frac{dz_2 - dz_2'}{z_2 - z_2'} \prod_{\substack{4 \text{ points} \\ (t_*, z_*)}} \frac{\text{orientation of the knot}}{\text{orientation arising from } t}$$
$$+ \frac{1}{48}(\text{ number of critical points of } t\circ K) - \frac{1}{24},$$

where the sum is taken over all choices of two pairs of points (t_i, z_i), (t_i, z_i'), $i = 1, 2$, on $K(S^1)$ such that points of the first pair alternate with points of the second pair, is again a knot invariant. The proof uses Stokes' theorem and the identity

$$\frac{dz_1 - dz_2}{z_1 - z_2} \wedge \frac{dz_2 - dz_3}{z_2 - z_3} + \frac{dz_2 - dz_3}{z_2 - z_3} \wedge \frac{dz_3 - dz_1}{z_3 - z_1} + \frac{dz_3 - dz_1}{z_3 - z_1} \wedge \frac{dz_1 - dz_2}{z_1 - z_2} = 0.$$

Both formulas give the same integer number.

§1. Vassiliev's invariants

It is clear that knot invariants (with values in some abelian group) are the same as locally constant functions on the space of embeddings of the standard circle S^1 into the 3-dimensional Euclidean space $\mathbb{R}^3$, or, equivalently, zero-cohomology classes.

Let us consider the space of embeddings as the complement in the infinite-dimensional vector space of all mappings from S^1 to $\mathbb{R}^3$ to the closed subspace of maps with self-intersections or with singular image. We intersect both spaces, the space of knots and its complement, with an appropriate generic family of finite-dimensional vector spaces of growing dimensions. For example, the spaces of trigonometric polynomial maps of fixed degrees will do. Then we can apply the usual Alexander duality. Of course, we can generalize it to the case of embeddings of an arbitrary manifold into Euclidean space of arbitrary dimension.

The main techical invention of V. Vassiliev is a very simple simplicial resolution of singularities of the space of nonembeddings, which allows us to compute the homology groups with closed support. This technique can be applied to a very broad class of situations, and in good cases it gives a complete description of the weak homotopy type of some functional spaces. The case of knot invariants turns out to be marginal. The spectral sequence arising in Vassiliev's approach does not converge well. The zero-degree part of its limit is a certain countable-dimensional subspace in the continuum-dimensional space of all cohomology classes.

Let us fix a nonnegative integer n. We want to define invariants of degree strictly less than n.

For any knot $K : S^1 \hookrightarrow \mathbb{R}^3$ and any family of nonintersecting balls B_1, $B_2, \dots, B_n \subset \mathbb{R}^3$ such that the intersection of any ball with $K(S^1)$ is the standard one (see Figure 1 on the next page), one can construct 2^n knots. These knots will be labeled by the sequences of $+1$ and -1 of length n. The knot $K_{\varepsilon_1, \dots, \varepsilon_n}$ is obtained from $K = K(S^1)$ by replacing, for each i, $1 \leq i \leq n$, such that $\varepsilon_i = -1$, the part of the knot in the interior of B_i by another standard sample (see Figure 2 on the next page). Of course, $K_{1,1,\dots,1}$ is the initial knot K.

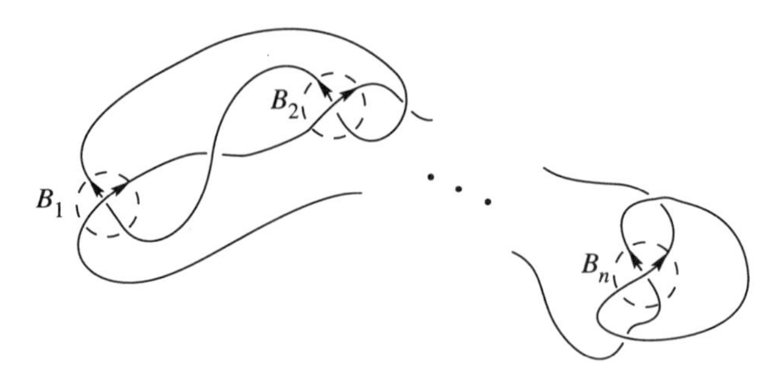

FIGURE 1

FIGURE 2

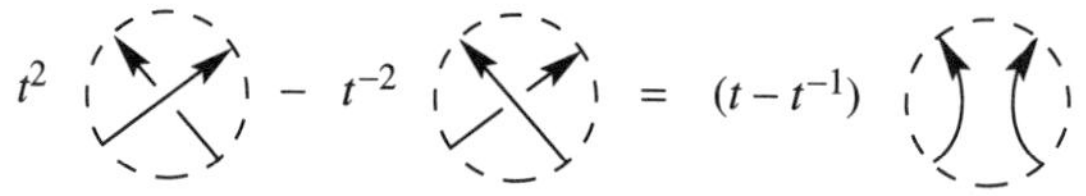

t^2 − t^{-2} = $(t - t^{-1})$

FIGURE 3

Let Φ be a knot invariant with the values in an abelian group A (for example, $A = \mathbb{Z}, \mathbb{Q}, \mathbb{C}, \dots$).

DEFINITION. Φ is an invariant of degree less than n if for all K and $B_1, \dots, B_n$ as above the following equality holds:

$$\sum_{\varepsilon_1, \dots, \varepsilon_n} \varepsilon_1 \cdots \varepsilon_n \Phi(K_{\varepsilon_1, \dots, \varepsilon_n}) = 0.$$

Now we show that, for example, Jones invariants are contained in Vassiliev's invariants. Recall that the Jones invariant

(1) takes values in the group $\mathbb{Z}[t, t^{-1}]$,
(2) is defined for links, and
(3) satisfies the skein relation (see Figure 3).

THEOREM 1.1 (Birman-Lin). *The kth coefficient in the Taylor expansion of the Jones invariant at $t = 1$ is a Vassiliev's invariant of degree less than $(k+1)$.*

PROOF. The skein relation degenerates at $t = 1$ to the relation $\Phi(K_0) = \Phi(K_1)$. This fact easily implies that

$$\sum_{\varepsilon_1, \dots, \varepsilon_{k+1}} \varepsilon_1 \cdots \varepsilon_{k+1} \Phi(K_{\varepsilon_1, \dots, \varepsilon_{k+1}})$$

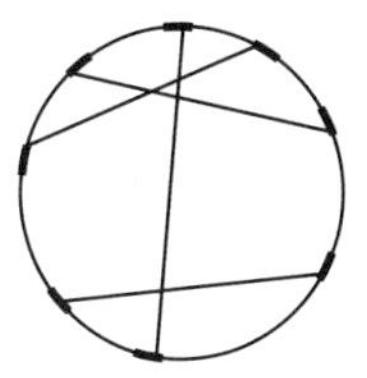

FIGURE 4

belongs to the $(k+1)$st degree of the ideal $(t-1)\mathbb{Z}[t, t^{-1}]$. Hence the kth Taylor coefficient of this alternating sum is zero. □

This proof can be generalized immediately to an arbitrary complex analytic family of solutions of the quantum Yang-Baxter equation containing the trivial solution. So, Vassiliev invariants are at least as strong as different kinds of polynomial invariants.

For $k = 0, 1, \dots$ denote by V_k the vector space of $\mathbb{Q}$-valued invariants of degree less than $k+1$. Denote by $V = \bigcup_k V_k$ the space of all invariants of finite degree. It is clear that $V_0 \subset V_1 \subset V_2 \subset \cdots$ is a growing family of vector spaces. First of all, let us prove that all V_k are finite-dimensional.

LEMMA 1.1. $\dim(V_k/V_{k-1}) \leq (2k-1)!! = 1 \cdot 3 \cdot 5 \cdots (2k-1)$.

PROOF. For an invariant $\Phi \in V_k$ and any data K, $B_1, \dots, B_k$ as above consider the value of the alternating sum

$$\nabla^k(\Phi) := \sum_{\varepsilon_1, \dots, \varepsilon_k} \varepsilon_1 \cdots \varepsilon_k \Phi(K_{\varepsilon_1, \dots, \varepsilon_k}).$$

We claim that this number does not change if we change the knot allowing self-intersections of the knot outside the union of balls B_i. It follows directly from the equation for invariants of degree less than $(k+1)$.

Hence these numbers depend only on k pairs of intervals on S^1, which are preimages of balls B_i under the map K. It is clear that these numbers are invariant under permutations of indices $1, \dots, k$ and under the homotopy of families of intervals. We can connect two points on both components of $K^{-1}(B_i)$ by lines and obtain a finite family of chords (see Figure 4). Each family of k chords can be obtained in this way for some K, $B_1, \dots, B_n$. So, we obtain a function $\nabla^k(\Phi)$ on families of k chords with all ends distinct. By definition, $\nabla^k(\Phi) = 0$ if and only if $\Phi \in V_{k-1}$. A simple computation shows that if we replace S^1 by $\mathbb{R}^1$, there are $(2k-1)!!$ topologically distinct families of pairs of points. It is clear that it gives an upper bound for the circle. □

Moreover, there are some linear relations between values of $\nabla^k(\Phi)$.

For a family S of $k-1$ chords, a point p on the circle S^1 different from all endpoints of chords in S, and a chord $a \in S$ we can construct four new familes S_1, S_2, S_3, S_4 of k chords. They are obtained by adding to S

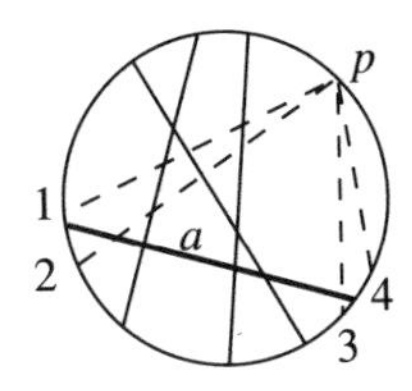

FIGURE 5

a new chord connecting p with a point on S^1 from the left and from the right side near both endpoints of a (see Figure 5).

LEMMA 1.2. *For any invariant $\Phi \in V_k$ its higher derivative $\nabla^k(\Phi)$ satisfies the following relations:*

(1) *for any family S of $k-1$ arcs, p, and a as above,*

$$\sum_{i=1}^{4}(-1)^i \nabla^k(\Phi)(S_i) = 0;$$

(2) *if the family S of k chords contains a chord that does not intersect other chords, then $\nabla^k(\Phi)(S) = 0$.*

PROOF. To obtain the first relation consider the situation when there exists a ball intersecting our knot in three intervals parallel to coordinate axes. Let the endpoints of a and p lie on these intervals. There are eight topologically different possible arrangements of such intervals in a ball. Our relation contains 16 terms. One can check easily that each configuration arises twice with opposite signs. The second relation is simpler: one can consider the situation when the knot consists of two spatially separated parts connected by two strings. The spinning of this pair of strings does not change the topological type of the knot. □

Vassiliev's invariants are closed under the multiplication $V_k \times V_l \subset V_{k+l}$. This fact follows easily from the definition and a generalization of the Leibniz formula to difference derivatives.

§2. Algebra of diagrams and the main theorem

It will be convenient for us to cut the circle at some point and to obtain a family of nonintersecting 2-element subsets of a (horizontal) line $\mathbb{R}^1$. We connect two points belonging to one subset by an arc above the line.

We shall use now notations from the previous lemma.

DEFINITION. For each $k = 0, 1, \ldots$, $\mathscr{A}_k$ is the vector space over $\mathbb{Q}$ generated by homotopy classes of families of k arcs with distinct ends on the line $\mathbb{R}^1$ modulo relations

(1) $\sum_{i=1}^4 (-1)^i S_i = 0$,

(2) if S contains an arc that does not intersect other arcs, then $S = 0$.

The first three spaces are: $\mathscr{A}_0 = \mathbb{Q}$, $\mathscr{A}_1 = 0$, $\mathscr{A}_2 = \mathbb{Q}$. Lemma 1.2 means that V_k/V_{k-1} is a subspace of the space $\mathscr{A}_k^* = Hom(\mathscr{A}_k, \mathbb{Q})$.

REMARK. D. Bar-Natan denotes our $\mathscr{A}$ by $\mathscr{A}^r$ and our $\hat{\mathscr{A}}$ (see the end of this section) by $\mathscr{A}$.

We shall prove the following

THEOREM 2.1. $V_k/V_{k-1} = \mathscr{A}_k^*$.

This theorem means that differentials for zero cohomology groups of the space of knots in higher terms of Vassiliev's spectral sequence are trivial up to the torsion.

The proof of Theorem 2.1 requires some preparations.

LEMMA-DEFINITION. *For two families of arcs such that all endpoints of S_1 are smaller than all endpoints of S_2 (as real numbers) the formula $S_1 \times S_2 = S_1 \cup S_2$ defines the structure of associative algebra on $\mathscr{A} := \bigoplus_{k=0}^{\infty} \mathscr{A}_k$.*

We have to verify that the product is compatible with the additive relations between families of arcs. It is almost evident.

A slightly less trivial fact is that any family of chords on the circle S^1 defines an element of $\mathscr{A}$.

LEMMA 2.1. *If S is a family of k chords with distinct ends on S^1 and $p_1, p_2 \in S^1$ are two points different from the endpoints of S, then two families of arcs S_1, S_2 on $\mathbb{R}^1$ obtained by deleting p_1 or p_2 from the circle, define the same element of $\mathscr{A}_k$.*

PROOF. We can proceed step by step. Let S' is a family of arcs on $\mathbb{R}^1$. Let $x \in \mathbb{R}^1$ be the minimal endpoint of S' and $\{x, y\} \in S'$ be the corresponding arc. Consider the sum of relations (1) over all arcs $a \in S' \backslash \{x, y\}$ for the triple

$$(\text{point } p, \ \text{arc } a, \ \text{family of arcs } S' \backslash \{x, y\}).$$

It is easy to see that this sum is equal to the difference $S' - S''$, where S'' is obtained from S' by replacing x with a point on $\mathbb{R}^1$ right to all endpoints of S'. □

COROLLARY. *$\mathscr{A}$ is a commutative algebra.*

Let us consider the standard Euclidean space $\mathbb{R}^3$ as the product of the real line $\mathbb{R}^1$ with the complex line $\mathbb{C}^1$. We denote the corresponding coordinates by t, z, $t \in \mathbb{R}$, $z \in \mathbb{C}$. A knot K is called a Morse knot if the function $(t \circ K)$ is a Morse function on S^1. As before, for any n we can define the notion of an invariant of Morse knots of degree less than n.

Define a sequence of invariants of Morse knots with values in $\mathscr{A}_n \otimes \mathbb{C}$.

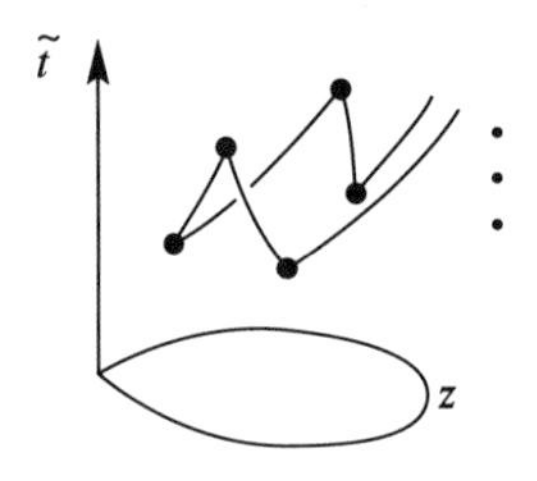

FIGURE 6

DEFINITION. $Z_0(K) = 1 \in \mathbb{Q} = \mathcal{A}_0 \subset \mathcal{A}_0 \otimes \mathbb{C}$. For $n > 0$

$$Z_n(K) = \int \cdots \int_{\substack{\text{noncritical} \\ t_1 < t_2 < \cdots < t_n}} \sum_{(z_i, z_i')} (\text{the corresponding element of } \mathcal{A}_n) \times \prod_{i=1}^{n} \frac{d \log(z_i - z_i')}{2\pi\sqrt{-1}} (-1)^{\varepsilon},$$

where the summation is taken over all choices of nonordered pairs of distinct points on the sections $t = t_i$ of the knot K, ε denotes the number of points (t_i, z_i), (t_i, z_i') with the orientation in the negative direction. The element of $\mathcal{A}_n$ corresponding to n pairs of points on $K(S^1)$ is well defined by Lemma 2.1.

THEOREM 2.2. (1) *The integral defining $Z_n(K)$ is absolutely convergent,*
(2) *$Z_n(K)$ is invariant under the homotopy in the class of Morse knots,*
(3) *$Z_n(K)$ is an invariant of degree less than $(n+1)$.*

PROOF. For any Morse knot there exists a homeomorphism $\tilde{t}(t)$ of $\mathbb{R}$ that is a diffeomorphism everywhere outside the set of critical values of the map $(t \circ K)$ and that transforms the image of the knot to the union of finitely many lines transversal to horizontal planes $\{t, z \mid t \text{ is fixed}\}$ (see Figure 6). One can see that all possible divergences arise at the domain where a short interval (z_i, z_i') arises. If there are no other endpoints of other intervals in the small part of the knot connecting these two points, then the integral is zero by the relation (2). In all other cases the integral can be estimated above by

$$\text{const} \int \cdots \int_{0 < x_1 < x_2 < \cdots < x_k < 1} dx_1 \frac{dx_2}{x_2} \frac{dx_3}{x_3} \cdots \frac{dx_k}{x_k}.$$

The last integral is absolutely convergent. Hence the first part of the theorem is proved.

Using the same estimates, one can verify that $Z_n(K)$ is a continuous function of Morse knot K. Now it is sufficient to verify that this function is locally constant for the variation of a Morse knot with all distinct critical values. In this situation we can use Stokes' formula since we integrate a

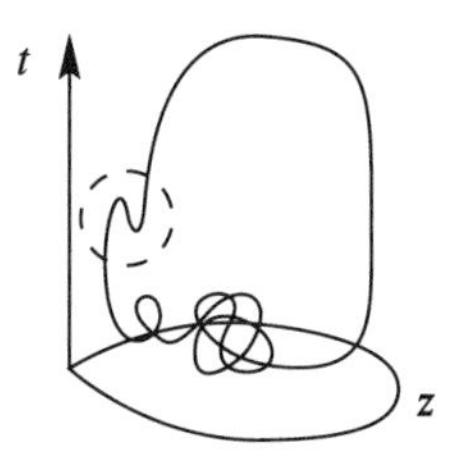

FIGURE 7

closed form over a manifold with the corners. The reader can check that the last relation from § 0 and relation (1) in the definition of $\mathscr{A}_n$ imply the homotopy invariance of $Z_n(K)$.

Part (3) is very simple. Using part (2) proved just now we can consider only the situation when all $(n+1)$ balls have nonintersecting images under the projection onto the vertical axis $\mathbb{R}$. Then the vanishing of the corresponding alternating sum is evident. □

The integral formula above arose from the attempts to understand the remarkable work of V. Drinfel'd [D1] on quasi-Hopf algebras. It follows more or less explicitly from another work of Drinfel'd [D2] that $Z_n(K) \in \mathscr{A}_n \subset \mathscr{A}_n \otimes \mathbb{C}$. We do not know any direct geometric proof of this fact.

It is easy to see that the set of complete invariants of Morse knot K consists of the topological type of the knot and the number of critical points of the map $(t \circ K)$. Let us consider the sequence of invariants for the special noncompact Morse knot K_0

$$K_0 : \mathbb{R}^1 \longrightarrow \mathbb{R} \times \mathbb{C}, \qquad K_0(x) = (x^3 - x, x).$$

Define an element $Z(K_0) \in \bar{\mathscr{A}} := \prod \mathscr{A}_n$ to be the sum $\sum_{k=0}^{\infty} Z_n(K_0)$. This element is invertible because the series for it starts from 1. One can replace here K_0 by any compact trivial Morse knot with four critical points.

THEOREM 2.3. *For any Morse knot K the element*

$$\tilde{Z}(K) := \sum_{k=0}^{\infty} Z_k(K) \times Z(K_0)^{-\frac{1}{2}\text{ number of critical points of } (t \circ K)} \in \bar{\mathscr{A}}$$

depends only on the topological type of K. For any n the first $(n+1)$ components $\tilde{Z}_0(K), \dots, \tilde{Z}_n(K)$ of $\tilde{Z}(K)$ together give the universal $\mathbb{Q}$-valued Vassiliev invariant of degree less than $(n+1)$.

PROOF. Consider the Morse knot such that its top part looks like the union of the middle part of K_0 with a vertical line far away from this curved part, and with a convex arc connecting these two parts (see Figure 7). The value of normalized invariant $\tilde{Z}(K)$ does not change when we replace the curved part on the top by a straight line. So, $\tilde{Z}(K)$ is a knot invariant. $\tilde{Z}_k(K)$, $k \leq n$, are invariants of degree less than $(n+1)$ because they are linear combinations of invariants $Z_k(K)$, $k \leq n$. The map $\nabla^n(\tilde{Z}_n) : \mathscr{A}_n \longrightarrow \mathscr{A}_n$ is the identity map. □

Theorem 2.1 follows immediately from Theorem 2.3. We have an explicit isomorphism

$$V_n \simeq \bigoplus_{k=0}^{n} \mathscr{A}_k^* .$$

Define a comultiplication $\Delta : \mathscr{A}_n \longrightarrow \bigoplus_{k+l=n} \mathscr{A}_k \otimes \mathscr{A}_l$ by the formula

$$\Delta(S) = \sum_{S' \subset S} S' \otimes (S \backslash S') .$$

A simple computation shows that this comultiplication is well defined and compatible with the algebra structure on $\mathscr{A}$. By the usual structure theorem (Milnor-Moore theorem) we conclude that $\mathscr{A}$ is canonically isomorphic to the symmetric algebra of the graded vector space of primitive elments

$$P = \mathrm{Prim}(\mathscr{A}) := \{a \in \mathscr{A} \mid \Delta(a) = 1 \otimes a + a \otimes 1\} .$$

The value of the universal knot invariant $\tilde{Z}(K)$ is always a group-like element of $\bar{\mathscr{A}}$, $\Delta(\tilde{Z}(K)) = \tilde{Z}(K) \otimes \tilde{Z}(K)$. Hence there is a space of algebraically independent invariants $\bigoplus P_k^*$ and all other rational-valued Vassiliev's invariants are polynomials in it.

D. Bar-Natan has calculated vector spaces P_k for $k \leq 8$ using a computer. The list of dimensions of these spaces is $0, 0, 1, 1, 2, 3, 5, 8, 12$ for $k = 0, 1, 2, 3, 4, 5, 6, 7, 8$ respectively. The computation time grows superexponentially, and the computation of P_9 seems to be quite hard.

We define a modified invariant of knot by the formula $\tilde{Z}_{\mathrm{mod}}(K) = \tilde{Z}(K) \times Z(K_0)$. One can check that again $\tilde{Z}_{\mathrm{mod}}(K)$ is a group-like element, and the modified primitive invariant $\log(\tilde{Z}_{\mathrm{mod}}(K)) \in \bar{P}$ is an additive invariant with the respect to the usual addition of knots. We can introduce the structure of the Hopf algebra on the space V of Vassiliev's invariants with the comultiplication coming from the addition of knots. Hopf algebra $\mathscr{A}$ is in a sense dual to V.

There are two obvious involutions on the set of knot types, corresponding to the change of the orientation of the Euclidean space $\mathbb{R}^3$, and of the knot S^1. The corresponding involutions on $\mathscr{A}$ are multiplication by $(-1)^{\mathrm{degree}}$ and the change of the orientation of the line $\mathbb{R}^1$.

One can extend the definition of Vassiliev invariants and all constructions above to the case of framed knots (= embeddings of the standard solid torus into $\mathbb{R}^3$). We define the group $\hat{\mathscr{A}_k}$ in the same way as $\mathscr{A}_k$ with relation (2) omitted.

The universal $\mathbb{Q}$-valued Vassilev's invariant for framed knots takes values in the space $\bar{\hat{\mathscr{A}}} := \prod \hat{\mathscr{A}_k}$. The integral has to be regularized near critical points. The algebra $\hat{\mathscr{A}}$ is a commutative cocommutative Hopf algebra canonically isomorphic to $\mathscr{A}[t]$, where t is the variable of degree 1 corresponding to the unique family of single arcs.

§3. Algebra $\hat{\mathscr{A}}$ and universal enveloping algebras

Let $\mathfrak{g}$ be a Lie algebra over $\mathbb{C}$, and $t \in \mathfrak{g} \otimes \mathfrak{g}$ be a symmetric element invariant under the adjoint action, i.e.,

$$t \in S^2(\mathfrak{g})^{\mathfrak{g}} \subset S^2(\mathfrak{g}) \subset \mathfrak{g} \otimes \mathfrak{g}.$$

Choose any decomposition of the tensor t into the sum of squares of elements of $\mathfrak{g}$,

$$t = \sum g_i \otimes g_i, \qquad g_i \in \mathfrak{g}.$$

Then for any finite sequence of letters $i, j, k, \dots$ such that every letter appears twice we can construct an element of the universal enveloping algebra $U\mathfrak{g}$. For example, the element corresponding to the word $ijkjlilk$ is

$$\sum_{i,j,k,l} g_i g_j g_k g_j g_l g_i g_l g_k.$$

With such a word one can associate also a family of arcs on the line. The invariance of t implies that the relation (1) in the definition of the group $\hat{\mathscr{A}}$ maps to zero. Our construction gives a morphism of algebras $\hat{\mathscr{A}} \to U\mathfrak{g}$, which does not depend on the choice of the decomposition of t. In general this map is not compatible with coproduct structures on $\hat{\mathscr{A}}$ and $U\mathfrak{g}$. Since the tensor t is invariant, the image of this map belongs to the subalgebra $(U\mathfrak{g})^{\mathfrak{g}}$ of invariants of $U\mathfrak{g}$ under the adjoint action of $\mathfrak{g}$. It is clear that $(U\mathfrak{g})^{\mathfrak{g}}$ is the center $ZU\mathfrak{g}$ of the algebra $U\mathfrak{g}$.

As a corollary, we obtain a map $\mathscr{A} \to ZU\mathfrak{g}/\hat{t}ZU\mathfrak{g}$, where $\hat{t} = \sum g_i g_i$.

Any linear functional $\chi : ZU\mathfrak{g} \to \mathbb{C}$ gives an infinite sequence of $\mathbb{C}$-valued Vassiliev's invariants for ordinary and framed knots. For example, any finite-dimensional representation ρ of the Lie algebra $\mathfrak{g}$ gives the trace functional

$$\chi_\rho(x) = \operatorname{tr}(\rho(x)), \qquad x \in (U\mathfrak{g})^{\mathfrak{g}} \subset U\mathfrak{g}.$$

All invariants up to degree 7 computed by D. Bar-Natan are linear combinations of invariants arising from simple Lie algebras of types A, B, C, D. Using this construction one can obtain the estimate

$$\dim(V_n) > e^{c\sqrt{n}}, \qquad n \to +\infty,$$

for any positive constant $c < \pi\sqrt{2/3}$ (see [BN, Exercise 6.14]).

§4. Problem of orientation

The construction described above gives a lot of Vassiliev's invariants. Unfortunately, all such invariants constructed from semisimple Lie algebras cannot detect the change of orientation of knots.

The change of orientation corresponds to the passing to the dual representation. Any simple Lie algebra admits an automorphism (Cartan involution) acting as the conjugation on the set of irreducible representations.

At the moment it is not clear whether or not ($\mathbb{Q}$-valued) Vassiliev invariants can detect the change of orientation. All the standard polynomial invariants also cannot do it. Nevertheless, I hope that there are such invariants. Probably, the first one has degree 9. Recall that the first noninvertible knot was discovered only in 1964 and the simplest diagram for noninvertible knot has eight self-intersections. D. Bar-Natan conjectured that all invariants arise from the classical Lie algebras. His conjecture implies that Vassiliev invariants cannot detect noninvertibility.

§5. Relation with the perturbative Chern-Simons theory

Three-dimensional topological field theory with Chern-Simons action is now known to be solvable (see [W]). The corresponding Feynman integral is defined for integral values k of some parameter called level, and the value of this integral is an algebraic number. From the physical point of view there must exist an asymptotic formula for the value of the Feynman integral for large k in terms of the set of representations of the fundamental group of 3-manifolds into a compact Lie group (=critical points for the Chern-Simons action).

The perturbation theory was discussed in [GMM] for two-loop diagrams and was worked out completely by S. Axelrod and I. Singer [AS] and by myself. It turns out that gauge fields, ghost fields, and anti-ghost fields can be considered as differential forms of degrees $0, 1, 2$. Feynman diagrams (for knots in $\mathbb{R}^3$) are connected 3-valent graphs Γ with a fixed oriented simple closed cycle C in it and an *orientation* that could be defined in several ways (see below).

Feynman integral associated with a diagram Γ and a knot $K(S^1) \subset \mathbb{R}^3$ is

$$I_\Gamma(K) = \int_{\substack{\text{straight embeddings} \\ (\Gamma, C) \hookrightarrow (\mathbb{R}^3, K(S^1))}} \prod_{\text{edges } \in \Gamma \setminus C} \omega(\text{edge}) ,$$

where the words "straight embeddings" denote embeddings of the graph Γ into $\mathbb{R}^3$ such that edges from C are parts of $K(S^1)$ with compatible orientations and other edges are straight line intervals in $\mathbb{R}^3$. The orientation of the graph Γ is a way to fix a sign in the definition of the corresponding integral. One can describe it as

1) the ordering on the set of vertices plus the choice of orientation of each edge from $\Gamma \setminus C$ modulo even permutations and an even number of changes, or, equivalently, as

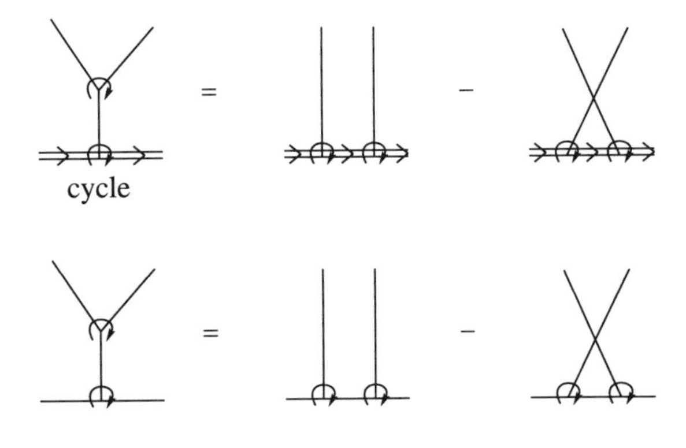

FIGURE 8

2) the choice of cyclic order on 3-element sets of edges coming to each vertex, modulo an even number of changes.

Of course, chord diagrams from Vassiliev's theory are examples of Feynman diagrams for CS theory. One can define a vector space generated by equivalence classes of Feynman diagrams with orientation modulo relations $(\Gamma, -\mathrm{or}) = -(\Gamma, \mathrm{or})$ and a kind of "Jacobi identity" (see Figure 8). D. Bar-Natan has proved that this space is canonically isomorphic to $\hat{\mathscr{A}}$.

The perturbative Chern-Simons theory for *all* compact Lie groups near the trivial representation gives the following series

$$1 + \sum_{\text{equivalence classes of } \Gamma} \frac{\Gamma}{\#\mathrm{Aut}(\Gamma)} I_\Gamma(K) \in \mathscr{A} \otimes \mathbb{R}.$$

The proof of the topological invariance of this formula is somewhat more technical than for the case with the Knizhnik-Zamolodchikov equation, but it uses only three properties of the form ω mentioned in Section 0.

Question. Are invariants arising from CS theory the same as invariants arising from the KZ equation?

The approach with the Gauss form has a straightforward generalization

1) to the case of higher-degree cohomology classes of the space of embeddings,

2) to the case of embeddings into $\mathbb{R}^n$ with $n \geq 3$. It implies collapsing of the spectral sequence of Vassiliev in the 3-dimensional case and "complete" computation of $H^*(\mathrm{Emb}(S^1, \mathbb{R}^n), \mathbb{Q})$ for $n \geq 4$.

Acknowledgments

I would like to thank V. Arnol'd, D. Bar-Natan, S. Piunikhin, and V. Vassiliev for useful discussions.

REFERENCES

[AS] S. Axelrod and I. M. Singer, *Chern-Simons perturabation theory*, MIT preprint, October 1991.

[BL] J. S. Birman and X.-S. Lin, *Knot polynomials and Vassiliev's invariants*, Preprint, Columbia University, 1991.

[BN] D. Bar-Natan, *On the Vassiliev knot invariants*, Preprint, Harvard University, August 1992.

[D1] V. G. Drinfel'd, *Quasi-Hopf algebras*, Algebra i Analiz **1** (1989), no. 6, 114–148; English transl., Leningrad Math. J. **1** (1990), 1419–1457.

[D2] ———, *On quasitriangular Quasi-Hopf algebras and a group closely connected with* Gal $(\overline{\mathbb{Q}}/\mathbb{Q})$, Algebra i Analiz **2** (1991), no. 4, 149–181; English transl., Leningrad Math. J. **2** (1991), 829–260.

[GMM] E. Guadagnini, M. Martinelli, and M. Mintchev, *Perturbative aspect of the Chern-Simons field theory*, Phys. Lett. **B227** (1989), 111.

[V1] V. A. Vassiliev, *Cohomology of knot spaces*, Theory of Singularities and its Applications, Advances in Soviet Mathematics, vol. 1, Amer. Math. Soc., Providence, RI, 1990, pp. 23–69.

[V2] ———, *Complements to discriminants of smooth maps: topology and applications*, Amer. Math. Soc., Providence, RI, 1992.

[W] E. Witten, *Quantum field theory and the Jones polynomial*, Comm. Math. Phys. **121** (1989), 351–399.

MAX-PLANCK-INSTITUT FÜR MATHEMATIK, GOTTFRIED-CLAREN-STR.26, 5300 BONN 3, GERMANY

E-mail address: maxim@mpim-bonn.mpg.de

ADVANCES IN SOVIET MATHEMATICS
Volume 16, Part 2, 1993

Kazhdan-Lusztig Polynomials for Lie Superalgebra $\mathfrak{gl}(m \mid n)$

VERA SERGANOVA

ABSTRACT. We define Kazhdan-Lusztig polynomials for Lie superalgebra $\mathfrak{gl}(m \mid n)$ and formulate the conjecture on characters of irreducible representations. We also discuss a superanalogue of the reflection functor.

Introduction

In this paper we present a new approach to a very unfortunate problem in the representation theory of Lie superalgebras. This is the problem of finding characters of all finite dimensional irreducible representations of a simple complex Lie superalgebra $\mathfrak{g}$. The question had arisen in 1977 after V. Kac generalized in [4] the highest weight representation theory to Lie superalgebras and showed that for a generic value of the highest weight the character formula can be written in a very simple manner. But for a general highest weight the problem appeared to be unexpectedly difficult. There were several conjectures as to what such a formula should be ([1, 8, 3]), but only in very particular cases was anything proven.

Here we try to develop Kazhdan-Lusztig theory [5] for Lie superalgebra $\mathfrak{g} = \mathfrak{gl}(m \mid n)$. This theory is not trivial even for finite dimensional $\mathfrak{g}$-modules. There are some similarities with the case of finite characteristic. We define Kazhdan-Lusztig polynomials as some generating function for characters and then present a conjecture on how to calculate these polynomials. In order to do this we follow D. Vogan's approach from [2]. Namely, we construct a derived functor in the category of $\mathfrak{g}$-modules and (assuming its semisimplicity) produce recurrent relations for the above polynomials.

The paper is organized in the following way. At first we define Kazhdan-Lusztig polynomials for Lie superalgebras in terms of the homology ring of some subalgebra of $\mathfrak{g}$. Then we construct a functor on the category of $\mathfrak{g}$-modules and formulate the main conjecture: this functor is semisimple. Using this conjecture we evaluate Kazhdan-Lusztig polynomials and therefore characters of irreducible representations.

1991 *Mathematics Subject Classification*. Primary 17B70.

At the moment it is unclear how to prove the semisimplicity conjecture. This can probably be done using vector bundles on algebraic supermanifolds. That is why the geometric definition of the above functor which we give in §5 seems important.

One can easily generalize our conjecture to any classical simple Lie superalgebra, but the construction of the functor is clear only for $\mathfrak{gl}(m \mid n)$. Therefore this generalization remains unproved (if we consider the case of $\mathfrak{gl}$ as proved).

There should certainly be some connection with the conjecture in [3], but I know only that these conjectures are equivalent in the case of $\mathfrak{gl}(2 \mid n)$.

I am glad to thank J. Bernstein who suggested the main idea to generalize the Vogan functor for Lie superalgebras and without whom this paper could not appear. I am also thankful to M. Finkelberg and I. Penkov for useful discussions.

§1. Preliminaries

Let $\mathfrak{g} = \mathfrak{gl}(m \mid n)$, $\mathfrak{h} \subset \mathfrak{g}$ be a Cartan subalgebra, and $\mathfrak{g} = \mathfrak{g}_{-1} \oplus \mathfrak{g}_0 \oplus \mathfrak{g}_1$ be the natural consistent $\mathbb{Z}$-grading, where $\mathfrak{g}_0 = \mathfrak{gl}(m) \oplus \mathfrak{gl}(n)$, $\mathfrak{g}_0' = [\mathfrak{g}_0, \mathfrak{g}_0]$. We fix a triangular decomposition $\mathfrak{g} = \mathfrak{n}^- \oplus \mathfrak{h} \oplus \mathfrak{n}^+$ such that $\mathfrak{g}_1 \subset \mathfrak{n}^+$, $\mathfrak{g}_{-1} \subset \mathfrak{n}^-$, and a Borel subalgebra $\mathfrak{b} = \mathfrak{h} \oplus \mathfrak{n}^+$. Let Δ be the set of roots of $\mathfrak{g}$, Δ^+ the set of positive roots, Δ_0 the set of even roots, and Δ_1 the set of odd roots. If $\{\varepsilon_1, \dots, \varepsilon_m, \delta_1, \dots, \delta_n\}$ is the standard basis in $\mathfrak{h}^*$, then

$$\Delta_1^+ = \{\varepsilon_i - \delta_j \mid 1 \leqslant i \leqslant m,\ 1 \leqslant j \leqslant n\},$$

$$\Delta_0^+ = \{\varepsilon_i - \varepsilon_j \mid 1 \leqslant i, j \leqslant m\} \cup \{\delta_i - \delta_j \mid 1 \leqslant i, j \leqslant n\}.$$

Put $\rho = (1/2)(\sum_{\alpha \in \Delta_0^+} \alpha - \sum_{\alpha \in \Delta_1^+} \alpha)$. By $(\cdot, \cdot)$ we denote the invariant symmetric form on $\mathfrak{h}^*$. In the standard basis, $(\varepsilon_i, \varepsilon_j) = -(\delta_i, \delta_j) = \delta_{ij}$ (Kronecker delta), $(\varepsilon_i, \delta_j) = 0$.

Let us denote by L_λ the irreducible $\mathfrak{g}$-module with highest weight λ and by V_λ the module $U(\mathfrak{g}) \otimes_{U(\mathfrak{g}_1 \oplus \mathfrak{g}_0)} L_\lambda^0$, where L_λ^0 is the irreducible $\mathfrak{g}_0$-module with highest weight λ and trivial action of $\mathfrak{g}_1$. It is clear that V_λ is free as a $U(\mathfrak{g}_{-1})$-module.

For a subalgebra $\mathfrak{k} \subset \mathfrak{g}$ denote by $L_\lambda(\mathfrak{k})$ the irreducible $\mathfrak{k}$-module with highest weight λ when this makes sense for $\mathfrak{k}$ (for example when $\mathfrak{k}$ is a regular reductive subalgebra of $\mathfrak{g}$). In particular, $L_\lambda(\mathfrak{g}) = L_\lambda$.

The weight $\lambda \in \mathfrak{h}^*$ is said to be dominant and integral if it is such with respect to $\mathfrak{g}_0$. In this paper we consider only integral weights. The weight λ is integral and dominant if and only if $\dim L_\lambda < \infty$ (see [4] for details).

LEMMA 1.1. *Any finite dimensional $\mathfrak{g}$-module that is free over $U(\mathfrak{g}_{-1})$ has a filtration with factors isomorphic to V_μ for some dominant μ. It is isomorphic to the direct sum of the factors as a $U(\mathfrak{g}_{-1} \oplus \mathfrak{g}_0')$-module.*

Proof. Let V be such a module. Then $V = \bigoplus V(\lambda)$, where $V(\lambda) = \operatorname{Ker}(h - \lambda \operatorname{Id})^{\dim V}$ is a generalized eigenspace for all $h \in \mathfrak{h}$. Then one can find $V(\lambda)$ such that $\mathfrak{n}^+ V(\lambda) = 0$. Take any eigenvector $v \in V(\lambda)$. Then the submodule W generated by v is isomorphic to $V(\lambda)$, and V/W is again free over $U(\mathfrak{g}_{-1})$. Then one can apply induction on $\dim V$ in order to prove the first part of the lemma.

To prove the second part let us consider a filtration

$$V = V_0 \supset V_1 \supset \cdots \supset V_l = 0$$

with factors $W_i = V_i/V_{i+1}$. Since $\mathfrak{g}_0'$ is semisimple, $V = \bigoplus W_i$ as a $\mathfrak{g}_0'$-module and $V/\mathfrak{g}_{-1}V = \bigoplus W_i/\mathfrak{g}_{-1}W_i$. Then $V = \bigoplus U(\mathfrak{g}_{-1}) \otimes W_i/\mathfrak{g}_{-1}W_i \cong W_i$ as a $U(\mathfrak{g}_0 \oplus \mathfrak{g}_{-1})$-module.

Lemma 1.2. *Any finite dimensional $\mathfrak{g}$-module has a resolution by $\mathfrak{g}$-modules that are free $U(\mathfrak{g}_{-1})$-modules.* ([1])

Proof. We just repeat the construction of BGG resolution for the supercase. First, construct a resolution for a trivial module as dual to the de Rham complex on the supermanifold G/G^+, where G and G^+ are supergroups with algebras $\mathfrak{g}$ and $\mathfrak{g}_0 \oplus \mathfrak{g}_1$ respectively. For a definition of such supermanifolds see [6, 7].

Then one can take the tensor product of this resolution with any finite dimensional module V and get a resolution of V.

One can find the complete proof in [9].

Let $\chi(\lambda)$ be the central character of weight λ. We say that $\lambda \leqslant \mu$ if $\chi(\lambda) = \chi(\mu)$ and $\mu = \lambda + \sum n_i \alpha_i$ where α_i are from Δ_1^+.

Lemma 1.3. *If $\lambda \leqslant \mu$ then there is a sequence $\alpha_1, \dots, \alpha_k$ of odd positive roots such that $(\lambda + \rho, \alpha_1) = 0, \dots, (\lambda + \rho + \alpha_1 + \cdots + \alpha_{k-1}, \alpha_k) = 0$ and $\mu = \lambda + \alpha_1 + \cdots + \alpha_k$.*

Proof. You can find the proof in [9]. It is purely combinatoric and we will not repeat it here.

A resolution $M^\cdot$ of L_λ by free $U(\mathfrak{g}_{-1})$-modules is called minimal if each term M^i contains only components V_μ with $\mu \leqslant \lambda$.

Lemma 1.4. *Any finite dimensional irreducible $\mathfrak{g}$-module L_λ has a minimal resolution by free $U(\mathfrak{g}_{-1})$-modules.*

Proof. Just take the resolution constructed in the proof of Lemma 1.2 and consider the component with central character $\chi(\lambda)$.

([1]) By resolution of V here we mean a complex

$$0 \leftarrow M^0 \leftarrow M^1 \leftarrow M^2 \leftarrow \cdots$$

of $\mathfrak{g}$-modules free over $U(\mathfrak{g}_{-1})$ which has only zero cohomology group isomorphic to V. In contrast with the Lie algebra case there exists usually only infinite resolution.

§2. Definition of Kazhdan-Lusztig polynomials $K(\lambda, \mu)$

Let us consider $H_i(\mathfrak{g}_{-1}; L_\lambda)$. This space has a natural structure of a $\mathfrak{g}_0$-module. Denote the multiplicity $[L_\mu^0: H_i(\mathfrak{g}_{-1}; L_\lambda)]$ by $K_i(\lambda, \mu)$. Define the generating function

$$K(\lambda, \mu) = \sum K_i(\lambda, \mu) q^i.$$

LEMMA 2.1. *The character of* L_λ *can be evaluated as*

$$\operatorname{ch} L_\lambda = \sum_\mu K(\lambda, \mu)(-1) \operatorname{ch} L_\mu^0 \prod_{\alpha \in \Delta_1^+} (1 + \varepsilon e^{-\alpha}).$$

PROOF. Let us consider a resolution of L_λ

$$0 \leftarrow M^0 \leftarrow M^1 \leftarrow M^2 \leftarrow \cdots$$

by free $U(\mathfrak{g}_{-1})$-modules. Then

$$\operatorname{ch} L_\lambda = \sum_{i=0}^{\infty} (-1)^i \operatorname{ch} M^i.$$

The character of M^i is given by

$$\operatorname{ch} M^i = \operatorname{ch} H_0(\mathfrak{g}_{-1}; M^i) \prod_{\alpha \in \Delta_1^+} (1 + \varepsilon e^{-\alpha}).$$

Therefore

$$\operatorname{ch} L_\lambda = \sum_{i=0}^{\infty} (-1)^i \operatorname{ch} H_0(\mathfrak{g}_{-1}; M^i) \prod_{\alpha \in \Delta_1^+} (1 + \varepsilon e^{-\alpha}).$$

But

$$\sum_{i=0}^{\infty} (-1)^i \operatorname{ch} H_0(\mathfrak{g}_{-1}; M^i) = \sum_{i=0}^{\infty} (-1)^i \operatorname{ch} H_i(\mathfrak{g}_{-1}; L_\lambda),$$

$$\sum_{\substack{i=0 \\ \mu}}^{\infty} (-1)^i K_i(\lambda, \mu) \operatorname{ch} L_\mu^0 = \sum_\mu K(\lambda, \mu)(-1) \operatorname{ch} L_\mu^0.$$

From the last two formulas we can easily obtain the formula of Lemma 2.1.

The next lemma is a standard fact from homological algebra.

LEMMA 2.2. *Let* V *be a* $\mathfrak{g}$*-module and*

$$0 \leftarrow M^0 \xleftarrow{d^1} M^1 \xleftarrow{d^2} M^2 \xleftarrow{d^3} c \cdots$$

be a resolution of V *by free* $U(\mathfrak{g}_{-1})$*-modules. Then*

$$H_i(\mathfrak{g}_{-1}; V) = H^i(H_0(\mathfrak{g}_{-1}; M^{\cdot})).$$

LEMMA 2.3. *The coefficients* $K(\lambda, \mu)$ *have the following properties*:

1) $K(\lambda, \lambda) = 1$;
2) *If* $K(\lambda, \mu) \neq 0$ *then* $\mu \leqslant \lambda$.

PROOF. The first property is obvious and the second one follows from the existence of a minimal resolution and Lemma 2.2.

§3. Definition of an analogue of the reflection functor $U_{t,h}$

Let $h \in \mathfrak{h}_{\mathbb{R}}$, $t \in \mathbb{R}$, and V be any finite dimensional $\mathfrak{g}$-module. Put $V^s = (h - s \cdot \mathrm{id})^n$ for sufficiently large n, $V^{\geqslant t} = \sum_{t \geqslant s} V^s$.

Let h be such that $\langle \alpha, h \rangle \geqslant 0$ for all positive roots α. Then $V^{\geqslant t}$ is $\mathfrak{b}$-invariant. Therefore $V_t = U(\mathfrak{n}^-)V^{\geqslant t}$ is a $\mathfrak{g}$-submodule in V.

Define a functor $\varphi_{t,h}$ on the category of finite dimensional $\mathfrak{g}$-modules as:

$$\varphi_{t,h}(V) = V/V_t.$$

LEMMA 3.1. a) *If* V *is free as a* $U(\mathfrak{g}_{-1})$*-module, then* $\varphi_{t,h}(V)$ *is also free as a* $U(\mathfrak{g}_{-1})$*-module*;

b) $\varphi_{t,h}$ *is exact on the category of free* $U(\mathfrak{g}_{-1})\mathfrak{g}$*-modules.*

PROOF. a) By Lemma 1.1, V has a filtration $V = V_0 \supset V_1 \supset \cdots \supset V_l$ such that $V_i/V_{i+1} = V_{\mu_i}$. It is easy to see that $\varphi_{t,h}(V_\mu) = 0$ if $\langle \mu, h \rangle \geqslant t$, $\varphi_{t,h}(V_\mu) = V_\mu$ if $\langle \mu, h \rangle < t$. As a $U(n^-)$-module, $V = \bigoplus V_{\mu_i}$ and therefore

$$\varphi_{t,h}(V_i/V_{i+1}) = \varphi_{t,h}(V_i)/\varphi_{t,h}(V_{i+1}) = V_{\mu_i} \quad \text{or} \quad 0.$$

Then it is clear that V is free over $U(\mathfrak{g}_{-1})$.

b) Let us consider a short exact sequence

$$0 \to U \to W \to V \to 0$$

of free $U(\mathfrak{g}_{-1})$-modules. By Lemma 1.1 it splits over $U(\mathfrak{n}^-)$. Then $W_t = U_t \oplus V_t$ as a $U(\mathfrak{n}^-)$-module and therefore

$$\varphi_{t,h}(W) = \varphi_{t,h}(U) \oplus \varphi_{t,h}(V)$$

as a $U(\mathfrak{n}^-)$-module and the sequence

$$0 \to \varphi_{t,h}(U) \to \varphi_{t,h}(W) \to \varphi_{t,h}(V) \to 0$$

is exact.

Let $\mathscr{A}$ be the category of finite dimensional $\mathfrak{g}$-modules that are free over $U(\mathfrak{g}_{-1})$ and $\mathscr{C}(\mathscr{A})$ be the derived category of left bounded complexes. We will construct a functor $U_{t,h}$ from the category of finite dimensional $\mathfrak{g}$-modules to $\mathscr{C}(\mathscr{A})$ as the derived functor for $\varphi_{t,h}$. In other words let

$$0 \leftarrow M^0 \leftarrow M^1 \leftarrow \cdots$$

be a resolution of V by modules from $\mathscr{A}$. Then $\varphi_{t,h}(V) = \varphi_{t,h}(M^\cdot)$, which is uniquely defined up to quasi-isomorphism.

§4. The main conjecture and its corollaries

CONJECTURE. *Let* $h = h_1 = \operatorname{diag}(1, 0, \dots, 0)$, $t = \langle\lambda, h\rangle$. *Then* $U_{t,h}(L_\lambda)$ *is semisimple, i.e., the complex* $U_{t,h}(L_\lambda)$ *is isomorphic to the direct sum of its cohomology.*

Below instead of $U_{\langle\lambda, h_1\rangle, h_1}(L_\lambda)$ we write $U_{h_1}(\lambda)$. Let $U^i_{h_1}(\lambda)$ be the ith cohomology of $U_{h_1}(\lambda)$ and $U^i_{h_1}(\lambda, \mu) = [L_\mu : U^i_{h_1}(\lambda)]$. Let us consider the generating function $U_{h_1}(\lambda, \mu) = \sum_{i=0}^{\infty} U^i_{h_1}(\lambda, \mu) q^i$.

THEOREM 4.1. *If the previous conjecture is true then for all* $\mu \in \mathfrak{h}^*$ *such that* $\langle\mu, h_1\rangle < \langle\lambda, h_1\rangle$ *the following relation is true:*

$$K(\lambda, \mu) = \sum_{\mu \leqslant \nu < \lambda} U_{h_1}(\lambda, \nu) K(\nu, \mu).$$

PROOF. Let us consider a resolution $M^\cdot$ of L_λ by free $U(\mathfrak{g}_{-1})$-modules. Take $t = \langle\lambda, h_1\rangle$. For each i we have the short exact sequence:

$$0 \to M^i_t \to M^i \to \varphi_{t, h_1}(M^i) \to 0,$$

which gives the exact sequence of homology:

$$0 \to H_0(\mathfrak{g}_{-1}; M^i_t) \to H_0(\mathfrak{g}_{-1}; M^i) \to H_0(\mathfrak{g}_{-1}; \varphi_{t, h_1}(M^i)) \to 0.$$

Let us consider the long exact sequence:

$$\begin{aligned} \cdots \to H^j(H_0(\mathfrak{g}_{-1}; M^\cdot_t)) &\to H^j(H_0(\mathfrak{g}_{-1}; M^\cdot)) \\ &\to H^j(H_0(\mathfrak{g}_{-1}; \varphi_{t, h_1}(M^\cdot))) \to \cdots. \end{aligned}$$

The term $H^j(H_0(\mathfrak{g}_{-1}; M^\cdot_t))$ contains only components L^0_μ with $\langle\mu, h_1\rangle \geqslant \langle\lambda, h_1\rangle$ and the term $H^j(H_0(\mathfrak{g}_{-1}; \varphi_{t, h_1}(M^\cdot)))$ contains only components L^0_μ with $\langle\mu, h_1\rangle < \langle\lambda, h_1\rangle$. If the conjecture is true, then $\varphi_{t, h_1}(M^\cdot)$ is isomorphic to the direct sum of resolutions of $\bigoplus U^i_{h_1}(\lambda)$. So we have the following relation:

$$[L^0_\mu : H_j(\mathfrak{g}_{-1}; L_\lambda)] = \sum_{i=0}^{j} [L^0_\mu : H_i(\mathfrak{g}_{-1}; U^{j-i}_{h_1}(\lambda))],$$

for all μ such that $\langle\mu, h_1\rangle < \langle\lambda, h_1\rangle$.

After rewriting this relation in terms of generating functions we obtain the statement of Theorem 4.1.

Put $h_i = \operatorname{diag}(0, \dots, 0, 1, 0, \dots, 0)$ for $i \leqslant m$, $h_i = \operatorname{diag}(0, \dots, 0, -1, 0, \dots 0)$ for $i > m$. Let $\mathfrak{g}^k$ be the centralizer of h_k,

$$\mathfrak{g}^{(k)} = \mathfrak{g}^1 \cap \cdots \cap \mathfrak{g}^k \quad \text{for } k \leqslant m,$$
$$\mathfrak{g}^{(k)} = \mathfrak{g}^k \cap \cdots \cap \mathfrak{g}^{m+n} \quad \text{for } k > m.$$

LEMMA 4.1. *Let $\mu \in \mathfrak{h}^*$ be such that $\langle \mu, h_j \rangle = \langle \lambda, h_j \rangle$ for all $j = 1, \dots, k$, $k \leqslant m$. Then $[L^0_\mu : H_i(\mathfrak{g}_{-1}; L_\lambda)] = [L^0_\mu(\mathfrak{g}^{(k)}) : H_i(\mathfrak{g}^{(k)}_{-1}; L_\lambda(\mathfrak{g}^{(k)}))]$.*

PROOF. Natural embeddings $\mathfrak{g}_{-1} \subset \mathfrak{g}^{(k)}_{-1}$ and $L_\lambda(\mathfrak{g}^{(k)}) \subset L_\lambda$ induce a map of homology:

$$p: H_i(\mathfrak{g}^{(k)}_{-1}; L_\lambda(\mathfrak{g}^{(k)}_{-1})) \to H_i(\mathfrak{g}_{-1}; L_\lambda)$$

which is a homomorphism of $\mathfrak{g}^{(k)}_0$-modules.

Let us describe $\operatorname{Im} p$. It coincides with the eigenspace of $h_1, \dots, h_k$ with eigenvalues $\langle \lambda, h_1 \rangle, \dots, \langle \lambda, h_k \rangle$. Let $H_i(\mathfrak{g}_{-1}; L_\lambda) = \bigoplus L^0_\mu$. By Lemma 2.3, $\mu \leqslant \lambda$ and therefore $\langle \lambda, h_i \rangle \geqslant \langle \mu, h_i \rangle$. Then

$$L^0_\mu \cap \operatorname{Im} p = \begin{cases} L^0_\mu(\mathfrak{g}^{(k)}) & \text{if } \langle \lambda, h_i \rangle = \langle \mu, h_i \rangle,\ 1, \dots, k, \\ 0 & \text{otherwise.} \end{cases}$$

Since p does not have a kernel it gives a 1-1 correspondence between irreducible components of $H_i(\mathfrak{g}^{(k)}_{-1}; L_\lambda(\mathfrak{g}^{(k)}_{-1}))$ and irreducible components of $H_i(\mathfrak{g}_{-1}; L_\lambda)$ with highest weight μ satisfying the conditions of the lemma.

Let

$$K^i_k(\lambda, \mu) = [L^0_\mu(\mathfrak{g}^{(k-1)}) : H_i(\mathfrak{g}^{(k-1)}_{-1}; L_\lambda(\mathfrak{g}^{(k-1)}))],$$
$$K_k(\lambda, \mu) = \sum K^i_k(\lambda, \mu) q^i,$$
$$U^i_{h_k}(\lambda, \mu) = [L^0_\mu(\mathfrak{g}^{(k-1)}) : U^i_{h_k}(L_\lambda(\mathfrak{g}^{(k-1)}))],$$
$$U_{h_k}(\lambda, \mu) = \sum U^i_{h_k}(\lambda, \mu) q^i.$$

For $k = 0$ put $K_0(\lambda, \mu) = K(\lambda, \mu)$.

The following theorem is a simple corollary of the previous two statements.

THEOREM 4.2.

$$K_k(\lambda, \mu) = K_{k+1}(\lambda, \mu) + \sum_{i=1}^{\infty} \sum_{\mu_1, \dots, \mu_i} U_{h_{k+1}}(\lambda, \mu_1) U_{h_{k+1}}(\mu_1, \mu_2) \cdots U_{h_{k+1}}(\mu_{i-1}, \mu_i) K_{k+1}(\mu_i, \mu).$$

§5. Calculation of $U(\lambda, \mu)$

Here we write down recurrent relations on coefficients $U(\lambda, \mu)$. Before that we will give another description of $U_{h_1}(\lambda)$. Let us consider the parabolic subalgebra $\mathscr{P}^{(1)} = \mathfrak{g}^{(1)} + b$. Let $P^{(1)}$ and G be supergroups corresponding to Lie superalgebras $\mathscr{P}^{(1)}$ and $\mathfrak{g}$. One can define a supermanifold $X = G/P^{(1)}$, which is the analogue of projective space. The existence of this supermanifold has been proved in [6]. One can look at it as at the projective space $X_0 = P^{m-1}$ equipped with the sheaf $\mathscr{O}_X$ of sections of a vector bundle, whose fiber is isomorphic to $U(\mathfrak{g}_{-1}/\mathfrak{g}^{(1)}_{-1}) = \Lambda^{\cdot}(\mathfrak{g}_{-1}/\mathfrak{g}^{(1)}_{-1})$.

Let us introduce the structure of $P^{(1)}$-modules on $L_\lambda(\mathfrak{g}^{(1)})$ putting the radical action as trivial.

Let V be a $P^{(1)}$-module. Denote by $H^\cdot(G/P^{(1)};V)$ the cohomology of vector bundles on the supermanifold $G/P^{(1)}$ with the fiber V. Put $H^\cdot(V) = (H^\cdot(G/P^{(1)};V^*))^*$.

LEMMA 5.1. *For any $\mathfrak{g}$-module V*

$$H^\cdot(X,V) = H^\cdot(X_0, U(\mathscr{P}_0(1)\oplus\mathfrak{g}_{-1})^* \otimes_{U(\mathscr{P}_0^{(1)}\oplus\mathscr{P}_{-1}^{(1)})} V),$$

as a $\mathfrak{g}_0$-module.

PROOF. Let $\pi\colon E\to X$ be our vector bundle with the fiber V and $p\colon X\to X_0$ be the natural projection of a supermanifold to the underlying manifold. Then $p\circ\pi\colon E\to X_0$ is a usual bundle on the projective space X_0. For a point $x_0\in X_0$ denote by W the fiber over x_0. One can check that the stabilizer of $p^{-1}(x_0)$ in $\mathfrak{g}$ is $\mathscr{P}_0^{(1)}\oplus\mathfrak{g}_{-1}$. Therefore W has a structure of a $(\mathscr{P}_0^{(1)}\oplus\mathfrak{g}_{-1})$-module. The straightforward calculations show that

$$W = U(\mathscr{P}_0^{(1)}\oplus\mathfrak{g}_{-1})^* \underset{U(\mathscr{P}_0^{(1)}\oplus\mathscr{P}_{-1}^{(1)})}{\otimes} V.$$

LEMMA 5.2. *Let $M = V_\mu$ and $\langle\mu, h_1\rangle = t$. Then $H^\cdot(M^{\geqslant t}) = M^\cdot$.*

PROOF. First, let us notice that $M^{\geqslant t} = V_\mu(\mathfrak{g}^{(1)})$. So we are evaluating $H^\cdot(X, V_\mu(\mathfrak{g}^{(1)}))$. It is clear that

$$V_\mu^*(\mathfrak{g}^{(1)}) = U(\mathfrak{g}^{(1)})^* \underset{U(\mathfrak{g}_0^{(1)}\oplus\mathfrak{g}_1^{(1)})}{\otimes} L_\mu^*(\mathfrak{g}_0^{(1)}).$$

Let us evaluate the fiber W as in the proof of Lemma 5.1.

$$W = U(\mathscr{P}^{(1)}\oplus\mathfrak{g}_{-1})^* \underset{U(\mathscr{P}_0^{(1)})\oplus\mathscr{P}_{-1}^{(1)}}{\otimes} U(\mathfrak{g}^{(1)})^* \underset{U(\mathfrak{g}_0^{(1)}\oplus\mathfrak{g}_1^{(1)})}{\otimes} L_\mu^*(\mathfrak{g}_0^{(1)}).$$

The simple observation that $\mathscr{P}_{-1}^{(1)} = \mathfrak{g}_{-1}^{(1)}$ gives us:

$$\begin{aligned} W &= U(\mathscr{P}^{(1)}\oplus\mathfrak{g}_{-1})^* \otimes_{U(\mathscr{P}_0^{(1)})} U(\mathfrak{g}^{(1)}\oplus\mathfrak{g}_1^{(1)})^* \otimes_{U(\mathfrak{g}^{(1)}\oplus\mathfrak{g}_1^{(1)})} L_\mu^*(\mathfrak{g}_0^{(1)}) \\ &= U(\mathscr{P}_0^{(1)}\oplus\mathfrak{g}_{-1})^* \otimes_{U(\mathscr{P}_0^{(1)})} L_\mu^*(\mathfrak{g}_0^{(1)}). \end{aligned}$$

The last module is isomorphic to $U(\mathfrak{g}_{-1})^*\otimes L_\mu(\mathfrak{g}_0^{(1)})$ as a $(\mathscr{P}_0^{(1)}\oplus\mathfrak{g}_{-1})$-module. Since $U(\mathfrak{g}_{-1})^*$ includes a trivial vector bundle on X_0 we have

$$H^\cdot(X_0, U(\mathfrak{g}_{-1})^*\otimes L_\mu^*(\mathfrak{g}_0^{(1)})) = U(\mathfrak{g}_{-1})^*\otimes H^\cdot(X_0, L_\mu^*(\mathfrak{g}_0^{(1)}))$$

as a $(\mathfrak{g}_0\oplus\mathfrak{g}_{-1})$-module. The well-known results for projective space X_0 give us the relation:

$$H^i(X_0, L_\mu^*(\mathfrak{g}_0^{(1)})) = \begin{cases} 0, & i \geq 1, \\ L_\mu^*(\mathfrak{g}_0), & i = 0. \end{cases}$$

Therefore $H^\cdot(M^{\geqslant t}) \cong (U(\mathfrak{g}_{-1}) \otimes L_\mu(\mathfrak{g}_0))^\cdot \cong V_\mu^\cdot$ at least as a $(\mathfrak{g}_0 \oplus \mathfrak{g}_{-1})$-module. But then $H^\cdot(M^{\geqslant t}) \cong V_\mu^\cdot$ over $\mathfrak{g}$.

Now take two complexes $L_\lambda^\cdot$ and $H^\cdot(L_\lambda(\mathfrak{g}^{(1)}))$. The map $H^0(L_\lambda(\mathfrak{g}^{(1)})) \to L_\lambda$ induces the morphism $a: H^\cdot(L_\lambda(\mathfrak{g}^{(1)})) \to L_\lambda^\cdot$.

Lemma 5.3. $U_{h_1}(\lambda) = \operatorname{Cone} a$ *for any dominant weight* λ.

Proof. Let $M^\cdot$ be a minimal resolution of L_λ by free $U(\mathfrak{g}_{-1})$-modules. We need to show that $H^\cdot(L_\lambda(\mathfrak{g}^{(1)})) = H^\cdot(M_t^\cdot)$, where $t = \langle \lambda, h_1 \rangle$. In fact the complex $(M^\cdot)^{\geqslant t}$ is a resolution of $L_\lambda(\mathfrak{g}^{(1)})$. The functor $H^\cdot$ is exact and therefore (by Lemma 5.2) $H^\cdot((M^i)^{\geqslant t}) = M_t^i$. So we have $H^\cdot(L_\lambda(\mathfrak{g}^{(1)})) = H^\cdot(M_t^\cdot)$.

Remark. The functor $H^\cdot(L_2(\mathfrak{g}^{(1)}))$ can be defined now not only for dominant λ. The weight λ must be dominant only with respect to $\mathfrak{g}^{(1)}$.

If $\lambda \in \mathfrak{h}^*$ is dominant then there is at most one $\alpha \in \Delta_1^+ \backslash \Delta_1^+(\mathfrak{g}^{(1)})$ for which $(\lambda + \rho, \alpha) = 0$. If there is no such α, we call such λ typical with respect to $\mathfrak{g}^{(1)}$. Otherwise we call λ atypical with respect to $\mathfrak{g}^{(1)}$.

Theorem 5.1. *If λ is typical with respect to $\mathfrak{g}^{(1)}$, then $U_{h_1}(\lambda) = 0$.*

Proof. Take a minimal resolution $M^\cdot$ of L_λ. Then each V_μ occurring in M^i has the highest weight $\mu \leqslant \lambda$. Then $\langle \mu, h_1 \rangle = \langle \lambda, h_1 \rangle$ and $M_{\langle \lambda, h_1 \rangle}^\cdot = M^\cdot$. That immediately implies the theorem.

Let λ be dominant atypical with respect to $\mathfrak{g}^{(1)}$, $\langle \lambda + \rho, \alpha \rangle = 0$. Let $\alpha = \varepsilon_1 - \delta_k$. Then we will consider four different types of λ:

A) $\lambda - \alpha$ is dominant, i.e., $\langle \lambda, h_1 \rangle > \langle \lambda, h_2 \rangle$ and $\langle \lambda, h_{m+k-1} \rangle < \langle \lambda, h_{m+k} \rangle$;
B) $\langle \lambda, h_1 \rangle > \langle \lambda, h_2 \rangle$ and $\langle \lambda, h_{m+k-1} \rangle = \langle \lambda, h_{m+k} \rangle$;
C) $\langle \lambda, h_1 \rangle = \langle \lambda, h_2 \rangle$ and $\langle \lambda, h_{m+k-1} \rangle < \langle \lambda, h_{m+k} \rangle$;
D) $\langle \lambda, h_1 \rangle = \langle \lambda, h_2 \rangle$ and $\langle \lambda, h_{m+k-1} \rangle = \langle \lambda, h_{m+k} \rangle$.

Theorem 5.2. *Let λ be atypical with respect to $\mathfrak{g}^{(1)}$, $\alpha = \varepsilon_1 - \delta_k$, and $(\lambda + \rho, \alpha) = 0$. Then the following relations hold according to the type of λ:*

A) $U_{h_1}(\lambda, \mu) = [q^{-1} U_{h_1}(\lambda - \alpha, \mu)]_+$ *for* $\mu \neq \lambda - \alpha$, $U_{h_1}(\lambda, \lambda - \alpha) = 0$, *here P_+ means the part of a polynomial P without constant term;*
B) $U_{h_1}(\lambda, \mu) = U_{h_{m+k-1}}(\lambda - \alpha, \mu)$;
C) $U_{h_1}(\lambda, \mu) = U_{h_2}(\lambda - \alpha, \mu)$;
D) $U_{h_1}(\lambda, \mu) = q U_{h_2}(\lambda - \alpha, \mu)$, *where we consider $U_{h_2}(\lambda - \alpha)$ restricted to the subalgebra $\mathfrak{g}^{(m+k)}$.*

§6. Proof of Theorem 5.2

First we formulate several simple propositions.

LEMMA 6.1. *Let λ be dominant. Then*

$$H^i(L_\lambda(\mathfrak{g}^{(1)})) = U_{h_1}^{i+1}(\lambda) \quad \textit{for } i \geqslant 1;$$

for $i = 0$ we have the following exact sequence:

$$0 \to U_{h_1}^1(\lambda) \to H^0(L_\lambda(\mathfrak{g}^{(1)})) \to L_\lambda \to 0.$$

PROOF. It follows from Lemma 5.3.

LEMMA 6.2. *Let λ be dominant and atypical with respect to $\mathfrak{g}^{(1)}$, and $\langle \lambda + \rho, \varepsilon_1 - \delta_k \rangle = 0$. Then $U_{h_1}(\lambda, \mu) = U_{h_{m+k}}(\lambda, \mu)$.*

PROOF. It follows from the construction of U_h if we apply it to a minimal resolution.

LEMMA 6.3. *Let λ be dominant with respect to $\mathfrak{g}^{(1)}$ and $\langle \lambda, h_1 \rangle = \langle \lambda, h_2 \rangle - 1$. Then*

$$[L_\mu : H^i(L_\lambda(\mathfrak{g}^{(1)}))] = [L_\mu(\mathfrak{g}^{(1)}) : U_{h_2}^i(\lambda)]$$

for $i \geqslant 1$, and $H^0(L_\lambda(\mathfrak{g}^{(1)})) = 0$.

PROOF. Take a resolution $M^\cdot$ of $L_\lambda(\mathfrak{g}^{(1)})$ by free $U(\mathfrak{g}_{-1}^{(1)})$-modules. Then we have the exact triangle:

$$\cdots \to M^\cdot_{\langle \lambda, h_2 \rangle} \to M^\cdot \to U_{h_2}(\lambda) \to \cdots.$$

Let us apply the functor $H^\cdot$ to it. Then $H^\cdot(M^\cdot_{\langle \lambda, h_2 \rangle}) = 0$ because all of its components $V_\mu(\mathfrak{g}^{(1)})$ induce cyclic bundles on $G/P^{(1)}$ (see [7]). So

$$H^\cdot(M^\cdot) = H^\cdot(U_{h_2}(\lambda)).$$

Let $U_{h_2}(\lambda) = \bigoplus L_\mu(\mathfrak{g}^{(1)})$ (by conjecture). The condition $\mu < \lambda$ implies that $\langle \mu, h_2 \rangle < \langle \lambda, h_2 \rangle$ and by simple combinatorial reason all μ are typical with respect to $\mathfrak{g}^{(1)}$. By Theorem 5.1, $H^\cdot(L_\mu(\mathfrak{g}^{(1)})) = L_\mu$. This proves Lemma 6.3.

Let W be the Weyl group of $\mathfrak{g}_0$, which acts naturally on $\mathfrak{h}^*$. Introduce a shifted action of W on $\mathfrak{h}^*$ by the formula $\lambda^w = w(\lambda + \rho) - \rho$. For any integral weight $\lambda \in \mathfrak{h}^*$ there exists the unique dominant weight on its W-orbit for the shifted action. We denote this weight by λ'.

LEMMA 6.4. *If $U_{h_1}(\lambda, \mu)$ is nonzero, then $\mu = (\lambda - p\alpha)'$ for some $p \in \mathbf{Z}^+$.*

PROOF. We use the conjecture. Since U_{h_1} is semisimple we are looking for highest, i.e., $\mathfrak{n}^+$-invariant, vectors in $H^\cdot(L_\lambda(\mathfrak{g}^{(1)}))$. By Lemma 5.1 any such weight can be obtained as a sum of λ and some weight from $\Lambda^\cdot(\mathfrak{g}_{-1}/\mathfrak{g}_{-1}^{(1)})$ modulo the Weyl group action. Therefore any such weight μ is equal to $(\lambda - \alpha_1 - \cdots - \alpha_p)'$, where $\alpha_1, \dots, \alpha_p \in \Delta_1^+ \backslash \Delta_1^+(\mathfrak{g}^{(1)})$. But then it is equal to $(\lambda - p\alpha)'$.

LEMMA 6.5. *Let E be a finite-dimensional $\mathfrak{g}$-module with the set of weights $P(E)$. Then $E \otimes L_\lambda$ have only components with central characters $\chi(\lambda + \nu)$, where $\nu \in P(E)$.*

PROOF. Since $L_\lambda \otimes E$ is a factor module of $V_\lambda \otimes E$, it has the same central characters. The last module is filtered by V_μ, where μ are some weights from the set $\lambda + P(E)$, hence all possible central characters are $\chi(\mu)$.

Now we will prove Theorem 5.2.

Let us consider the exact sequence of $P^{(1)}$-modules:

$$0 \to L_\lambda(\mathfrak{g}^{(1)}) \to L_{\lambda-\varepsilon_1}(\mathfrak{g}^{(1)}) \otimes L_{\varepsilon_1} \to M \to 0.$$

The idea of the proof is to apply $H^\bullet$ to this sequence. Since at the end we will take the component with central character $\chi(\lambda)$, we have to take this component in the sequence:

$$0 \to L_\lambda(\mathfrak{g}^{(1)}) \to p_\lambda(L_{\lambda-\varepsilon_1}(\mathfrak{g}^{(1)}) \otimes L_{\varepsilon_1}) \to p_\lambda(M) \to 0.$$

Here p_λ is the projection onto the component with central character $\chi(\lambda)$.

A). LEMMA 6.6. *If λ is of type* A *then $p_\lambda(M) = L_{\lambda-\alpha}(\mathfrak{g}^{(1)})$.*

PROOF. It is clear that $M = L_{\varepsilon_2}(\mathfrak{g}^{(1)}) \otimes L_{\lambda-\varepsilon_1}(\mathfrak{g}^{(1)})$.

Take $\mathfrak{k} = \mathfrak{g}^{(1)} \cap \mathfrak{g}^{(m+k)}$. Then

$$p_\lambda(L_{\varepsilon_2}(\mathfrak{g}^{(1)}) \otimes L_{\lambda-\varepsilon_1}(\mathfrak{k})) = L_{\lambda-\varepsilon_1+\delta_k}(\mathfrak{k}) = L_{\lambda-\alpha}(\mathfrak{k}).$$

Take a parabolic subgroup K in supergroup $G^{(1)}$ that corresponds to Lie superalgebra $\mathfrak{k} + \mathfrak{b} \cap \mathfrak{g}^{(1)}$. Consider the functor $H^\bullet(G^{(1)}/K, \cdot)$ which can be defined in the same way as in §5. Apply this functor to both sides of the last equality. From the left-hand side we obtain

$$H^\bullet(G^{(1)}/K, p_\lambda(L_{\varepsilon_2}(\mathfrak{g}^{(1)}) \otimes L_{\lambda-\varepsilon_1}(\mathfrak{k}))) = p_\lambda(H^\bullet(G^{(1)}/K, L_{\lambda-\varepsilon_1}(\mathfrak{k})) \otimes L_{\varepsilon_2}(\mathfrak{g}^{(1)})).$$

Since $\lambda - \varepsilon_1$ is typical with respect to $\mathfrak{k}$ in $\mathfrak{g}^{(1)}$,

$$H^\bullet(G^{(1)}/K, L_{\lambda-\varepsilon_1}(\mathfrak{k})) = L_{\lambda-\varepsilon_1}(\mathfrak{g}^{(1)}).$$

By the same reason

$$H^\bullet(G^{(1)}/K, L_{\lambda-\alpha}(\mathfrak{k})) = L_{\lambda-\alpha}(\mathfrak{g}^{(1)}),$$

and this completes the proof.

Therefore we have

$$H^\bullet(p_\lambda(L_{\lambda-\varepsilon_1}(\mathfrak{g}^{(1)}) \otimes L_{\varepsilon_1})) = p_\lambda(H^\bullet(L_{\lambda-\varepsilon_1}(\mathfrak{g}^{(1)})) \otimes L_{\varepsilon_1})$$

and $H^\bullet(L_{\lambda-\varepsilon_1}(\mathfrak{g}^{(1)})) = L_{\lambda-\varepsilon_1}$, because $\lambda - \varepsilon_1$ is typical with respect to $\mathfrak{g}^{(1)}$.

So we obtain the following long exact sequence:

$$\cdots \to H^\bullet(L_\lambda(\mathfrak{g}^{(1)})) \to p_\lambda(L_{\lambda-\varepsilon_1} \otimes L_{\varepsilon_1}) \to H^\bullet(L_{\lambda-\alpha}(\mathfrak{g}^{(1)})) \to \cdots,$$

which splits into the following pieces:

$$0 \to H^{i+1}(L_{\lambda-\alpha}(\mathfrak{g}^{(1)})) \to H^{i}(L_{\lambda}(\mathfrak{g}^{(1)})) \to 0$$

for $i \geqslant 1$;

$$0 \to H^{1}(L_{\lambda-\alpha}(\mathfrak{g}^{(1)})) \to H^{0}(L_{\lambda}(\mathfrak{g}^{(1)})) \to p_{\lambda}(L_{\lambda-\varepsilon_1} \otimes L_{\varepsilon_1}) \to H^{0}(L_{\lambda-\alpha}(\mathfrak{g}^{(1)})) \to 0.$$

From Lemma 6.1, we have $U_{h_1}^{i+1}(\lambda-\alpha, \mu) = U_{h_1}^{i}(\lambda, \mu)$ for $i \geqslant 2$.

For $i = 1$ we use the last sequence. Write it in the following form:

$$0 \to H^{0}(L_{\lambda}(\mathfrak{g}^{(1)}))/H^{1}(L_{\lambda-\alpha}(\mathfrak{g}^{(1)})) \to p_{\lambda}(L_{\lambda-\varepsilon_1} \otimes L_{\varepsilon_1}) \to H^{0}(L_{\lambda-\alpha}(\mathfrak{g}^{(1)})) \to 0.$$

Let S be the left term. Then $S \subset p_{\lambda}(L_{\lambda-\varepsilon_1} \otimes L_{\varepsilon_1})$. By conjecture $U_{h_1}^{1}(\lambda) = \operatorname{Ker}(H^{0}(L_{\lambda}(\mathfrak{g}^{(1)})) \to L_{\lambda})$ is semisimple. Then $T = \operatorname{Ker}(S \to L_{\lambda})$ is also semisimple. Let $T = \bigoplus L_{\mu}$. Since T is a submodule in $p_{\lambda}(L_{\lambda-\varepsilon_1} \otimes L_{\varepsilon_1})$ (Lemma 6.5) each highest vector can have weight either $\lambda - \varepsilon_1 + \delta_i$ or $\lambda - \varepsilon_1 + \varepsilon_j$. But from the condition $\chi(\lambda) = \chi(\mu)$ there are only two possibilities: $\mu = \lambda$ or $\mu = \lambda - \alpha$. The case $\mu = \lambda$ is impossible in $\operatorname{Ker}(S \to L_{\lambda})$. Therefore $T = L_{\lambda-\alpha}$. By Lemma 6.1,

$$\begin{aligned} &\operatorname{Ker}[H^{0}(L_{\lambda}(\mathfrak{g}^{(1)}))/H^{1}(L_{\lambda-\alpha}(\mathfrak{g}^{(1)})) \to L_{\lambda}] \\ &\quad = \operatorname{Ker}[H^{0}(L_{\lambda}(\mathfrak{g}^{(1)})) \to L_{\lambda}]/H^{1}(L_{\lambda-\alpha}(\mathfrak{g}^{(1)})) = U_{h_1}^{1}(\lambda)/U_{h_1}^{2}(\lambda-\alpha) = T. \end{aligned}$$

This completes the proof of the theorem for the case A.

B). For λ of type B we have (by Lemma 6.5) $p_{\lambda}(M) = 0$. Proceeding in the same way as in case A, we obtain

$$p_{\lambda}(H^{\cdot}(L_{\lambda-\varepsilon_1}(\mathfrak{g}^{(1)})) \otimes L_{\varepsilon_1}) = H^{\cdot}(L_{\lambda}(\mathfrak{g}^{(1)})).$$

Lemma 6.1 implies that

$$p_{\lambda}(U_{h_1}^{i}(\lambda-\varepsilon_1) \otimes L_{\varepsilon_1}) = U_{h_1}^{i}(\lambda) \quad \text{for } i \geqslant 2.$$

Let $U_{h_1}^{i}(\lambda-\varepsilon_1) = \bigoplus L_{\mu}$. Then $\mu \leqslant \lambda$ for each μ. It is easy to see that $\langle \mu, h_{m+k} \rangle = \langle \lambda, h_{m+k} \rangle$. Then in $p_{\lambda}(L_{\mu} \otimes L_{\varepsilon_1})$ we can find only components of weight $\mu + \delta_k$. So for $i \geqslant 2$ we have:

$$U_{h_1}^{i}(\lambda, \mu + \delta_k) = U_{h_1}^{i}(\lambda - \varepsilon_1, \mu).$$

For $i = 1$

$$\begin{aligned} &\operatorname{Ker}[\varphi \colon p_{\lambda}(H^{0}(L_{\lambda-\varepsilon_1}(\mathfrak{g}^{(1)}) \otimes L_{\varepsilon_1})) \to p_{\lambda}(L_{\lambda-\varepsilon_1} \otimes L_{\varepsilon_1})] \\ &\quad = \operatorname{Ker}[H^{0}(L_{\lambda}(\mathfrak{g}^{(1)})) \to L_{\lambda}] = U_{h_1}^{1}(\lambda). \end{aligned}$$

Take the natural projection $p \colon p_{\lambda}(L_{\lambda-\varepsilon_1} \otimes L_{\varepsilon_1}) \to L_{\lambda}$. Let us show that p is an isomorphism. Indeed, $\operatorname{Ker} p\varphi = U_{h_1}^{1}(\lambda)$ is semisimple. Therefore $\operatorname{Ker} p\varphi = \operatorname{Ker} p \oplus \operatorname{Ker} \varphi$, and $\operatorname{Ker} p$ is also semisimple. Let $\operatorname{Ker} p = \bigoplus L_{\mu}$. All μ are equal to $\lambda - \varepsilon_1 + \varepsilon_i$ or $\lambda - \varepsilon_1 + \delta_j$. But there is no such dominant μ for which $\chi(\lambda) = \chi(\mu)$. Therefore $\operatorname{Ker} p = 0$. So we have $p_\lambda(L_{\lambda-\varepsilon_1} \otimes L_{\varepsilon_1}) = L_\lambda$. By Lemma 6.1, $U^1_{h_1}(\lambda, \mu + \delta_k) = U^1_{h_1}(\lambda - \varepsilon_1, \mu)$.

By Lemma 6.2 one can get $U_{h_1}(\lambda, \mu + \delta_k) = U_{h_{m+k}}(\lambda, \mu + \delta_k)$ and

$$U_{h_1}(\lambda - \varepsilon_1, \mu) = U_{h_{m+k-1}}(\lambda - \varepsilon_1, \mu).$$

But

$$U_{h_{m+k-1}}(\lambda - \varepsilon_1, \mu) = U_{h_{m+k-1}}(\lambda - \varepsilon_1 + \delta_k, \mu + \delta_k).$$

Therefore we have

$$U_{h_1}(\lambda, \mu) = U_{h_{m+k-1}}(\lambda - \alpha, \mu)$$

after substitution μ instead of $\mu + \delta_k$ and α instead of $\varepsilon_1 - \delta_k$.

C). We can reduce this case to the case B using Lemma 6.2. If we consider $U_{h_{m+k}}$ instead of U_{h_1} then λ of type C becomes a weight of type B.

D). Let us use the relation:

$$p_{\lambda-\varepsilon_1}(L_\lambda(\mathfrak{g}^{(1)}) \otimes L_{-\delta_n}) = L_{\lambda-\varepsilon_1}(\mathfrak{g}^{(1)}).$$

Applying to it the functor $H^\cdot$, we obtain

$$p_{\lambda-\varepsilon_1}(H^\cdot(L_\lambda(\mathfrak{g}^{(1)}) \otimes L_{-\delta_n})) = H^\cdot(L_{\lambda-\varepsilon_1}(\mathfrak{g}^{(1)})).$$

Since $p_{\lambda-\varepsilon_1}(L_\lambda \otimes L_{-\delta_n}) = 0$, Lemma 6.1 implies

$$p_{\lambda-\varepsilon_1}(U^{i+1}_{h_1}(\lambda) \otimes L_{-\delta_n}) = H^i(L_{\lambda-\varepsilon_1}(\mathfrak{g}^{(1)})).$$

Let $U^i_{h_1}(\lambda) = \bigoplus L_\mu$. One can easily get from Lemma 6.4 that

$$\langle \mu, h_{m+k} \rangle = \langle \lambda, h_{m+k} \rangle - 1.$$

Therefore $p_{\lambda-\varepsilon_1}(L_\mu \otimes L_{-\delta_n}) = L_{\mu-\delta_k}$ (Lemma 6.5 and conjecture). So we have the following relations:

$$U^{i+1}_{h_1}(\lambda, \mu) = [L_{\mu-\delta_k} : H^i(L_{\lambda-\varepsilon_1}(\mathfrak{g}^{(1)}))].$$

Now apply Lemma 6.3 to $H^i(L_{\lambda-\varepsilon_1}(\mathfrak{g}^{(1)}))$. We have

$$[L_{\mu-\delta_k} : H^i(L_{\lambda-\varepsilon_1}(\mathfrak{g}^{(1)}))] = [L_{\mu-\delta_k}(\mathfrak{g}^{(1)}) : U^i_{h_2}(\lambda - \varepsilon_1)].$$

Finally

$$[L_{\mu-\delta_k}(\mathfrak{g}^{(1)}) : U^i_{h_2}(\lambda - \varepsilon_1)] = U^i_{h_2}(\lambda - \alpha, \mu),$$

where the last expression we understand as in the statement of the theorem.

So we have proved the theorem.

§7. Formulas for $U(\lambda, \mu)$ and $K(\lambda, \mu)$

Now we are able to solve the recurrent relations from Theorem 5.2. Here we will give exact formulas for $U(\lambda, \mu)$, $K(\lambda, \mu)$.

Let us consider the free $\mathbf{C}[q, q^{-1}]$-module H with a basis T_λ, where λ runs over all integral regular dominant weights with given central character.

It is convenient to consider T_λ for all integral λ (not only dominant) subject to the following relations:

$$T_\lambda q^{l(w)} = T_{w(\lambda)},$$

where λ is dominant, $w \in W$, and $l(w)$ is the length of w. From this relation we have that $T_\lambda = 0$ for nonregular λ.

Let us introduce a linear operator σ_i $(i = 1, \dots, m)$ on H by:

$$\sigma_i(T_\lambda) = \sum_\nu U_{h_i}(\lambda - \rho, \nu - \rho) T_\nu.$$

For regular λ let us denote by $l(\lambda)$ the length of $w \in W$ for which $w(\lambda)$ is dominant. Let $\alpha \in \Delta_1^+$. We define the operator σ_α by:

$$\sigma_\alpha(T_\lambda) = \sum q^{2-k} T_{\lambda - k\alpha} \quad \text{if } (\lambda, \alpha) = 0,$$

where we take the summation over all $k \geqslant 1$ satisfying the condition

$$2 + j - k + l(\lambda - j\alpha) - l(\lambda - k\alpha) > 0 \quad \text{for any } j = 0, \dots, k.$$

For (λ, α) different from 0 we put $\sigma_\alpha(T_\lambda) = 0$.

THEOREM 7.1. $\sigma_i(T_\lambda) = \sum_{\alpha \in \Delta_1^+(\mathfrak{g}^{(i-1)}) \setminus \Delta_1^+(\mathfrak{g}^{(i)})} \sigma_\alpha T_\lambda$.

PROOF. We prove the statement for $i = 1$. One can obviously generalize the proof to any i. If λ is typical with respect to $\mathfrak{g}^{(1)}$, then all $\sigma_\alpha(T_\lambda) = 0$ and therefore $\sigma_i(T_\lambda) = 0$ which is true because of Theorem 5.1.

Let λ be atypical. Then there is only one α from $\Delta_1^+ \setminus \Delta_1^+(\mathfrak{g}^{(1)})$ with $(\lambda, \alpha) = 0$. Therefore we can omit the summation sign from the formula we need to prove. We can rewrite recurrent relations obtained in Theorem 5.2 in the following form:

$$\sigma_1(T_\lambda) = \sigma_j(T_{\lambda - k\alpha}) + q T_{\lambda - \alpha},$$

where k is the first number for which $\lambda - k\alpha$ is regular and $j = w(1)$ ([2]) where $w(\lambda)$ is dominant. One can check that $\sigma_1(T_\lambda)$ given in the theorem satisfies exactly this recurrent relation.

Now for each $i = 0, \dots, m$ define a basis $\{S_\lambda^i\}$ in H by the expression:

$$S_\lambda^i = \sum K_i(\lambda - \rho, \mu - \rho) T_\mu.$$

Put $S_\lambda^0 = S_\lambda$. Our main goal is the expression for S_λ.

THEOREM 7.2. $S_\lambda = \prod_{i=1}^m (1 - \sigma_i)^{-1} T_\lambda$.

PROOF. It follows from Theorem 4.2 by induction of i.

([2]) $W \cong S_n \times S_m$. Therefore W acts naturally on the set of indices $\{1, \dots, m + n\}$.

REFERENCES

1. Joseph N. Bernstein and D. A. Leites, *A formula for the characters of the irreducible finite-dimensional representations of Lie superalgebras of series Gl and sl*, C. R. Acad. Bulgare Sci. **33** (1980), no. 8, 1049–1051. (Russian)
2. David A. Vogan, Jr., *Irreducible characters of semisimple Lie algebras*, Duke Math. J. **49** (1979), 1081–1098.
3. J. Hughes, R. King, and J. Thierry-Mieg, *On characters of irreducible representations of $gl(m, n)$*, preprint, 1991.
4. V. Kac, *Lie superalgebras*, Adv. Math. **26** (1977), 8–96.
5. David Kazhdan and George Lusztig, *Representations of Coxeter groups and Hecke algebras*, Invent. Math. **53** (1979), no. 2, 165–184.
6. Yu. I. Manin, *Gauge fields and complex geometry*, "Nauka", Moscow, 1984; English transl., Springer-Verlag, Berlin and New York, 1988.
7. I. Penkov, *Borel-Weil-Bott theory for classical lie supergroups*, Itogi Nauki i Tekhniki. Sovremennye Problemy Mathematiki. Noveĭshie Dostizheniya, vol. 32, VINITI, Moscow, 1988, pp. 71–124; English transl. in J. Soviet Math. **51** (1990).
8. I. Penkov and V. Serganova, *Cohomology of g/p for classical complex Lie supergroups g and characters of some atypical g-modules*, Ann. Inst. Fourier (Grenoble) **39** (1989), no. 4, 845–873.
9. ———, *On irreducible representations of classical Lie superalgebras*, Indag. Math. **2** (1992), no. 3.

DEPARTMENT OF MATHEMATICS, UNIVERSITY OF CALIFORNIA AT BERKELEY, BERKELEY, CALIFORNIA 94720

E-mail address: serganov@math.berkeley.edu

ADVANCES IN SOVIET MATHEMATICS
Volume 16, Part 2, 1993

Hodge Filtration of Hypergeometric Integrals Associated with an Affine Configuration of General Position and a Local Torelli Theorem

A. N. VARCHENKO

To I. M. Gelfand on his 80th birthday

§1. Hypergeometric differential forms

Let $\mathscr{C} = \{A_0, \dots, A_N\}$ be a configuration of hyperplanes in general position in $\mathbb{CP}^n$, $\alpha = (\alpha_0, \dots, \alpha_N)$ a collection of complex numbers, $\alpha_0 + \dots + \alpha_N = 0$. We may assume that A_0 is the hyperplane at infinity and $\mathbb{C}^n = \mathbb{CP}^n \setminus A_0$. Let $f_1, \dots, f_N$ be linear functions on $\mathbb{C}^n$ whose zeroes define the remaining hyperplanes of the configuration:

$$A_j: f_j = 0.$$

Let $I = \{i_1, \dots, i_n\} \subset \{1, \dots, N\}$ be a subset of n elements. The differential form

$$\omega(I) = \Big(\prod_{i \notin I} f_i^{\alpha_i}\Big)\, df_{i_1}^{\alpha_{i_1}} \wedge \dots \wedge df_{i_n}^{\alpha_{i_n}}$$

is called the *hypergeometric form* of the configuration $\mathscr{C}$. It is a closed multivalued form.

Let L be the coefficient system of rank 1 on $\mathbb{P}^n \setminus \mathscr{C}$, whose monodromy around the hyperplane A_j is equal to $\exp(-2\pi i\alpha_j)$. Then $\omega\langle I\rangle$ defines a single-valued section of $\Omega^n(\mathbb{P}^n \setminus \mathscr{C}; L)$. Let L^* be the local coefficient system of rank 1 on $\mathbb{P}^n \setminus \mathscr{C}$, whose monodromy around the hyperplane A_j is equal to $\exp(2\pi i\alpha_j)$, $H_n^{\mathrm{lf}}(\mathbb{CP}^n \setminus \mathscr{C}, L^*)$ the group of locally finite n-homology which is defined using locally finite cycles, where each of them is situated in a bounded part of $\mathbb{C}^n$.

EXAMPLE. Let all hyperplanes of the configuration be invariant with respect to complex conjugation, i.e., the configuration is real. Then each bounded component of $\mathbb{R}^n - \mathscr{C}$ is a cycle from $H_n^{\mathrm{lf}}(\mathbb{CP}^n \setminus \mathscr{C}, L^*)$, and,

1991 *Mathematics Subject Classification*. Primary 33C20, 14C34, 32S35; Secondary 32S40, 14D05.

1051-8037/93 $1.00 + $.25 per page

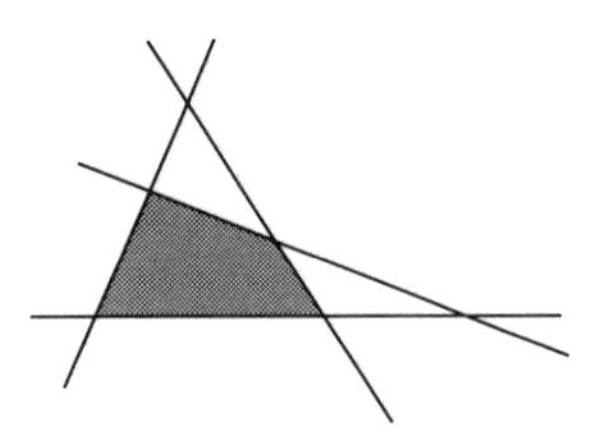

FIGURE 1

moreover, the set of such components is a basis in $H_n^{\mathrm{lf}}(\mathbb{CP}^n \setminus \mathscr{Q}, L^*)$. The pair $(\mathbb{C}^n, \bigcup A_j)$ for a configuration in general position is smoothly isotopic to the pair $(\mathbb{C}^n, \bigcup A'_j)$ for any other configuration $\mathscr{Q}'$ in general position with the same number of hyperplanes. This implies that

$$\dim H_n^{\mathrm{lf}}(\mathbb{CP}^n \setminus \mathscr{Q}, L^*) = \binom{N-1}{n},$$

and H_n^{lf} is spanned by curvilinear n-dimensional polyhedra with boundary in $\bigcup A_j$.

Assume that real parts of $\alpha_1, \dots, \alpha_N$ are positive, $\mathrm{Re}(\alpha_j) > 0$. Then the integral

$$\int_\gamma \omega\langle I\rangle, \qquad \gamma \in H_n^{\mathrm{lf}}(\mathbb{CP}^n \setminus \mathscr{H}, L^*),$$

converges. If $\alpha_1, \dots, \alpha_N$ are not nonnegative integers, then this integral is defined by analytic continuation with respect to $\alpha_1, \dots, \alpha_N$. In this case it is shown in [1] that

1. The forms $\{\omega(i_1, \dots, i_n): 1 < i_1 < \cdots \leqslant N\}$ are linearly independent and if $-\alpha_0$ is not a nonnegative integer, they form a basis in $(H_n^{\mathrm{lf}})^*$.
2. For any $I_0 = \{i_2, \dots, i_n\} \subset \{1, \dots, N\}$ the relation

$$\sum_{j \notin I_0} \omega(j, I_0) = 0$$

holds.

Therefore, the hypergeometric forms correspond to the vertices $A_{i_1} \cap \cdots \cap A_{i_n}$ of the configuration $A_1, \dots, A_N$ in $\mathbb{C}^n$. The forms attached to the vertices outside a fixed hyperplane form a basis in the cohomology. The sum of the forms (with appropriate signs) corresponding to the vertices on a fixed line $A_{i_2} \cap \cdots \cap A_{i_n}$ is equal to zero.

EXAMPLE. Let $n = 2$, $N = 4$. Then

$$\omega\langle 1, 3\rangle + \omega\langle 2, 3\rangle + \omega\langle 4, 3\rangle = 0,$$
$$\omega\langle 2, 4\rangle, \omega\langle 2, 3\rangle, \omega\langle 3, 4\rangle \quad \text{form a basis.}$$

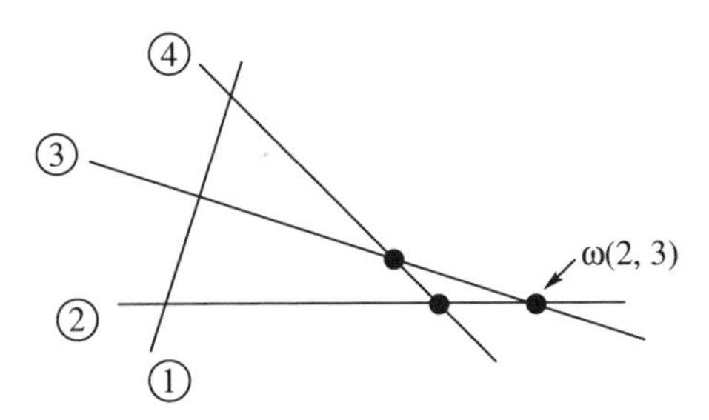

FIGURE 2

§2. Hodge filtration in the cohomology $H_n^{\mathrm{lf}}(\mathbb{CP} \setminus \mathscr{C}, L^*)^*$

An *edge* of a configuration is any nonempty intersection of some of its hyperplanes. For example, a hyperplane belonging to a configuration is an $(n-1)$-edge. A vertex is a zero-dimensional edge.

A vertex $v = A_{i_1} \cap \cdots \cap A_{i_n}$ lies on n one-dimensional edges $\{A_{i_2} \cap \cdots \cap A_{i_n}, \dots, A_{i_1} \cap \cdots \cap A_{i_n}\}$, $\binom{n}{2}$ two-dimensional edges, $\binom{n}{k}$ k-dimensional edges, and so on.

The *k-dimensional fan* of a vertex v is the union of all $\binom{n}{k}$ k-dimensional edges of the configuration $\mathscr{C}$ that contain v.

We define the *Hodge filtration*

$$0 \subset F^n \subset F^{n-1} \subset \cdots \subset F^0 = H^n$$

as follows.

Let F^k be the subspace spanned by hypergeometric forms corresponding to the vertices of the configuration lying on the $(n-k)$-dimensional fan of the vertex $A_1 \cap \cdots \cap A_n$. In other words,

$$F^k = \operatorname{span}\{\omega\langle i_1, \dots, i_n\rangle : (i_1, \dots, i_k) \subset (1, \dots, n)\}.$$

EXAMPLE. The subspace F^n is spanned by a single form $\omega(1, \dots, n)$. The subspace F^{n-1} is spanned by the forms

$$\omega(1, \dots, n), \qquad \omega(1, \dots, \widehat{k}, \dots, n, j),$$

where $j = n+1, \dots, N$, $k = 1, \dots, n$. The subspace F^{n-2} is spanned by F^{n-1} and the forms $\omega(1, \dots, \widehat{i}, \dots, \widehat{j}, \dots, n, k_1, k_2)$, where $i, j \in (1, \dots, n)$, $k_1, k_2 \in (n+1, \dots, N)$, and so on.

EXAMPLE. The subspace F^2 is spanned by $\omega(1, 2)$, the subspace F^1 is spanned by $\omega(1, 2)$, $\omega(3, 1)$, $\omega(4, 1)$, $\omega(2, 3)$, $\omega(2, 4)$, and the subspace F^0 is spanned by all forms (see Figure 3 on the next page).

THEOREM 1. *If $\alpha_1, \dots, \alpha_N, -\alpha_0$ are not nonnegative integers, then*

$$\dim F^n = 1,$$

$$\dim F^i - \dim F^{i+1} = \binom{n}{i}\binom{N-n-1}{n-i}, \qquad i = 0, \dots, n-1.$$

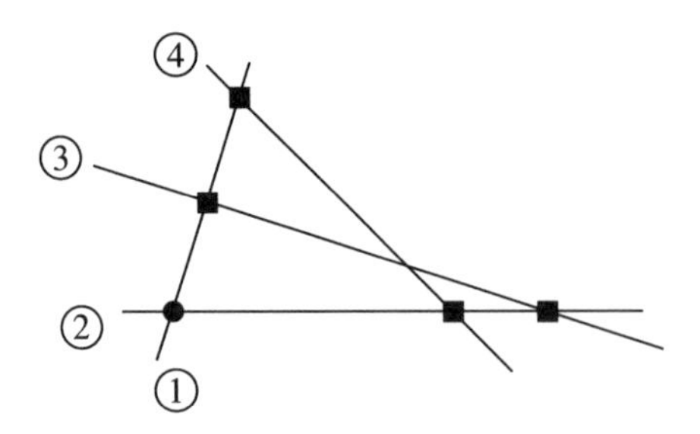

FIGURE 3

PROOF. The number $\dim F^i - \dim F^{i+1}$ is equal to the number of those vertices outside A_{n+1} that lie on $\binom{n}{i}$ $(n-i)$-dimensional edges containing the vertex $v_0 = A_1 \cap \dots \cap A_n$ and do not lie on $\binom{n}{i+1}$ $(n-i-1)$-dimensional edges containing the same vertex. It suffices to make computations for just one such $(n-i)$-dimensional edge.

LEMMA 2. *The number of vertices of the configuration $\mathscr{Q}$ that do not lie on the hyperplanes $A_1, \dots, A_{n+1}$ is equal to $\binom{N-n-1}{n}$.*

PROOF. Obvious.

Let us complete the proof of the theorem. The edge $A_1 \cap A_2 \cap \dots \cap A_i$, $i \leqslant n$, contains $\binom{N-n-1}{n-i}$ vertices not belonging to the hyperplanes $A_{i+1}, \dots, A_n, A_{n+1}$ (write $N-n-1 = (N-i)-(n-i)-1$ and apply Lemma 2).

§3. The space of ordered configurations of $N+1$ hyperplanes of general position in $\mathbb{CP}^n$

We denote this space by Q_{N+1}. Its points are configurations $\mathscr{Q} = (A_0, \dots, A_N)$ of general position. Each configuration $\mathscr{Q}$ defines the space $H_n^{\mathrm{lf}}(\mathbb{CP}^n \backslash \mathscr{Q}, L^*)$ and its dual space H^n. The spaces H_n^{lf} and H^n form vector bundles $\mathscr{H}_n \to Q_{N+1}$, $\mathscr{H}^n \to Q_{N+1}$ over Q_{N+1}. These vector bundles are holomorphic, mutually dual, and are equipped with the Gauss-Manin connection. The bundle $\mathscr{H}^n$ is also equipped with the Hodge filtration defined in Section 2.

LEMMA 3. *The fiberwise Hodge filtration defines a filtration of the bundle $\mathscr{H}^n$ by holomorphic subbundles*

$$0 \subset \mathscr{F}^n \subset \mathscr{F}^{n-1} \subset \dots \subset \mathscr{F}^0 = \mathscr{H}^n.$$

PROOF. Each hypergeometric differential form $\omega(I) = \omega(i_1, \dots, i_n)$ defines a section $s(i_1, \dots, i_n)$ of the bundle $\mathscr{H}^n$. For any family of cycles $\delta(\mathscr{Q}) \in H_n^{\mathrm{lf}}(\mathbb{P}^n \setminus \mathscr{Q}, L^*)$, $\mathscr{Q} \in Q_N$, defining a covariantly constant section of the bundle $\mathscr{H}_n$, the integral $\int_{\delta(\mathscr{Q})} \omega(I)$ depends holomorphically on $\mathscr{Q}$. Hence, the section $s(i_1, \dots, i_n)$ is holomorphic. This proves the lemma.

The bundle $\mathscr{F}^n$ is one-dimensional and is spanned by the form $\omega(1, \dots, n)$, that is, by the section $s(1, \dots, n)$.

§4. Derivation of sections

Let s be the germ at a point $\mathscr{Q} \in Q$ of a holomorphic section of the vector bundle $\mathscr{H}$ on Q, v the germ at the same point of a holomorphic vector field on Q, ∇ a connection on the bundle $\mathscr{H}$. Then $\nabla_v s$ is a germ of a holomorphic section of $\mathscr{H}$.

Let $X \subset \mathscr{H}$ be a subset of $\mathscr{H}$. Define a subset $\mathscr{D}(X) \subset \mathscr{H}$ as follows. Consider germs s of holomorphic sections of $\mathscr{H}^n$ at an arbitrary point $\mathscr{Q} \in Q$ such that their values belong to X. Denote the set of such germs by $\sec(X)$. Next consider an arbitrary germ v at $\mathscr{Q}$ of a holonomic vector field on Q, and take the value $\nabla_v s(\mathscr{Q}) \in \mathscr{H}$ of the derivative at the point $\mathscr{Q}$. Now we define $\mathscr{D}(X)$ as the subset in $\mathscr{H}$, whose intersection with the fiber of $\mathscr{H}$ over a point $\mathscr{Q} \in Q$ is equal to the linear span of the set $\{\nabla_v s(\mathscr{Q}),$ $s \in \sec(X)$, v is a germ of a vector field$\}$. We set

$$\mathscr{D}^2(X) = \mathscr{D}(\mathscr{D}(X)), \quad \mathscr{D}^i(X) = \mathscr{D}(\mathscr{D}^{i-1}(X)), \quad k = 2, 3, \dots.$$

§5. Back to the situation described in Section 3

THEOREM 4. *If $\alpha_1, \dots, \alpha_N, -\alpha_0$ are not nonnegative integers, then*

$$\mathscr{D}(\mathscr{F}^i) = \mathscr{F}^{i-1}, \qquad i = 1, \dots, n.$$

COROLLARY. *We have*

$$\mathscr{F}^i = \mathscr{D}^{n-i}(\mathscr{F}^n).$$

PROOF OF THE THEOREM. It is sufficient to show that for any $I = \{i_1, \dots, i_n\}$, the set $\mathscr{D}(s(I))$ is equal to the total space of the subbundle in $\mathscr{H}$ spanned by the sections

$$\{s(I), s(i_1, \dots, \widehat{i_k}, \dots, i_n, j) \colon 1 \leqslant k \leqslant n,\ j \notin I\},$$

or, equivalently, that the derivatives of the form $\omega(I)$ are spanned by the form itself and the forms in which one of the indices $i_1, \dots, i_n$ is replaced by any index from $\{1, \dots, N\} \setminus I$. In other words, the derivatives of the form corresponding to the vertex $A_{i_1} \cap \cdots \cap A_{i_n}$ are spanned by the form itself and by the forms corresponding to the vertices lying on n one-dimensional edges of the configuration passing through $A_{i_1} \cap \cdots \cap A_{i_n}$.

We prove the assertion for the form $\omega(1, \dots, n)$. The general case will follow by renumbering the hyperplanes.

LEMMA 5. *We have*

$$\mathscr{D}(s(1, \dots, n)) = \mathscr{F}^{n-1} \overset{\text{def}}{=} \{s(1, \dots, n), s(1, \dots, \widehat{k}, \dots, n, j) \colon 1 \leqslant k \leqslant n,\ j > n\}.$$

First we make the following

REMARK. Let $A \in Q_{N+1}$, $T\colon \mathbb{C}^n \to \mathbb{C}^n$ be a linear transformation, $T(\mathscr{Q})$ be the transformed configuration. Then we have $T^*\omega(I)_{T(\mathscr{Q})} = c\omega(I)_{\mathscr{Q}}$, where $\omega(I)_{T(\mathscr{Q})}$ is the hypergeometric form with parameters I for the configuration $T(\mathscr{Q})$, $\omega(I)_{\mathscr{Q}}$ is the corresponding form for the configuration $\mathscr{Q}$, the constant c depends on the choice of linear equations defining the hyperplanes of $\mathscr{Q}$ and $T(\mathscr{Q})$. In particular, the Hodge filtration of the configurations $\mathscr{Q}$ and $T(\mathscr{Q})$ are canonically isomorphic.

By the previous remark, to prove the lemma it suffices to consider the set Q^v_{N+1} of configurations in general position that can be defined by the equations $f_j = 0$, where

$$f_1 = x_1, \dots, f_n = x_n, \qquad f_{n+1} = x_1 + \cdots + x_n - 1,$$
$$f_i = k_i^1 x_1 + \cdots + k_i^n x_n + 1, \quad i > n+1.$$

For such configurations,

$$\begin{aligned}\omega = \omega(1, \dots, n) &= \alpha_1 x_1^{\alpha_1 - 1} \cdots \alpha_n x_n^{\alpha_n - 1}(x_1 + \cdots + x_n - 1)^{\alpha_{n+1}} \\ &\times (k^1_{n+1}x_1 + \cdots + k^n_{n+1}x_n + 1)^{\alpha_{n+2}} \cdots \\ &\times (k^1_N x_1 + \cdots + k^n_N x_n + 1)^{\alpha_N} dx_1 \wedge dx_2 \wedge \cdots \wedge dx_n.\end{aligned}$$

LEMMA 6. *We have*

$$\frac{d\omega}{dk_i^j} = \frac{(-1)^{j-n}}{k_i^j}\alpha_j \omega(I_{ij}),$$

where $I_{ij} = \{1, \dots, \hat{j}, \dots, n, i\}$ *for any* $1 \leqslant j \leqslant n$, $i > n+1$.

The proof is obvious.

EXAMPLE. Let $n = 2$. Then

$$\omega = \alpha_1 x_1^{\alpha_1 - 1}\alpha_2 x_2^{\alpha_2 - 1}(x_1 + x_2 - 1)^{\alpha_3}(k_4^1 x_1 + k_4^2 x_2 + 1)^{\alpha_4} dx_1 \wedge dx_2,$$

$$\begin{aligned}\frac{d\omega}{dk_4^2} &= \alpha_1 x_1^{\alpha_1 - 1}\alpha_2 x_2^{\alpha_2 - 1}(x_1 + x_2 - 1)^{\alpha_3}\alpha_4 x_2 (k_4^1 x_1 + k_4^2 x_2 + 1)^{\alpha_4 - 1} dx_1 \wedge dx_2 \\ &= \alpha_2 x_2^{\alpha_2}(x_1 + x_2 - 1)^{\alpha_3} dx_1^{\alpha_1} \wedge \frac{d(k_4^1 x_1 + k_4^2 x_2 + 1)^{\alpha_4}}{k_4^2} = \frac{\alpha_2}{k_4^2}\omega(1, 4).\end{aligned}$$

§6. The period map

Let $S_{N+1} = Q_{N+1}/\operatorname{PGL}(n+1)$. Then the bundles $\mathscr{H}_n$ and $\mathscr{H}^n$, as well as the Hodge filtration $\mathscr{F}^n \subset \cdots \subset \mathscr{F}^0 = \mathscr{H}^n$, are well defined on S_{N+1}. Both bundles have a Gauss-Manin connection. The bundle $\mathscr{F}^n$ is of rank 1, hence it defines a section in the projectivization $\mathbb{P}(\mathscr{H}^n)$:

$$\sigma\colon S_{N+1} \to \mathbb{P}(\mathscr{H}^n).$$

Since the projective bundle $\mathbb{P}(\mathscr{H}^n)$ is equipped with a Gauss-Manin connection, the section σ defines a holomorphic map $\tilde{s}$ from the universal covering $\widetilde{S}_{N+1}$ of S_{N+1} to the projective space $\mathbb{CP}^{M-1}$, $M = \binom{N-1}{n}$, which is equal to some fiber of the bundle $\mathbb{P}(\mathscr{H}^n)$. We call this map the *period map.*

THEOREM 7 (Local Torelli Theorem). *The period map*

$$\widetilde{s}\colon \widetilde{S}_{N+1} \to \mathbb{CP}^{M-1}$$

is an immersion.

PROOF. By Theorem 1 we know that

$$\dim S_{N+1} = n(N-n-1),$$
$$\dim \mathscr{F}^{n-1} = n(N-n-1)+1.$$

It remains to apply the equality $\mathscr{D}(\mathscr{F}^n) = \mathscr{F}^{n-1}$ (Theorem 4).

§7. The case of rational exponents

Let $\beta_0 = k_0/d, \ldots, \beta_N = k_N/d$ be rational positive numbers with the properties

$$0 < \beta_j < 1, \qquad \beta_0 + \cdots + \beta_N = n+1.$$

Let $\mathscr{Q} = (A_0, \ldots, A_N)$ be a configuration in general position. Consider a local coefficient system L of rank 1 of $\mathbb{CP}^n \setminus \mathscr{Q}$ defined by the monodromies $\exp(2\pi i \beta_j)$ around the hyperplanes A_j. The cohomology $H^n(\mathbb{CP}^n \setminus \mathscr{Q}, L)$ have the following Hodge filtration.

Let $(x_0, \ldots, x_n)$ be homogeneous coordinates in $\mathbb{CP}^n$ such that our hyperplanes are given by the equations

$$A_0\colon x_0 = 0, \ldots, A_n\colon x_n = 0,$$
$$A_j\colon \overset{0}{a}_j x_0 + \cdots + \overset{n}{a}_j x_n = 0, \qquad j = n+1, \ldots, N.$$

The subvariety X of $\mathbb{CP}^N$ given by the equations

$$y_j^d = \overset{0}{a}_j y_0^d + \cdots + \overset{n}{a}_j y_j^d, \qquad j = n+1, \ldots, N,$$

is a nonsingular projective variety (recall that $\mathscr{Q}$ is in general position). The group $(\mathbb{Z}_d)^{N+1}$ acts naturally on X by multiplying the coordinates by dth roots of unity.

THEOREM 8. *The space $H^n(\mathbb{CP}^n \setminus \mathscr{Q}, L)$ is naturally isomorphic to the eigensubspace $H^n(X, \mathbb{C})_\mu$ corresponding to the character of $(\mathbb{Z}_d)^{N+1}$ defined by the numbers β_j.*

PROOF. This is a straightforward generalization of the corresponding result for $n = 1$ from [2, 2.23].

The cohomology space $H^*(X, C)$ has a pure Hodge structure. In particular,

$$H^n(X, \mathbb{C})_\mu = \bigoplus_p H_\mu^{p, n-p}.$$

THEOREM 9 (Deligne [3], Dolgachev). *We have*

$$\dim H_\mu^{p,n-p} = \binom{n}{p}\binom{M-n-1}{n-p}.$$

Under the same assumptions as in the beginning of Section 1, let us consider the hypergeometric differential form:

$$\eta = f_1^{-\beta_1}\cdots f_N^{-\beta_N}\,dx_1\wedge\cdots\wedge dx_n.$$

Let $F\colon X\to\mathbb{CP}^n$ be the map given by the formula:

$$(y_0, \dots, y_N)\to(x_0, \dots, x_n) = (y_0^d, \dots, y_n^d).$$

PROPOSITION 10. $F^*(n)$ *is a single-valued holomorphic form on* X.

PROOF. Trivial verification.

Define new exponents $\alpha_0, \dots, \alpha_N$ by the formulas

$$\alpha_0 = 1-\beta_0, \qquad \alpha_1 = 1-\beta_1, \dots, \alpha_n = 1-\beta_n,$$
$$\alpha_j = -\beta_j, \qquad j = n+1, \dots, N.$$

Then $\alpha_0+\cdots+\alpha_N = 0$, $\alpha_1, \dots, \alpha_N, -\alpha_0$ are not nonpositive integers. For the configuration $\mathscr{Q}$ with attached numbers α, the following equality holds:

$$\omega(1, \dots, n) = \operatorname{const}\eta.$$

In the described situation there are two Hodge filtrations in $H^n(\mathbb{CP}^n\setminus\mathscr{Q}, L)$: the first one is the hypergeometric filtration $\{\mathscr{F}^i\}$ from Section 2, the second one is induced from the Hodge filtration in $H^n(X, \mathbb{C})_\mu$, and is denoted by $\{F_\mu^i\}$.

THEOREM 11. *The two Hodge filtrations in* $H^n(\mathbb{CP}^n\setminus\mathscr{Q}, L)$ *coincide.*

PROOF. By Proposition 10 we get $F_\mu^n = \mathscr{F}^n$. It is well known that

$$\mathscr{D}(F_\mu^{i+1})\subset F_\mu^i.$$

Therefore, by Theorem 4,

$$\mathscr{F}^i\subset F_\mu^i.$$

Applying Theorem 1 and Theorem 9, we see that

$$\dim F_\mu^i = \dim\mathscr{F}^i$$

for all i. Hence, the two filtrations coincide.

REMARK. In [3] Deligne defined a Hodge filtration in $H^n(\mathbb{CP}^n\setminus\mathscr{Q}, L)$ by using logarithmic multivalued differential forms and proved the formula from Theorem 9 in the case of not necessary rational exponents. Our proof of Theorem 11 shows that our filtration and Deligne's filtration coincide.

§8. Simplest degenerations of configurations in general position

Let Q_{N+1}^0 denote the space of all stable ordered configurations of $N+1$ hyperplanes $(A_0, \dots, A_N)$ in $\mathbb{CP}^n$. (Here we consider a configuration as an ordered set of points in the dual projective space, and use the corresponding notion of stability for the diagonal action of $\mathrm{PGL}(n+1)$, see [4].) Let $Q_{N+1} \subset Q^0_{N+1}$ be the subspace of all configurations of general position,

$$S^0_{N+1} = Q^0_{N+1}/\,\mathrm{PGL}(n+1),$$
$$S_{N+1} = Q_{N+1}/\,\mathrm{PGL}(n+1).$$

Consider the divisor $D \subset S^0_{N+1}$ that consists of all configurations in which the hyperplanes $A_2, \dots, A_{n+2}$ have exactly one common point (we assume here that $N > n+1$ to ensure the stability of such configurations). Let d be a natural number. Consider a d-sheet covering $\pi\colon V \to S^0_{N+1}$ ramified along D. Let $\alpha_0, \dots, \alpha_N$ be complex numbers with the properties that $\alpha_1, \dots, \alpha_N, -\alpha_0$ are not nonnegative integers, and $\alpha_2+\dots+\alpha_{n+2}+1 = 1/d$.

Let $s\colon S_{N+1} \to \mathbb{P}(\mathscr{H}^n)$ be the section introduced in Section 6; it is defined by the subbundle $\mathscr{F}^n$, or, equivalently, by the hypergeometric form $\omega(1, \dots, n)$. If we write this section in covariantly constant projective coordinates, then s can be viewed as a multivalued holomorphic map $\tilde{s}$ from the base S_{N+1} to a fixed fiber $\mathbb{CP}^{M-1}$. Explicitly, if $\delta_1(\mathscr{Q}), \dots, \delta_M(\mathscr{Q}) \in H^{\mathrm{lf}}_n(\mathbb{CP}^n \setminus \mathscr{Q}, L^*)$ is a covariantly constant basis, then

$$\tilde{s}(a) = \Big(\int_{\delta_1(a)} \omega(1, \dots, n), \dots, \int_{\delta_M(a)} \omega(1, \dots, n)\Big).$$

Let us consider the multivalued holomorphic map:

$$\tilde{s}\circ\pi\colon V \setminus \pi^{-1}(D) \to \mathbb{CP}^{M-1}.$$

Fix a nonsingular point v of the divisor $\pi^{-1}(D)$ and a small neighborhood U of the point v in V. Fix some branch $(\tilde{s}\circ\pi)_0$ of $\tilde{s}\circ\pi$ at a point $u \in U \setminus \pi^{-1}(D)$.

THEOREM 12. (i) *The branch* $(\tilde{s}\circ\pi)_0$ *can be analytically extended to a single-valued holomorphic map of* $U \setminus \pi^{-1}(D)$.

(ii) *It can be analytically extended to* $\pi^{-1}(D) \cap U$.

(iii) *The analytic continuation defines an immersion* $U \to \mathbb{CP}^{M-1}$.

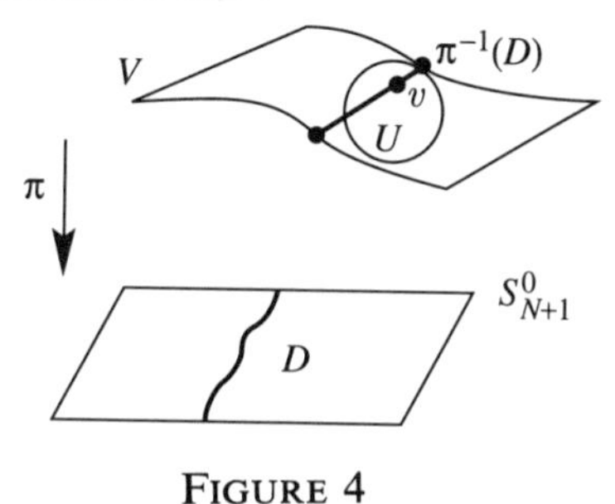

FIGURE 4

PROOF. The monodromy operator in $H_n^{\mathrm{lf}}(\mathbb{CP}^n \setminus \mathscr{C}, L^*)$ corresponding to the path around D can be diagonalized. Its eigenvalues are $\exp(2\pi i/d)$ (simple eigenvalue) and 1 (eigenvalue of multiplicity $M-1$). Let $\delta_1(\mathscr{C}), \dots, \delta_M(\mathscr{C}) \in H_n^{\mathrm{lf}}(\mathbb{CP}^n \setminus \mathscr{C}, L^*)$ be the corresponding covariantly constant basis over $\pi(U) \setminus D$. We may take for $\delta_1(\mathscr{C})$ the "vanishing real n-dimensional tetrahedron" bounded by the hyperplanes $A_2, \dots, A_{n+2}$. The multivalued functions

$$\int_{\delta_j(\mathscr{C})} \omega(1, \dots, n), \qquad j = 1, \dots, M,$$

defined on $\pi(U)$ will be denoted by H_j, $j = 1, \dots, M$.

Let $t_1, \dots, t_q$, $q = n(N-n-1)$, be local coordinates on U, in which $\pi^{-1}(D)$ is given by the equation $t_1 = 0$. An easy local computation shows that

$$\pi^* H_1 = t_1 h_1(t_2, \dots, t_q) + o(t_1),$$

$$\pi^* H_j = h_j(t_1^d, t_2, \dots, t_q) + o(t_1), \quad j = 2, \dots, M,$$

where $h_1, \dots, h_M$ are suitable single-valued holonomic functions; moreover, h_1 does not vanish. In particular, this implies the first assertion of the theorem.

Now let us choose special coordinates in a neighborhood of the point $\pi(v)$. If the hyperplanes of the configuration are given by equations (1) from Section 5, then the functions k_j^i define local coordinates on S_{N+1}^0. In these coordinates the divisor D is described by the equation $k_{n+2}^1 = 1$.

Now let $k_{n+2}^1 = 1$, that is, the intersection of the hyperplanes

$$f_1 = 0, \dots, f_{n-1} = 0, f_{n+1} = 0, f_{n+2} = 0$$

is the point $(0, \dots, 0, 1)$. Assume that in the rest our hyperplanes are in general position. As is shown in [1, Part II], for this configuration one has

$$\dim H_n^{\mathrm{lf}}(\mathbb{CP}^n \setminus \mathscr{C}, L^*) = \binom{N-1}{n} - 1.$$

Moreover, the hypergeometric forms corresponding to the vertices of the configuration not lying on the hyperplane A_{n+1} (which contains the singular vertex of the configuration) form a basis in the dual space H^n. In particular, the form $\omega(1, \dots, n)$ defines a nonzero cohomology class. As a corollary we obtain that the vector $(0, h_2, \dots, h_M)$ does not vanish at a generic point of the divisor D. This proves the second assertion of the theorem.

The third assertion of the theorem is proved similarly to the proof of Theorem 7 by using Lemma 6 and the abovementioned results from [1, Part II].

§9. Acknowledgments

The hypergeometric Hodge filtrations and the local Torelli theorem might be useful for higher-dimensional analogs of the results of Deligne and Mostov [2] on the uniformization of the spaces of configurations of points on the line. I was attracted to this kind of problems by I. Dolgachev. I am grateful to him for numerous stimulating discussions during our common visit to the Mathematics Institute of the University of Warwick in May of 1989. I would like to thank the faculty of the Institute for their hospitality.

§10. Appendix

In connection with higher-dimensional generalizations of [2] a solution of the following arithmetical problem could be useful:

Find all rational numbers $a_1, \dots, a_{n+k}$ with the following properties:

1. Each a_i satisfies $0 < a_i < 1$.
2. $a_1 + \dots + a_{n+k} = n + 1$.
3. For any $i_1, \dots, i_{n+1}$, $1 \leqslant i_1 < \dots < i_{n+1} \leqslant n+k$,

$$a_{i_1} + \dots + a_{i_{n+1}} < n \quad \text{implies} \quad n - (a_{i_1} + \dots + a_{i_{n+1}}) = 1/N,$$

where N is a natural number depending, in general, on $i_1, \dots, i_{n+1}$.

As was observed independently by Deligne and Dolgachev, if $a_1, \dots, a_{n+k}$ is a solution of this problem for the pair (n, k), then the numbers $1 - a_1, \dots, 1 - a_{n+k}$ give a solution of this problem for the pair $(k-2, n+2)$. For $n = 1$ all solutions of the problem are listed in [2].

PROPOSITION 12. *In addition to the solutions from* [2] *and the dual solutions, there are only the following six solutions*:

$$n = 2, \quad k = 7, \quad a_1 = \dots = a_9 = 1/3;$$
$$n = 2, \quad k = 4, \quad a_1 = \dots = a_6 = 1/2;$$
$$n = 2, \quad k = 4, \quad a_1 = a_2 = 1/6; \quad a_3 = a_4 = a_5 = a_6 = 2/3;$$
$$n = 2, \quad k = 4, \quad a_1 = a_2 = a_3 = a_4 = 1/3; \quad a_5 = a_6 = 5/6;$$
$$n = 3, \quad k = 5, \quad a_1 = \dots = a_8 = 1/2;$$
$$n = 5, \quad k = 4, \quad a_1 = \dots = a_9 = 2/3.$$

REFERENCES

1. A. N. Varchenko, *Euler beta-functions, Vandermonde determinant, Legendre equations and critical values of linear functions on configurations of hyperplanes.* I, Izv. Akad. Nauk SSSR Ser. Mat. **53** (1989), no. 6, 1206–1235; II, **54** (1990), no. 1, 146–158; English transl. in Math. USSR-Izv. **35** (1990); **36** (1991).
2. P. Deligne and G. Mostov, *Monodromy of hypergeometric functions and non-lattice integral monodromy groups*, Inst. Hautes Etudes Sci. Publ. Math. **63** (1986), 5–90.
3. P. Deligne, A letter to I. M. Gelfand, 1989.
4. D. Mumford, *Geometric invariant theory*, Springer-Verlag, Berlin and New York, 1969.

DEPARTMENT OF MATHEMATICS, UNIVERSITY OF NORTH CAROLINA, CHAPEL HILL, NORTH CAROLINA 27599

ADVANCES IN SOVIET MATHEMATICS
Volume 16, Part 2, 1993

The Second Gelfand-Dickey Bracket as a Bracket on a Poisson-Lie Grassmannian

ILYA ZAKHAREVICH

To my teacher Israel M. Gelfand on his 80th birthday

ABSTRACT. We introduce a Poisson structure on the Grassmannian $\mathrm{Gr}_k(V)$, on which the Poisson-Lie group $\mathrm{GL}(V)$ acts in a Poisson-Lie way. We discuss the analytic complications connected with the infinite-dimensional case $V = C^\infty(\mathbb{R})$ and show that an open subset of $\mathrm{Gr}_k(V)$ with this Poisson structure is isomorphic to the Gelfand-Dickey manifold of differential operators of order k with the second Gelfand-Dickey bracket. As a consequence, we introduce a Poisson-Lie action of an enormous group on the Gelfand-Dickey manifold generalizing (on the semiclassical level) the Sugavara inclusion.

§ 0. Introduction

In this paper we consider the manifold of differential operators of order n with the leading term ∂^n. There is a remarkable Poisson structure on this manifold, introduced by Adler [1] (the proof that this structure satisfies the Jacobi relation is due to Gelfand and Dickey [5]). This Poisson bracket on this manifold is usually called the *second Gelfand-Dickey structure.* ([1]) The usual geometrical interpretation comes from identification of this Poisson manifold with the Hamiltonian reduction from the dual space of an affine Lie coalgebra [3].

A new topic in the theory of this Poisson manifold was introduced by Radul, who invented an action of the Lie algebra of differential operators on this manifold. This action does not preserve the Poisson structure on the manifold, but changes it in a quite regular way. This kind of action was introduced first in the theory of quantum groups (on the semiclassical level) and is called a *Poisson-Lie action.* For this notion to be correct, we should first determine a structure of bialgebra on this Lie algebra, or, equivalently, a *compatible* structure of Lie algebra on the dual space to this Lie algebra.

1991 *Mathematics Subject Classification.* Primary 22E65, 58E07; Secondary 58E05.

([1]) The *first Gelfand-Dickey* structure is another topic.

However, the dual space to the Lie algebra of differential operators is the space of pseudodifferential operators and this space carries a natural structure of Lie algebra. In fact this structure is compatible with the Lie bracket on the algebra. That means that we can define a notion of a Poisson-Lie action of this algebra on a Poisson manifold.

It was a great surprise when (inspired by A. Givental) Radul discovered that the above action is actually a Poisson-Lie action. It seems that it is not easy to explain this fact using the Poisson reduction from the dual space to affine coalgebra.

In this paper we introduce another geometrical interpretation of the second Gelfand-Dickey structure. We introduce an enormous group $\mathrm{GL}(C^{\infty}(\mathbb{R}))$, consider the usual GL-type Poisson-Lie structure on this group, and consider a homogeneous space for this group, the Grassmannian of n-dimensional subspaces in $C^{\infty}(\mathbb{R})$. The Poisson structure on the group induces a Poisson structure on this quotient by a parabolic subgroup.

After that it is a good idea to try to write down this Poisson structure in some coordinate system on the Grassmannian. We use the following coordinate system: a differential operator of order n with leading term ∂^n is identified with the (n-dimensional) space of its solutions in $C^{\infty}(\mathbb{R})$. It is easy to see that this correspondence is 1-1 on an open subset of a Grassmannian (for a given n-dimensional subspace to be a space of solutions of a differential equation, some Wronskian should be nonzero everywhere). We compute the Poisson structure in this coordinate system (indeed, the space of differential operators with the leading term ∂^n is an affine space, so we can consider this identification as a coordinate system). What we get is the second Gelfand-Dickey structure!

Therefore we got the following fact: *there is a Poisson-Lie action of an enormous group* $\mathrm{GL}(C^{\infty}(\mathbb{R}))$ *on the second Gelfand-Dickey structure*! (2) Now to explain the Radul action we should only note that the Lie algebra of differential operators is in fact a *Poisson-Lie subalgebra* of the Lie bialgebra $\mathfrak{gl}(C^{\infty}(\mathbb{R}))$. Therefore the restriction of the action on this subalgebra is (essentially by definition!) also a Poisson-Lie action.

Actually, the last fact is not *completely* obvious. On first sight the bialgebra structure on the space of differential operators has nothing to do with the decomposition of an operator into a sum of a *raising* and a *lowering* Volterra operator, that is used in the definition of the Poisson-Lie structure on $\mathrm{GL}(C^{\infty}(\mathbb{R}))$. However, the former structure is indeed a restriction of the latter, and the proof of this fact is given in 2.2.

In fact what is announced here is just a simple calculation. The main inspiration for this calculation was the Radul action. To go from Radul's

(2) As often happens with actions of Poisson-Lie groups, it is a *local action*: the action of "big" elements of the group can force a point "to get out of the manifold it lives on". Say, we identify the Gelfand-Dickey manifold with an open subset of the Grassmannian $\mathrm{Gr}(C^{\infty}(\mathbb{R}))$, but this open subset is not stable with respect to the action of $\mathrm{GL}(C^{\infty}(\mathbb{R}))$.

result to those discussed here we should only note that if we introduce the identification of the Grassmannian with the set of differential operators, then the Radul action can be extended from the Lie algebra of differential operators to the Lie algebra of all operators in $C^{\infty}(\mathbb{R})$. After that we noted that this action is evidently (if the calculation in 2.1 can be called a calculation in two rows, and it is, if we drop all the preliminary definitions) a Poisson-Lie action, and the Lie subalgebra of differential operators is a Poisson-Lie subalgebra.

Another point of view on the identification of the Grassmannian with the set of differential operators is given in the paper [6]. It is shown there that this mapping is just a *nonlinear momentum mapping* for the action of the bialgebra of differential operators. Let us recall that the usual momentum mapping is defined in the case when the action of the Lie algebra preserves the symplectic structure. It sends a manifold into the dual space of some central extension of the Lie algebra. In the case of the Poisson-Lie action such construction is possible, at least in some cases, but the mapping sends the manifold to the *dual Lie group* to some central extension. We do not want to discuss this notion now, and want only to note that in the case of the Lie algebra of differential operators this dual group is an extension of the group corresponding to the Lie algebra of integral operators by the one-parametric group $\partial^s = e^{s\cdot\log\partial}$, $s \in \mathbb{R}$. Therefore, an element of this group looks like

$$(1 + u_1\partial^{-1} + u_2\partial^{-2} + \cdots)\partial^s.$$

The main result of [6] is that the momentum mapping sends an element of the Grassmannian into the corresponding *differential* operator, i.e., $s = k$ and $u_l = 0$ for $l > k$. Therefore the above identification is just a momentum mapping.

This construction explains in particular why we can choose a coordinate system on the Grassmannian $\mathrm{Gr}_k(C^{\infty}(\mathbb{R}))$ in such a way that the Poisson bracket becomes quadratic. Indeed, such a coordinate system is given by the nonlinear momentum mapping, at least if it is "almost surjective".

I am indebted to a lot of people for fruitful discussions and inestimable help, among them I. M. Gelfand, A. Givental, A. Goncharov, D. Kazhdan, B. Khesin, M. Kontsevich, O. Kravchenko, J.-H. Lu, H. McKean, A. Radul, N. Reshetikhin. This article could not appear without the hospitality of Harvard University, the MSRI, and the MIT during the (extended) period of preparation of this paper.

The structure of this paper is the following: in § 1 we discuss the usual notions from the theory of Poisson-Lie groups. This discussion is superseded by any text on the theory of Poisson-Lie groups and affine Poisson-Lie groups (such as [2, 9, 10, 7]). In 2.0 we introduce the notion of pseudodifferential symbols. In 2.1 we prove the main formula that identifies the Poisson Grassmannian with the second Gelfand-Dickey structure. In 2.2 we prove the fact that the Lie algebra of differential operators is a Poisson-Lie subalgebra and find the corresponding bialgebra structure. In the next two subsections we discuss the periodic and matrix cases of this construction. In 2.5 we formulate a conjecture on quantization of the described action in the simplest possible case, when differential operators are of order 1, and in 2.6 we give some hints about what could be the rigorous topological justification of what we do here.

In this paper we are concerned with two different questions. Therefore all the discussion is divided into two parts: the first part is purely algebraic, the second part is concerned with the "topologizations" of the results in the infinite-dimensional case. A reader interested only in the purely algebraic picture can skip any discussion of the topology on the manifolds in question. Moreover, we fully understand that the discussion of topology is only sketchy here and hope to continue it elsewhere. The only reason we include the topological discussions in this (otherwise purely algebraic) paper is that the topological methods we use are apparantly quite simple, and they help to add some sense to the formulae we write.

We remark also that one of the interpretations of the results obtained here is that the enormous Lie group $\mathrm{GL}(C^\infty(\mathbb{R}))$ acts on the Gelfand-Dickey Poisson manifold in a Poisson-Lie way. From this point of view this is a generalization of the fact that was observed first by M. Semenov: the dressing action of Volterra operators is of Poisson-Lie nature [10]. Indeed, Volterra operators form a Poisson-Lie subgroup in $\mathrm{GL}(C^\infty(\mathbb{R}))$, therefore the action of this subgroup is also a Poisson-Lie action.

§ 1. Classical r-matrices and Poisson-Lie structures

1.1. The classical Yang-Baxter equation. Let us recall what the classical Yang-Baxter equation is. Let $r \in \mathfrak{g} \otimes \mathfrak{g}$, where $\mathfrak{g}$ is a Lie algebra. We call r a classical r-matrix if r satisfies the equation

$$[r^{12}, r^{13}] + [r^{12}, r^{23}] + [r^{13}, r^{23}] = 0. \tag{1.1}$$

Here for a tensor $x \in \mathfrak{g} \otimes \mathfrak{g}$ we denote by x^{ij} the tensor $\sigma_{\binom{12}{ij}}(x \otimes \mathrm{id}) \in U\mathfrak{g} \otimes U\mathfrak{g} \otimes U\mathfrak{g}$, $\binom{12}{ij} \in \mathfrak{S}_3$ being the permutation of three elements sending 1 to i and 2 to j, σ being the standard action of $\mathfrak{S}_n$ on $\underbrace{V \otimes V \otimes \cdots \otimes V}_{n \text{ times}}$, and $U\mathfrak{g}$ being the universal enveloping algebra. The bracket $[\ ,\ ]$ denotes the Lie bracket in the Lie algebra $U\mathfrak{g} \otimes U\mathfrak{g} \otimes U\mathfrak{g}$ (of course, the image of this bracket in the above formula lies in $\mathfrak{g} \otimes \mathfrak{g} \otimes \mathfrak{g}$).

There are several different objects called the Yang-Baxter equation. We use this name for the equation (1.1) and call a solution of this equation a classical r-matrix. However, below r-matrices with a particular symmetry property will play a special role, so we need to study the symmetry properties for the left-hand side of the Yang-Baxter equation. Let $YB(r)$ denote the left-hand side of the Yang-Baxter equation. This is a quadratic form in an argument r; let us denote the corresponding symmetric bilinear form by $[r, s]$,

$$[\ ,\ ]: (\mathfrak{g}\otimes\mathfrak{g})\otimes(\mathfrak{g}\otimes\mathfrak{g})\to\mathfrak{g}\otimes\mathfrak{g}\otimes\mathfrak{g}, \qquad [r, r] = YB(r).$$

CLAIM 1.1. *If r is skewsymmetric ($r\in\Lambda^2\mathfrak{g}$), then $YB(r)$ is also skewsymmetric ($YB(r)\in\Lambda^3\mathfrak{g}$).*

PROOF. First of all, we note that if $\sigma_{(12)}r = \alpha_1 r$, $\sigma_{(12)}s = \alpha_2 s$ (here $\alpha_{1,2}\in\{\pm 1\}$), then $\sigma_{(13)}[r, s] = -\alpha_1\alpha_2[r, s]$. Indeed, $\sigma_{(13)}$ changes the first and the last summand of the formula (1.1) (this, of course, does not change any signs), changes the order of terms inside brackets $[\ ,\ ]$ (this changes the sign of the bracket), and changes the order of indices for r (this multiplies one term by α_1, another by α_2).

Second, the action of $\sigma_{(12)}$ on $[r, r]$ can be described in a similar way: on the first two summands it acts by interchanging them and interchanging the indices for the first occurrence of r; on the last summand it acts by interchanging the arguments of the bracket. Both operations result in the change of the sign.

Hence, two generators of $\mathfrak{S}_3$ act on $YB(r)$ in the same way as they act on $\Lambda^3\mathfrak{g}$. Therefore, $YB(r)\in\Lambda^3\mathfrak{g}$.

Now we can compare the Yang-Baxter equation with another quadratic equation on an element of $\Lambda^2\mathfrak{g}$, the Jacobi equation on Poisson brackets.

1.2. Poisson manifolds. Let η be a 2-vector field on a manifold X, i.e., a section of Λ^2TX. Then we can define a skewsymmetric bracket on functions on the manifold X:

$$(f, g)\mapsto\{f, g\}, \qquad \{f, g\}|_x = \langle\eta|_x, df|_x\wedge dg|_x\rangle.$$

Here $\langle\ ,\ \rangle$ denotes the pairing between Λ^2T_xX and $\Lambda^2T_x^*X$. Consider the mapping

$$\mathrm{Jac}(\eta): (f, g, h)\mapsto \operatorname*{Alt}_{f,g,h}\{\{f, g\}, h\},$$

where Alt denotes the alternation operation. This mapping measures the degree of the non-Jacobi property for the bracket $\{\ ,\ \}$.

It is easy to see that $\mathrm{Jac}(\eta)(f, g, h)$ depends only on the 1-jet of the function f and vanishes if f is a constant function. Hence $\mathrm{Jac}(\eta)|_x$ is, in fact, a mapping

$$\Lambda^3T_x^*X\to\mathbb{C},$$

therefore $\mathrm{Jac}(\eta)$ corresponds to some section $SN(\eta)$ of Λ^3TX:

$$\mathrm{Jac}(\eta)(f, g, h)|_x = \langle SN(\eta), df|_x\wedge dg|_x\wedge dh|_x\rangle.$$

Therefore we have defined a quadratic mapping

$$SN: \Gamma(\Lambda^2TX)\to\Gamma(\Lambda^3TX): \eta\mapsto SN(\eta). \tag{1.2}$$

This mapping ([3]) is called the Schouten-Nijenhuis bracket. It is quadratic in η and is a differential operator of the first order in η (this means that the value of $SN(\eta)$ at x depends only on the 1-jet of η at x).

REMARK 1.1. Let us extend the commutator operator from vector fields to the multivector fields in a term-by-term way:

$$[v_1 \wedge v_2 \wedge \cdots \wedge v_n, w_1 \wedge w_2 \wedge \cdots \wedge w_m] = \sum_{i,j} (-1)^{i+j} [v_i, w_j] \wedge v_1 \wedge \cdots \wedge \widehat{v}_i \wedge \cdots \wedge v_n \wedge w_1 \wedge \cdots \wedge \widehat{w}_j \wedge \cdots \wedge w_m.$$

This operation becomes a supercommutator if we define the degree of $v_1 \wedge v_2 \wedge \cdots \wedge v_n$ to be $n-1$. The commutator of a vector field and a multivector field is the usual Lie derivative action, so we get a generalization of this notion. Now it is easy to see that the Schouten-Nijenhuis bracket is a square of a bivector field:

$$SN(\eta) = [\eta, \eta].$$

In a local coordinate frame we can express $SN(\eta)$ as

$$SN(\eta)^{ijk} = \operatorname*{Alt}_{ijk} \eta^{il} \eta^{jk}_{,l}. \tag{1.3}$$

We use here the tensor notation: if t is a component of a tensor, then $t_{,i}$ denotes the derivative of this component in the direction of the index i.

DEFINITION 1.1. A *Poisson manifold* is a pair consisting of a manifold X and a bivector field η on X such that the corresponding bracket $\{\ ,\ \}$ on functions satisfies the Jacobi identity.

Later in this paper we will often describe Poisson brackets by the corresponding *Hamiltonian mappings* $T_x^* X \to T_x X$. This is just the image of the bivector under the identification $\Lambda^2 T_x X$ with a subspace in $\operatorname{Hom}(T_x^* X \to T_x X)$.

1.3. Skewsymmetric *r*-matrices. Let $\eta \in \Lambda^2 \mathfrak{g}$. Then η determines a left-invariant section $\widetilde{\eta}$ of $\Lambda^2 TG$, G being the group corresponding to the Lie algebra $\mathfrak{g}$. The 3-vector $SN(\widetilde{\eta})$ is, evidently, left-invariant, hence it corresponds to an element $\widetilde{SN}(\eta) \in \Lambda^3 \mathfrak{g}$. A natural question is to compare two elements of $\Lambda^3 \mathfrak{g}$, $YB(\eta)$ and $\widetilde{SN}(\eta)$. For this we can use equation (1.3) in the exponential coordinate system. It is easy to see that the derivative (in this coordinate system) at the unity of a left-invariant tensor t in the direction of a vector $X \in \mathfrak{g}$ is proportional to $\operatorname{ad} X \cdot t$. Let c^i_{jk} be the structure constants for $\mathfrak{g}$:

$$[X_j, X_k] = c^i_{jk} X_i.$$

Then we can describe $\widetilde{SN}(\eta)$ up to a multiplicative constant as

$$\operatorname*{Alt}_{ijk} \eta^{il} \eta^{jk}_{,l} = \operatorname*{Alt}_{ijk} (\eta^{il} c^j_{lm} \eta^{mk} + \eta^{il} c^k_{lm} \eta^{jm}). \tag{1.4}$$

([3]) More explicitly, the symmetrical bilinear mapping that corresponds to this quadratic mapping.

Now it is easy to see that the first summand in parentheses coincides with the term $[r^{12}, r^{23}]$, while the second one coincides with the term $[r^{13}, r^{23}]$ from formula (1.1) at $r = \eta$. (4) Hence the alternation of the expression (1.1) is proportional to the expression (1.4). However, the expression (1.1) is already skewsymmetric if η is skewsymmetric. Therefore equation (1.1) is equivalent to the bivector η being a Poisson bracket.

Now we can see that skewsymmetric classical r-matrices for Lie algebra $\mathfrak{g}$ are "the same" as left-invariant Poisson brackets on the corresponding group G. Let us consider one such bracket $\widetilde{\eta}$. The action of the group G by left multiplication $g \mapsto L_g$ preserves the tensor field $\widetilde{\eta}$. On the contrary, the action of the group G by right multiplications $g \mapsto R_g$ does not preserve this tensor field. We want to describe *the change* of this bivector field under the action of the right multiplication.

Let us consider the bivector field $\widetilde{\eta}^g = \widetilde{\eta} - R_{g*}\widetilde{\eta}$ on G, were $g \in G$ is fixed. Of course, one possible description is that $\widetilde{\eta}^g$ is a left-invariant bivector field corresponding to a bivector $\eta - \mathrm{Ad}(g^{-1})\eta$ at $e \in G$. However, we can look on this field in another way. A left-invariant vector field on G corresponds to the right action of the Lie algebra $\mathfrak{g}$. Therefore the left-invariant bivector field corresponds to the "action" of a bivector on G. On the other hand, the bivector field we consider is the result of an action of the element $g \in G$. This can motivate a consideration of the bivector which is defined as the value of this bivector field at $g \in G$.

Therefore, we can identify the Lie algebra with a tangent space to G at g (let us recall that g is fixed) in a left-invariant way. As before, we can consider a value $\eta^{(g)} \in \Lambda^2 T_g G$ of the field $\widetilde{\eta}^g$. Now we claim the compatibility condition between the right action of $g \in G$ and the bivector $\eta^{(g)}$:

$$\Lambda^2\left(\frac{\partial R_g}{\partial g}\right)\Big|_{h,g} \cdot \eta^{(g)} = \widetilde{\eta}^g|_{hg}.$$

Here we consider $R_g: h \mapsto hg$ as a function of two variables. Since the tangent map $(dR_g/dg)|_{h,g}$ sends $X \in T_g G$ to

$$\frac{d}{d\varepsilon} h(g + \varepsilon X) \in T_{hg} G,$$

the mapping $\Lambda^2(\partial R_g/\partial g)|_{h,g}$ sends $\Lambda^2 T_g G$ to $\Lambda^2 T_{hg} G$.

Now we can describe the identity

$$\Lambda^2\left(\frac{dR_g}{dg}\right) \cdot \eta^{(g)} = \widetilde{\eta}^g = \widetilde{\eta} - R_{g*}\widetilde{\eta} \tag{1.5}$$

as saying that the map $m: G \times G \to G: (g, h) \mapsto gh$ is a *Poisson map* with respect to the bivector fields $\widetilde{\eta}$ on the first and the third occurrences of G and the bivector field $g \mapsto \eta^{(g)} \in \Lambda^2 T_g G$ on the second occurrence. That

(4) Unfortunately, the indices denote two different operations in (1.4) and (1.1).

means that the adjoint map m^* agrees with the Poisson brackets on the image and the preimage, i.e.,

$$\{m^*f, m^*g\}_1 = m^*\{f, g\}_2.$$

Here the bracket $\{\ ,\ \}_1$ on $G\times G$ is the sum of brackets on the components (the bracket $\widetilde{\eta}$ on the first and the bracket $\eta^{(g)}$ on the second), and the bracket $\{\ ,\ \}_2$ is a bracket $\widetilde{\eta}$ on G.

This is usually expressed by saying that *the group G acts in a Lie-Poisson way* by right multiplication. These words mean that there is a remarkable relation on the bivector field $\eta^{(g)}$ itself (it determines the structure of a Poisson-Lie group), and the above compatibility condition between $\eta^{(g)}$ and a change of $\widetilde{\eta}$.

DEFINITION 1.2. A group G with a Poisson structure η is called a *Poisson-Lie group*, if the multiplication map $G\times G\to G$ is compatible with the Poisson brackets and the inversion map $G\to G$ multiplies the Poisson bracket by -1.

DEFINITION 1.3. A Poisson-Lie group G acts on a Poisson manifold X *in a Poisson-Lie way* if the action map $X\times G\to X$ preserves the Poisson brackets.

REMARK 1.2. An example of such objects can be found in 2.1.

Formally, to use this expression we should show first that the bivector field $g\mapsto \eta^{(g)}\in \Lambda^2 T_g G$ on G determines a Poisson structure on G, which is odd with respect to inversion on the group and that the map $G\times G\to G$ is a Poisson map with respect to structures $\eta^{(g)}$ on any occurrence of G. As before, we can rewrite the last condition as the condition of the right action of G on itself being a Poisson-Lie action, or that the change of the bivector field with respect to the right action is compatible with the bivector field. But this is a trivial consequence of what is already proven. Really,

$$\eta^{(g)} = (\Lambda^2 L_g)\eta - (\Lambda^2 R_g)\eta, \quad L_g, R_g\colon T_eG\to T_gG. \tag{1.6}$$

However, the first summand is just the bivector field $\widetilde{\eta}$. On the other hand, the action by right multiplication preserves the second summand. Therefore a change of the vector field $g\mapsto \eta^{(g)}$ under the right multiplication coincides with the change of the first summand. However, this change is already computed and coincides with the term coming from the second component of $G\times G$.

To prove that the bracket corresponding to $g\mapsto \eta^{(g)}$ is a Poisson bracket, we can note that by (1.6) this bracket is a difference of two Poisson brackets. ([5]) So it is sufficient to show that

$$[(\Lambda^2 L_g)\eta, (\Lambda^2 R_g)\eta] = 0 \quad \text{or} \quad [\widetilde{\eta}, \operatorname{Inv}\widetilde{\eta}] = 0$$

([5]) The second summand is symmetrical to the first under the mapping $g\mapsto g^{-1}$, so should also be Poisson.

where Inv is the inversion map on the group, and this is easy. Indeed, since the commutator of a left-invariant and a right-invariant vector field on G is 0 and the Schouten-Nijenhuis bracket is an extension of the commutator on vector fields, the Schouten-Nijenhuis bracket of a left-invariant and a right-invariant bivector field is also 0.

1.4. The modified Yang-Baxter equation. As we have already seen, the skewsymmetric r-matrix makes it possible to construct a Poisson-Lie bracket on the group G and a left-invariant Poisson structure on G that is a Poisson torsor ([6]) with respect to the action by right multiplication. However, if we are ready to drop some of these properties, we can weaken the restrictions on an r-matrix. As we will see at the end of 1.5, the amount we need to drop is quite small.

Let $r = \eta + t$ be a decomposition of a (nonskewsymmetric) matrix from $\mathfrak{g} \otimes \mathfrak{g}$ into skewsymmetric and symmetric components. Then

$$[r, r] = [\eta, \eta] + 2[\eta, t] + [t, t],$$

and decomposition of a 3-tensor with respect to the action of $\sigma_{(13)}$ is, as we have shown in 1.3, $([\eta, \eta] + [t, t]) + 2[\eta, t]$, where the second summand is invariant and the first is multiplied by -1.

The restriction we want to impose is the $\operatorname{ad} G$-invariance of the 2-vector t. In this case

$$2[\eta, t] = [\eta^{12}, t^{13}] + [\eta^{12}, t^{23}] + [\eta^{13}, t^{23}] + [t^{12}, \eta^{13}] + [t^{12}, \eta^{23}] + [t^{13}, \eta^{23}].$$

The invariance of t implies that $[\eta^{ij}, t^{jk}] + [\eta^{ik}, t^{jk}] = 0$ for any i, j, k. Hence the sum of the second and the third summands, as well as the sum of the fourth and the fifth summands, is 0. However,

$$[\eta^{12}, t^{13}] + [t^{13}, \eta^{23}] = -[\eta^{21}, t^{13}] - [\eta^{23}, t^{13}] = 0.$$

Hence in this case the 3-tensor is also skewsymmetric with respect to $\sigma_{(13)}$:

$$[r, r] = [\eta, \eta] + [t, t].$$

The same arguments show that $[t, t]$ is skewsymmetric, hence $[r, r]$ is skewsymmetric. If r is an r-matrix, then $[\eta, \eta] = -[t, t]$.

DEFINITION 1.4. The equation

$$[\eta, \eta] = -[t, t] \tag{1.7}$$

on a skewsymmetric matrix $\eta \in \Lambda^2\mathfrak{g}$ for a fixed invariant symmetric matrix $t \in S^2\mathfrak{g}$ is called the *modified Yang-Baxter equation.* Usually, slightly abusing notation, a solution of this equation is also called an r-matrix.

We conclude that the Schouten-Nijenhuis bracket $SN(\tilde{\eta})$ for $\tilde{\eta}$ can be computed based on the invariant symmetric bilinear form t. Though the bivector field $\tilde{\eta}$ does not determine the Poisson bracket, the discussion above concerning the compatibility of brackets with the map of multiplication remains applicable to this case. Moreover, we have the following claim.

([6]) I.e., the principal homogeneous space.

CLAIM 1.2. *The bivector field* $g \mapsto \eta^{(g)}$ *on the group* G *determines a Poisson-Lie structure on* G.

REMARK 1.3. We can note here that for a bivector field of the form (1.6) to be a Poisson structure, the Schouten-Nijenhuis brackets $SN((\Lambda^2 L_g)\eta)$ and $SN((\Lambda^2 R_g)\eta)$ should coincide. Since one is left-invariant and the other is right-invariant, they should both be $\operatorname{Ad} G$-invariant. That means that $[\eta, \eta]$ should be $\operatorname{ad} G$-invariant. If all $\operatorname{ad} G$-invariant elements of $\Lambda^3\mathfrak{g}$ can be expressed as $[t, t]$ for some invariant symmetric element $t \in S^2\mathfrak{g}$, then the modified Yang-Baxter equation is equivalent to the definition of the Poisson-Lie group with the restriction (1.6).

Therefore, for a given solution of the modified Yang-Baxter equation we obtained a Poisson-Lie structure on the group G and a compatibile left-invariant bivector field on the torsor for G with respect to right multiplication. The Schouten-Nijenhuis bracket of this bivector with itself is a left-invariant (and right-invariant) 3-vector field determined by the given symmetric bilinear form.

The usual way to construct such a solution is to consider a decomposition of a Lie algebra with an invariant nondegenerate symmetric pairing t^* into a direct sum $\mathfrak{g} = \mathfrak{a} \oplus \mathfrak{b}$ of two Lie subalgebras that are both isotropic with respect to this form. Since the skew pairing

$$r^*((a_1, b_1), (a_2, b_2)) \stackrel{\text{def}}{=} ((-a_1, b_1), (a_2, b_2)) \tag{1.8}$$

(where outer parentheses in the right-hand side denote the scalar multiplication) is skewsymmetric, it defines (by t^*-duality) a skewsymmetric r-matrix.

A very similar way to do it is to consider a decomposition of a semisimple Lie algebra

$$\mathfrak{g} = \mathfrak{n}_- \oplus \mathfrak{h} \oplus \mathfrak{n},$$

and a bracket

$$[(a_1, b_1, c_1), (a_2, b_2, c_2)] \stackrel{\text{def}}{=} ((-a_1, c_1), (a_2, c_2)). \tag{1.9}$$

1.5. Homogeneous spaces with Poisson-Lie action. Let us return first to the situation in 1.3. There we had a Poisson structure on a torsor for G with a Poisson-Lie action. Since this Poisson structure is left-invariant, we can push it back to a structure on any quotient space $H \backslash G$ with respect to the action by left multiplication. Indeed, consider the projection $G \xrightarrow{\pi} H \backslash G$. If f and g are two functions on $H \backslash G$, then their pull-backs to G are H-invariant with respect to the left action. Therefore the Poisson bracket of these functions is also H-invariant, i.e., is a push-back of a function on $H \backslash G$. Hence we can define a bracket $\eta_{H\backslash G}$ on $H \backslash G$, which evidently is a Poisson bracket.

The same considerations applied to the right action show that the action of G on $H \backslash G$ by right multiplication remains the Poisson-Lie action: the map

$$(H \setminus G \times G, \eta_{H\setminus G} \times \eta^g) \to (H \setminus G, \eta_{H\setminus G})$$

is a map of Poisson manifolds.

Hence on any right homogeneous space X for G we can construct a Poisson structure such that this Poisson-Lie group acts in a Poisson-Lie way. (It is easy to see that this Poisson structure does not depend on the choice of the basic point on X used for the identification of X and $H \setminus G$. At the point $x \in X$ the bivector $\eta|_x$ is the image of r under the map

$$\Lambda^2\mathfrak{g} \to \Lambda^2(\mathfrak{g}/\mathfrak{g}_x) = \Lambda^2\mathcal{T}_x X,$$

where $\mathfrak{g}_x = \operatorname{Stab} x \subset \mathfrak{g}$.)

Let us return now to the situation in 1.4. Let r be a solution of the modified Yang-Baxter equation. Then we can, as above, define on any right G-homogeneous space a bivector field satisfying the condition of compatibility with the action of the Poisson-Lie group. However, this bivector field should not necessarily correspond to a Poisson structure.

Nevertheless, in many important cases this form *is* a Poisson structure. To investigate these cases, we can note that the descent to the space $H\setminus G$ results in restriction of the 3-vector $[\eta, \eta] \in \Lambda^3\mathfrak{g}$ to the quotient space $\Lambda^3(\mathfrak{g}/\mathfrak{h})$. Hence if a Lie subalgebra $\mathfrak{h} \subset \mathfrak{g}$ satisfies the condition

$$\operatorname{Im}([t, t]) \text{ in } \Lambda^3(\mathfrak{g}/\mathfrak{h}) \quad \textit{is} \quad 0,$$

then any r-matrix associated with t results in the *true* Poisson structure on the corresponding homogeneous space $H \setminus G$ which is compatible with the Poisson-Lie action of G.

In particular, if G is semisimple, then any homogeneous space with a parabolic stabilizer of a point has a Poisson structure associated with any solution of the modified Yang-Baxter equation. In fact, it is sufficient that the stabilizer contains a maximal nilpotent subgroup.

EXAMPLE 1.1 (A Grassmannian manifold). Let us consider a Grassmannian manifold $\operatorname{Gr}_k(V)$ considered as $\operatorname{GL}(V)/M$, where $M = \operatorname{Stab} W_0$, $W_0 \subset V$. The pairing

$$(\alpha, \beta) \mapsto \operatorname{Tr} \alpha \circ \beta, \qquad \alpha \in \operatorname{Hom}(V/W, W), \ \beta \in \operatorname{Hom}(W, V/W),$$

allows us to identify the tangent space $T_W \operatorname{Gr}_k(\mathbb{C}^n)$ (here $W \subset \mathbb{C}^n$) with $\operatorname{Hom}(W, V/W)$ and the cotangent space $T^*_W \operatorname{Gr}_k(\mathbb{C}^n)$ with $\operatorname{Hom}(V/W, W)$. Under these identifications the Poisson structure η_{Gr} associated with an r-matrix $r \in \Lambda^2\mathfrak{gl}(V)$ can be expressed via the corresponding Hamiltonian mapping

$$\begin{aligned} &\widetilde{\eta}_{\mathrm{Gr}}: T^*_W \operatorname{Gr}_k(\mathbb{C}^n) \to T_W \operatorname{Gr}_k(\mathbb{C}^n), \\ &\widetilde{\eta}_{\mathrm{Gr}}: \alpha \mapsto \widetilde{r}(\alpha)_{|_{W/W}} \in \operatorname{Hom}(W, V/W), \qquad \alpha \in \operatorname{Hom}(V/W, W). \end{aligned}$$

Here $\widetilde{r}$ is the image of r under the standard identification $\mathfrak{gl}(V)^{\otimes 2} \supset \Lambda^2\mathfrak{gl}(V)$ with $\mathrm{End}(\mathfrak{gl}(V))$; for $\beta \in \mathrm{End}\, V$ we denote the corresponding mapping $W \to V/W$ by $\beta_{|_{W/W}}$.

PROOF. Consider a tangent vector X at $e \in \mathrm{GL}(V)$. We can pull the corresponding right-invariant vector field to $\mathrm{Gr}_k(V)$. It is easy to see that the corresponding tangent vector at $W \in \mathrm{Gr}_k(V)$ is $X_{|_{W/W}}: W \to V/W$. (Here we consider X as a mapping $V \to V$.)

Now we can consider $\alpha \in \mathrm{Hom}(V/W, W) = T^*_W \mathrm{Gr}_k(\mathbb{C}^n) = T^*_W \mathrm{GL}(V)/M$ and the corresponding mapping $\alpha': V \to V$. The above arguments show that the right translation $R^*_{g^{-1}}\pi^*\alpha$ of the corresponding cotangent vector $\pi^*\alpha \in T^*_g \mathrm{GL}(V)$ (here $gW_0 = W$) is exactly α' (if we identify $\mathfrak{gl}^*$ with $\mathfrak{gl}$).

Now the Hamiltonian mapping on the GL sends the right-invariant covector field corresponding to α to the right-invariant vector field corresponding to $\widetilde{r}(\alpha)$.

A different approach to a definition of a Poisson structure on a parabolic quotient can be found in [7]. We should note that this approach results in a *different* Poisson structure.

REMARK 1.4. Poisson Grassmannians satisfy a remarkable property. A choice of r-matrix for $\mathfrak{gl}(V)$ in the standard form (1.9) is given by a Cartan subgroup H, i.e., a basis in V, and an ordering of this basis (that gives us two opposite Borel subgroups). It is easy to check that the Poisson structure on, say, the flag variety essentially depends on the choice of this ordering. However, *the Poisson structures on the Grassmannians are the same for two cyclically permuted orderings.*

PROOF. Indeed, if we consider an r-matrix for $\mathfrak{gl}$ as a mapping $\widetilde{r}: \mathfrak{gl} \to \mathfrak{gl}$, two r-matrices corresponding to two cyclically permuted orderings differ by a commutation mapping

$$\widetilde{r}_1 - \widetilde{r}_2: \mathfrak{gl}(V) \to \mathfrak{gl}(V): X \mapsto [z, X]$$

for an appropriate $z \in \mathrm{End}\, V$. Therefore, if $\alpha, \beta \in \mathrm{Hom}(V/W, W)$ are two covectors on $\mathrm{Gr}_k(V)$ at $W \in \mathrm{Gr}_k(V)$, the difference of values of two Poisson structures on the Grassmannian can be expressed as

$$\langle \widetilde{\eta}_{\mathrm{Gr},1} - \widetilde{\eta}_{\mathrm{Gr},2}, \alpha \wedge \beta \rangle = \mathrm{Tr}\, \beta \circ [z, \alpha] = \mathrm{Tr}[\alpha, \beta] z.$$

However, $[\alpha, \beta] = 0$ (this means, in fact, that the radical of the stabilizer is commutative), since already $\alpha\beta = \beta\alpha = 0$.

This shows also that other parabolic spaces with the same properties could be isotropic Grassmannians of middle dimension in the symplectic or even-dimensional orthogonal space or the isotropic conic in a projectivization of an orthogonal space (dropping some exceptional examples). However, we do not know exact results in these cases.

1.5.1. *Poisson-Lie subgroups.* The notion of a Poisson-Lie group is so tightly connected with the notion of a Poisson-Lie action that the usual notions of a Lie subgroup and a Lie subalgebra are useless in this context: the restriction of a Poisson-Lie action to a subgroup usually is not a Poisson-Lie action. To define the additional condition on a subalgebra we need the definition of the Lie algebra structure on the dual space to a Poisson-Lie algebra (i.e., a tangent space to a Poisson-Lie group at $e \in G$).

DEFINITION 1.5. Consider a Poisson-Lie group (G, η). A Poisson bracket $\{f, g\}$ of any two functions on G vanishes at e, and the linear part of this bracket in e is uniquely determined by linear parts of f and g at e, if $f(e) = g(e) = 0$. This determines a Lie algebra structure on $\mathfrak{g}^*$.

DEFINITION 1.6. A Lie subalgebra $\mathfrak{h} \subset \mathfrak{g}$ is called a Poisson-Lie subalgebra if its orthogonal complement $\mathfrak{h}^\perp \subset \mathfrak{g}^*$ is an ideal in $\mathfrak{g}^*$. A connected Lie subgroup $H \subset G$ is called a Poisson-Lie subgroup if its Lie algebra is a Poisson-Lie subalgebra.

LEMMA 1.1. *A Poisson-Lie subgroup is a Poisson submanifold* (*i.e., a union of symplectic leaves*). *The restricted Poisson structure makes it a Poisson-Lie group.*

CLAIM 1.3. *A restriction of a Poisson-Lie action on a Poisson-Lie subgroup is a Poisson-Lie action.*

§ 2. Gelfand-Dickey brackets

The topics of this section are inspired by the paper [8], where an action of the Lie algebra $\mathscr{D}$ of differential operators of an arbitrary order on the space $\mathscr{D}_n$ of differential operators of order n with the leading term d^n/dx^n (i.e., on the Gelfand-Dickey space) is defined.

DEFINITION 2.1. Denote by $\mathscr{D}$ the Lie algebra of differential operators

$$\sum_{k=0}^{K} f_k(x)\left(\frac{d}{dx}\right)^k$$

on the line.

DEFINITION 2.2. Let $\mathscr{D}_n$ denote the set of differential operators of order n with the leading coefficient 1:

$$\left(\frac{d}{dx}\right)^n + \sum_{k=0}^{n-1} f_k(x)\left(\frac{d}{dx}\right)^k.$$

In the above definitions we did not specify restrictions on the coefficient of differential operators. In what follows we will consider two important cases: in the first (the case of a line) we allow any (smooth) coefficients, in the second (the case of a circle) we consider only periodic coefficients.

As was shown by Radul, if we consider the second Gelfand-Dickey Poisson structure on the space $\mathscr{D}_n$ and the standard bialgebra structure on the Lie algebra $\mathscr{D}$, then the mentioned action becomes a Poisson-Lie action. This bialgebra structure is associated with the structure of a Lie algebra on the linear space $\mathscr{D}^*$ arising from the isomorphism of this space and the space of pseudodifferential symbols of negative order.

2.0. The second Gelfand-Dickey bracket on the set of differential operators. To define the isomorphism above let as define the space of pseudodifferential symbols (ΨDS) as an algebra of formal sums

$$f(x, \xi) = \sum_{-\infty < k \leqslant N} f_k(x) \xi^k$$

with the composition rule

$$f(x, \xi) \circ g(x, \xi) = \sum_{n=0}^{\infty} \frac{1}{n!} \left(\frac{\partial}{\partial \xi}\right)^n f(x, \xi) \left(\frac{\partial}{\partial x}\right)^n g(x, \xi)$$

and a function $\widetilde{\operatorname{Tr}}$ on 1-periodic pseudodifferential symbols as

$$\widetilde{\operatorname{Tr}}\Big(\sum_{k=-\infty}^{N} f_k(x)\xi^k\Big) = \int f_{-1}(x)\, dx.$$

The definition of $\widetilde{\operatorname{Tr}}$ makes sense in the periodic case (where the integral is taken over the period) or if the coefficients of the pseudodifferential operator are rapidly decreasing. The basic property of $\widetilde{\operatorname{Tr}}$,

$$\widetilde{\operatorname{Tr}}(f \circ g - g \circ f) = 0,$$

is the main reason for the notation we use. (7)

It is easy to see that the mapping $d/dx \mapsto \xi$ extends to the mapping from the algebra of differential operators into the algebra of pseudodifferential symbols. Hence we can change the letter ξ in the notation of a pseudodifferential symbol to d/dx without a risk of misinterpretation.

Now the formula

$$(f, g) \mapsto \widetilde{\operatorname{Tr}}(f \circ g)$$

defines a symmetric bilinear form on the vector space of pseudodifferential symbols. It is easy to see that this form is nondegenerate. Using this form we can identify the dual space to the space of differential operators with the space of pseudodifferential operators of degree $\leqslant -1$. Here again there is no problem in the periodic case, and in the case of a line we define a pairing between the vector space of all ΨDS and the vector space of ΨDS with compact support.

However, the usual dual space to a space of smooth functions is a space of generalized functions. Therefore the above formula identifies the dual space to differential operators with the space of ΨDS of negative order with generalized functions as coefficients. We are going to give a sketch of an explanation of how "to put this difficulty under the carpet" later, in 2.6.

(7) Another connection between the usual trace and the function $\widetilde{\operatorname{Tr}}$ is given by (2.7).

Here we consider only the linear functionals that correspond to ΨDS with smooth coefficients. Moreover, we pretend that any linear function on the space of ΨDS corresponds to such a functional. This is a usual difficulty in trying to define what an infinite-dimensional smooth manifold is.

Now consider a cotangent vector to $\mathscr{D}_n \subset \mathscr{D}$. The dual space to a subspace is a quotient of the space dual to the ambient space, and we can easily see that this covector corresponds to an element of $\Psi DS_{\leqslant -1}/\Psi DS_{\leqslant -n-1}$. Below we abuse our notations and call this element of the quotient a symbol too.

Let $L \in \mathscr{D}_n$, A and B be two cotangent covectors to $\mathscr{D}_n$ at the point L, represented by pseudodifferential symbols of negative order. Then the formula

$$(A, B) \mapsto \widetilde{\mathrm{Tr}}(A((LB)_+L - L(BL)_+)) \tag{2.1}$$

(where the subscript $+$ denotes taking the purely differential part of a ΨDS) determines a bivector field on $\mathscr{D}_n$ that induces (as a very cumbersome calculation shows) a Poisson structure on the manifold $\mathscr{D}_n$. In fact in what follows we will anyway give another description of this bivector field that will prove that the defined structure is indeed Poisson.

In the periodic case the existence of the right-hand side is clear, in the case of *all* functions on a line we can note that the symbols A and B are of finite support (since they represent covectors), hence the right-hand side is of finite support.

REMARK 2.1. Let us note that the structure (2.1) can be easily rewritten as

$$(A, B) \mapsto -\widetilde{\mathrm{Tr}}(A((LB)_-L - L(BL)_-)), \tag{2.2}$$

where $M_- = M - M_+$ is an "integral" part of a pseudodifferential operator M. In fact *to define* the second Gelfand-Dickey structure we need both these descriptions: the first one shows that the Hamiltonian mapping is given by the formula

$$B \mapsto (LB)_+L - L(BL)_+,$$

therefore the image is a differential operator, the second gives that it is

$$B \mapsto -((LB)_-L - L(BL)_-),$$

therefore it is a pseudodifferential operator of order $< n$. (And therefore a tangent vector to $\mathscr{D}_n$.)

In fact in what follows we use only the following property of this Hamiltonian mapping: it sends B to a differential operator of order $< n$ of the form

$$L(BL)_- - \text{something} \circ L$$

(i.e., it is a remainder of $L(BL)_-$ in right division by L).

Our task here is to give a description of this Poisson structure using classical r-matrices.

2.1. The identification with a Grassmannian. Here we define some Poisson structure on a Grassmannian $\mathrm{Gr}_k(C^\infty(\mathbb{R}))$ and show the relation between this Poisson structure and the Gelfand-Dickey structure. In turn, to define a Poisson structure on the Grassmannian we must define an r-matrix for the Lie algebra $\mathfrak{gl}(C^\infty(\mathbb{R}))$. Let us recall that an r-matrix for $\mathfrak{g}$ is an element of $\mathfrak{g} \otimes \mathfrak{g}$, therefore we need a definition of tensor square of an infinite-dimensional vector space. As usual in topology, there are some imminent difficulties in such a definition. We postpone the discussion of how we can deal with it until 2.6, and use for a while some ad hoc definitions of objects in question.

So let V be a space of smooth functions on $\mathbb{R}$. We will "take" as V^* the space of smooth 1-forms on $\mathbb{R}$ with a compact support. Hence as $\mathfrak{gl}(V)$ we "can" consider a space of smooth forms $f(x, y)\,dy$, such that

$$\forall M\, \exists N \quad f(x, y) = 0 \quad \text{if} \quad |x| \leqslant M,\, |y| \geqslant N.$$

As $\mathfrak{gl}(V)^*$ we "consider" a space of smooth forms $g(x, y)\,dx$, such that there exists $M > 0$ satisfying $f(x, y) = 0$ for $|x| \geqslant M$. The pairing between these two spaces is

$$(f(x, y)\,dy,\, g(x, y)\,dx) \mapsto \int_{\mathbb{R}\times\mathbb{R}} f(x, y) g(x, y)\,dx\,dy.$$

Similarly to (1.9) we can define a symmetric invariant bilinear form on $\mathfrak{gl}(V)^*$ as

$$(g_1(x, y)\,dy,\, g_2(x, y)\,dy) \overset{t}{\mapsto} \int_{\mathbb{R}\times\mathbb{R}} g_1(x, y) g_2(y, x)\,dx\,dy, \qquad (*)$$

and a skewsymmetric bilinear form on $\mathfrak{gl}(V)^*$ as

$$(g_1(x, y)\,dy,\, g_2(x, y)\,dy) \overset{r}{\mapsto} \int_{\mathbb{R}\times\mathbb{R}} g_1(x, y) g_2(y, x)\,\mathrm{sgn}(x-y)\,dx\,dy. \quad (2.3)$$

The same calculations as in the finite-dimensional case show that $t + r$ is a solution of the classical Yang-Baxter equation. Now the form (2.3) (being an r-matrix) determines a Poisson structure on the space $\mathrm{Gr}_k(V)$ and a Poisson-Lie structure on the corresponding group of invertible operators.

Let us emphasize that the above conditions on the supports are absolutely natural and arise from the operators being continuous in the corresponding spaces. The only tricks we did are tricks with smoothness conditions on the kernels.

REMARK 2.2. Now, when we have defined the r-matrix, we can explain why it cannot be defined as an element of the tensor square of some Lie algebra. Indeed, let us try to extend the definition (2.3) to g_1, g_2 being elements of the space dual to some algebra. Here we consider two cases: the algebras of continuous operators in two spaces of functions on the line, $C^\infty(\mathbb{R})$ and $C^{-\infty}(\mathbb{R})$. (The definitions of these topological vector spaces can be found in 2.6.)

The Lie algebra $\mathfrak{g}_1 = \mathfrak{gl}(C^\infty(\mathbb{R}))$ consists of operators with a kernel of the form $f(x, y)\,dy$ such that the wavefront of f is nonvertical on the (x, y)-plane and with the same condition on the support as before. Therefore the dual space contains operators with kernels $g(x, y)\,dy$ such that the wavefront of g is vertical and with the same condition on the support as above. Now in the formula $(*)$ we have a product $g_1 g_2^\perp$, with the first factor having a vertical wavefront, and the second factor having a horizontal one. Therefore the product is well defined as a generalized function and has a compact support, therefore the integral $(*)$ and, hence, tensor t is correctly defined. However, in the case of formula (2.3) we cannot pair this product with $\operatorname{sgn}(x - y)$, since $\operatorname{sgn}(x - y)$ is not a smooth function on a plane!

Similarly, if we consider $\mathfrak{g}_2 = \mathfrak{gl}(C^{-\infty}(\mathbb{R}))$, then we get the same result with a change of vertical and horizontal directions. However, if we consider $\mathfrak{g}_1^* \cap \mathfrak{g}_2^*$, as in 2.6, then the wavefront should be both horizontal and vertical, therefore any operator from this set has a smooth kernel, and the product with $\operatorname{sgn}(x - y)$ in formula (2.3) is well defined.

Theorem 2.1. *The above Poisson structure on* $\mathrm{Gr}_k(V)$ *coincides (on an open subset of* $\mathrm{Gr}_k(V)$*) with twice the second Gelfand-Dickey Poisson structure on the set* $\mathcal{D}_k$ *of differential operators of degree* k *with the leading term* $(d/dx)^k$ *under the identification*

$$\mathcal{D}_k \ni L \overset{s}{\mapsto} W = \{f \mid Lf = 0\}. \tag{2.4}$$

Proof. The two Poisson structures we want to compare are both given by their Hamiltonian mappings. However, these Hamiltonian mappings are described in different languages. In the ΨDS description we use the identification of the tangent space with the set of differential operators and of the cotangent space with pseudodifferential operators; in the Grassmannian description we identify them with the space of linear mappings $W \to V/W$ and $V/W \to W$ respectively. So the first two steps are to relate these two pairs of identifications. These steps take the longest part of the proof.

First of all we describe which mapping $W \to V/W$ corresponds to a variation δL of a differential operator L, $W = \operatorname{Ker} L$.

Lemma 2.1. *Consider the above identification of* $\mathcal{D}_k$ *with the subset of the Grassmannian* $\mathrm{Gr}_k(V)$. *Consider a tangent vector to* $\mathcal{D}_k$ *at* $L \in \mathcal{D}_k$. *Let* $\operatorname{Ker} L = W$. *We have two different representations of this vector: a differential operator* δL *and a mapping* $M : W \to V/W$. *Then*

$$M = -\pi \circ L^{-1} \delta L_{|_W},$$

where π *is the projection* $V \to V/W$.

Proof. Let us consider the equation

$$(L + \delta L)(f + \delta f) = 0.$$

It means that

$$\delta f = -L^{-1}\delta L \cdot f. \tag{2.5}$$

Hence the map $W \to V/W$ corresponding to δL is $-\pi \circ L^{-1}\delta L_{|_W}$, where π is the projection $V \to V/W$. Actually, the map L^{-1} is undefined, since the map L has a kernel. However, in our case this kernel is exactly W! Hence the composition $\pi \circ L^{-1}$ *is* correctly defined. (The map L is obviously a surjection.) Without abusing the notations we can denote $\pi \circ L^{-1}$ as L^{-1}. Now we can interpret (2.5) as two descriptions of the tangent space being connected by

$$\delta L \mapsto -L^{-1}\delta L_{|_{W/W}}.$$

Here for a mapping $B\colon V \to V$ we denote by $B_{|_{W/W}}$ the corresponding mapping $W \to V/W$.

Second, let us describe the relations on the cotangent space.

LEMMA 2.2. *Consider the identification of $\mathscr{D}_k$ with the subset of the Grassmannian $\mathrm{Gr}_k(V)$. Consider a cotangent vector to $\mathscr{D}_k$ at $L \in \mathscr{D}_k$. Let $\operatorname{Ker} L = W$. We have two different representations of this covector: a pseudodifferential symbol A and a mapping $A'\colon V/W \to W$. Consider A' as a finite-dimensional mapping $V \to V$. Then*

$$A = \textit{the symbol of } -\tfrac{1}{2}\tilde{r}(A' \circ L^{-1}).$$

PROOF. The identification above shows that

$$\widetilde{\operatorname{Tr}} A \circ \delta L = -\operatorname{Tr} A' \circ L^{-1}\delta L \tag{2.6}$$

for any δL. We are going to use this equation to express A in terms of A', therefore we need here a relation between functionals $\widetilde{\operatorname{Tr}}$ and Tr. However, this relation is very simple:

$$\operatorname{Tr} K(x, y)\,dy = \widetilde{\operatorname{Tr}}\left(\tfrac{1}{2}\operatorname{sgn}(x-y)K(x, y)\,dy\right) \tag{2.7}$$

for a smooth kernel $K(x, y)\,dy$. Some comments about what is in the right-hand side follow.

So far we defined only the notion of a pseudodifferential *symbol*. However, in a similar way we could define a notion of a pseudodifferential operator. It is the sum of a differential operator and of an operator of negative order. In turn, an operator of negative order is an operator with a piecewise-smooth kernel $L(x, y)\,dy$ with the only jump (of the first kind) on the diagonal $x = y$ (as the operator in the right-hand side of (2.7)). It is easy to construct a homomorphism from operators to symbols (this homomorphism is uniquely determined by the fact that it kills the set of operators with smooth kernels).

Now we prove that the symbol corresponding to the kernel $\frac{1}{2}\operatorname{sgn}(x-y)\cdot K(x, y)\,dy$ is $K(x, x)\partial^{-1} + O(\partial^{-2})$, proving the above formula. Indeed, if $K(x, y)\,dy$ is a kernel of a finite-dimensional operator $K\colon V \to V$, we can write $\operatorname{Tr} K \stackrel{\text{def}}{=} \operatorname{Tr}(K|_{\operatorname{Im} K})$ as $\int K(x, x)\,dx$. This formula extends naturally to operators with smooth kernel and satisfies any natural relation that is valid in the finite-dimensional case. In fact we can rewrite (2.7) as

$$\operatorname{Tr} K = \tfrac{1}{2}\widetilde{\operatorname{Tr}}\,\widetilde{r}(K).$$

In what follows we use formula (2.7) not only for smooth K, but also for operators of finite rank. The only difference in the proof of this modification is the reference to the wavefront of K being horizontal.

Hence we can rewrite (2.6) as

$$\widetilde{\operatorname{Tr}} A \circ \delta L = -\tfrac{1}{2}\widetilde{\operatorname{Tr}}\,\widetilde{r}(A' \circ L^{-1}\delta L).$$

(Since the operator A' is of finite rank, the operator in brackets also is of finite rank, therefore we can apply the formula (2.7) to it.)

I claim that A is the *symbol* of the *operator* $-\frac{1}{2}\widetilde{r}(A' \circ L^{-1})$. Indeed, if B is an operator with a smooth kernel and M is a differential operator, we can note that the kernels of $\widetilde{r}(B \circ M)$ and $\widetilde{r}(B) \circ M$ coincide outside of the diagonal, therefore the former is an integral part of the latter:

$$\widetilde{r}(B \circ M) = (\widetilde{r}(B) \circ M)_{-}. \tag{2.8}$$

Therefore values of $\widetilde{\operatorname{Tr}}$ on $\widetilde{r}(B \circ M)$ and on $\widetilde{r}(B) \circ M$ are the same. We obtain

$$\widetilde{\operatorname{Tr}} A \circ \delta L = -\tfrac{1}{2}\widetilde{\operatorname{Tr}}\,\widetilde{r}(A' \circ L^{-1}) \circ \delta L,$$

which proves the assertion. Now we can write the second translation formula connecting two descriptions of the cotangent space: let a cotangent vector to $\mathcal{D}_n$ correspond both to a pseudodifferential symbol A of negative order and to an operator $A' : V/W \to W$. Then ([8])

$$A = \text{the symbol of } -\tfrac{1}{2}\widetilde{r}(A' \circ L^{-1}).$$

Now we can begin the comparison of the Hamiltonian mappings themselves, since we know the translation formulas between the languages they are defined in. The Hamiltonian mapping defined in terms of a Grassmannian sends a cotangent vector represented by $A' : V/W \to W$ to the tangent vector to the Grassmannian represented by $\widetilde{r}(A')_{|_{W/W}} : W \to V/W$. From the other side, the pseudodifferential *symbol* A corresponding to the cotangent vector is already expressed in terms of the pseudodifferential *operator* $\mathbf{A} = \widetilde{r}(A' \circ L^{-1})$. Therefore we can write $\widetilde{r}(A')$ in terms of $\mathbf{A}$:

$$\widetilde{r}(A') = \widetilde{r}(A' \circ L^{-1} \circ L) = (\widetilde{r}(A' \circ L^{-1}) \circ L)_{-} = -2(\mathbf{A} \circ L)_{-}$$

(we used (2.8)). Therefore to find the differential operator δL corresponding to $\widetilde{r}(A')_{|_{W/W}}$ we need to solve the equation

$$-L^{-1}\delta L_{|_W} = -2(\mathbf{A} \circ L)_{-|_{W/W}}.$$

([8]) In fact we cannot claim that two sides of this formula are *exactly* equal. They are equal mod $\Psi DS_{\leqslant -n-1}$, since a cotangent vector to $\mathcal{D}_n$ is an element of $\Psi DS_{\leqslant -1}/\Psi DS_{\leqslant -n-1}$.

It is equivalent to

$$\delta L_{|_W} = 2L \circ (\mathbf{A} \circ L)_{-|_W}.$$

Therefore to find δL we need to find a differential operator of order $< n$ that coincides with $2L \circ (\mathbf{A} \circ L)_-$ when considered on W. Now the same arguments as in Remark 2.1 show that the operator

$$-2((L\mathbf{A})_- L - L(\mathbf{A}L)_-)$$

is a differential operator of order $< n$ and that it coincides with $2L \circ (\mathbf{A} \circ L)_-$ when restricted on $W = \operatorname{Ker} L$. Moreover, we can change operators to symbols in this formula, since the result is a differential operator anyway. Therefore the Hamiltonian mapping associated with the Poisson structure on the Grassmannian is indeed a multiple of the Hamiltonian mapping for the second Gelfand-Dickey structure under the above identification.

REMARK 2.3. It is easy to see that the identification of the theorem sends the space $\mathscr{D}_k$ of differential operators to an open subset of the Grassmannian. In fact we have extended the second Gelfand-Dickey structure to a "compactification" of the space $\mathscr{D}_k$.

COROLLARY 2.1. *On the space $\mathscr{D}_k$ of differential operators with the second Gelfand-Dickey Poisson structure a (local) action of a Poisson-Lie group* $\mathrm{GL}(C^\infty(\mathbb{R}))$ *is defined.*

COROLLARY 2.2. *The bracket* (2.1) *satisfies the Jacobi identity.*

PROOF. Indeed, it coincides with the Poisson bracket on the Grassmannian.

2.2. A Poisson-Lie algebra of differential operators as a Poisson-Lie subalgebra of $\mathfrak{gl}$. The previous corollary shows that there is a Poisson-Lie action of an enormous group GL(functions on $\mathbb{R}$) on (an extension of) the second Gelfand-Dickey structure. However, as we have already said, Radul defined a Poisson-Lie action of a Lie algebra of *differential* operators on the same space. Since these two actions are compatible, the natural conjecture is that the latter algebra is a Poisson-Lie subalgebra of the Lie algebra of GL(functions on $\mathbb{R}$).

THEOREM 2.2. *The subspace $\mathscr{D}$ of differential operators is a Lie-Poisson subalgebra of the Poisson-Lie algebra* $\mathfrak{gl}$*(functions on* $\mathbb{R}$*). The Poisson-Lie structure on this space coincides with the structure defined in* 2.0.

PROOF. We want to prove that the Lie algebra $\mathscr{D}$ of differential operators is a Lie-Poisson subalgebra (in the sense of 1.5.1) of the Lie algebra of continuous operators in the space of functions on $\mathbb{R}$. It is easy to see that $\mathscr{D}^\perp \subset \mathfrak{gl}(\text{functions on } \mathbb{R})^* = \{f(x, y)\,dy\}$ coincides with the space of forms $f(x, y)\,dy$ that vanish together with any derivative on the diagonal $x = y$.

On the other hand, it is easy to see that the Lie-algebraic structure on the coalgebra $\mathfrak{gl}(\text{functions on } \mathbb{R})^*$ is connected with a decomposition of an element f of this coalgebra into a sum of "increasing" and "decreasing" Volterra operators:

$$\begin{aligned} f(x, y)\,dy &= \frac{(\operatorname{sgn}(x-y)+1)}{2} f(x, y)\,dy + \frac{(-\operatorname{sgn}(x-y)+1)}{2} f(x, y)\,dy \\ &= f_+(x, y)\,dy + f_-(x, y)\,dy. \end{aligned}$$

Using this decomposition we can describe the commutator of two elements $f, g \in \mathfrak{gl}(\text{functions on } \mathbb{R})^*$ as

$$[f, g] = [f_+, g_+] - [f_-, g_-]$$

and to determine the bracket on the spaces of increasing (and decreasing) Volterra operators as a usual Lie bracket:

$$[f_\pm, g_\pm](x, y)\,dy = \left(\int (f_\pm(x, z)g_\pm(z, y) - g_\pm(x, z)f_\pm(z, y))\,dz\right) dy.$$

A simple calculation shows that these two formulae indeed determine a bracket with values in smooth functions on $\mathbb{R} \times \mathbb{R}$:

LEMMA 2.3. *The form* $[f, g]$ *is smooth.*

PROOF. It is obvious that the forms $[f_+, g_+]$ and $[f_-, g_-]$ are smooth outside of the diagonal and up to it as compositions of Volterra operators with smooth kernels. Hence the form $[f, g]$ can have a jump of the first kind on the diagonal maximum. To prove that it has no jump let as consider *a symbol* of this kernel considered as a pseudodifferential operator. It is sufficient to show that this symbol is 0.

Again, for this it is sufficient to show that the operators $[f_+, g_+]$ and $[f_-, g_-] = [-f_-, -g_-]$ have equal symbols. However, the kernels of operators f_+ and $-f_-$ differ on a smooth function f, hence have equal symbols. Therefore the commutators in question also have equal symbols.

Now it is almost clear that the space of forms $f(x, y)\,dy$ with the zero ∞-jet on the diagonal is an ideal with respect to this bracket. Indeed, it is sufficient to show that the bracket with, say, a lowering Volterra operator g_- satisfies this condition. Now the corresponding kernels $f_\pm$ are smooth on the whole $\mathbb{R} \times \mathbb{R}$ including diagonal, hence have zero symbol, hence their brackets with an arbitrary Volterra operator (of the same direction) also have zero symbols, hence are smooth on the whole $\mathbb{R} \times \mathbb{R}$ including the diagonal. Moreover, $[f_+, g_-] = 0$, and the kernel $[f_-, g_-]$ is again Volterra, therefore the symbol corresponds to the jet on the diagonal. Therefore the kernel of $[f, g_-]$ has a vanishing ∞-jet on the diagonal.

Therefore the Lie subalgebra $\mathscr{D}$ of differential operators is indeed a Poisson-Lie subalgebra. To determine the corresponding Poisson bracket on the dual space we should consider a structure of the quotient-algebra on

$$\mathscr{D}^* = \mathfrak{gl}(\text{functions on } \mathbb{R})^*/\mathscr{D}^\perp.$$

The latter space is a space of ∞-jets of functions along the diagonal, and the corresponding Lie-algebraic structure is the quotient structure of the Lie algebra of (say, increasing) Volterra operators with kernels that are smooth outside the diagonal by the ideal of Volterra operators with kernels smooth everywhere. However, this is nothing else but the algebra of pseudodifferential symbols of an order $\leqslant -1$.

PROPOSITION 2.1. *The action of the differential operator $X \in \mathscr{D}$ on the differential operator $L \in \mathscr{D}_k$ is given by the left remainder of $-LX$ modulo L:*

$$X \cdot L = -LX + YL,$$

where $Y \in \mathscr{D}$ is chosen in such a way that the right-hand side is of order $< k$:

$$Y = (LXL^{-1})_+.$$

Here the subscript $+$ denotes the differential part.

PROOF. By definition, if $Lf = 0$, then $(L + \varepsilon X \cdot L)(f + \varepsilon Xf) = O(\varepsilon^2)$. This means that $X \cdot L + L \cdot X$ is 0 on any function f such that $Lf = 0$. Therefore $X \cdot L + L \cdot X = YL$ for some Y.

2.3. The periodic case. We have considered a Poisson structure on the Gelfand-Dickey manifold in the case of operators on the line. However, it is known that the absolutely parallel theory exists in the case of periodic differential operators. Here we discuss the changes we should include in the definition of the Grassmannian and the group $\mathrm{GL}(C^\infty(\mathbb{R}))$ to use the natural identification of the set of periodic differential operators with a subset of the Grassmannian. Such an identification obviously exists, since the periodic differential operator is automatically a differential operator on a line. We get a natural subset of the Grassmannian corresponding to operators with periodic coefficients, that consists of invariant subspaces with respect to the action of $\mathbb{Z}$ by translations on the line. However, we want to consider this space as a space with an action of some analogue of $\mathrm{GL}(C^\infty(\mathbb{R}))$.

It is easy to see that in this way we can generalize the above construction to the periodic case. The main difficulty is to define the corresponding version of the group of matrices $\mathrm{GL}(C^\infty(\mathbb{R}))$. However, consider a natural action of the group $\mathbb{Z}$ on $C^\infty(\mathbb{R})$ by translations. Denote by $\mathrm{GL}(C^\infty(\mathbb{R}))^{\mathbb{Z}}$ the subgroup of operators that commute with $\mathbb{Z}$. Now if $f(x)$ is a function satisfying the condition $f(x+1) = \lambda f(x)$, $\lambda \neq 0$, then Af for $A \in \mathrm{GL}(C^\infty(\mathbb{R}))^{\mathbb{Z}}$ also satisfies this condition. Denote by $\mathrm{Gr}_k(C^\infty(\mathbb{R}))^{\mathbb{Z}}$ the subspace of $\mathrm{Gr}_k(C^\infty(\mathbb{R}))$ consisting of $\mathbb{Z}$-invariant subspaces in $C^\infty(\mathbb{R})$. For example, the Grassmannian Gr_1 consists of spaces spanned by functions f satisfying the above condition with an arbitrary λ. If $V \in \mathrm{Gr}_k(C^\infty(\mathbb{R}))^{\mathbb{Z}}$, then the translation operator can be restricted to V. This determines an "invariant" of V, which is the conjugacy class of this restriction. In the case of Gr_1 this invariant is equal to λ.

Now the group $\mathrm{GL}(C^\infty(\mathbb{R}))^{\mathbb{Z}}$ acts on the space $\mathrm{Gr}_k(C^\infty(\mathbb{R}))^{\mathbb{Z}}$ and this action preserves the above "invariant". It is clear that the orbits of this action are numerated by this invariant.

The dual space to the Lie algebra $(\mathfrak{gl}(C^\infty(\mathbb{R}))^{\mathbb{Z}})^*$ (almost—see above) has a natural decomposition into the sum of the space of upper-triangular (lowering Volterra) operators and lower-triangular (raising Volterra) operators. The corresponding r-matrix is nevertheless well defined. Therefore, we can define a structure of the Poisson-Lie group on $\mathrm{GL}(C^\infty(\mathbb{R}))^{\mathbb{Z}}$ and the corresponding Poisson structures on the homogeneous spaces for this group, say, on orbits of action on $\mathrm{Gr}_k(C^\infty(\mathbb{R}))^{\mathbb{Z}}$.

Taking the union of these structures, we get a Poisson structure on the entire manifold $\mathrm{Gr}_k(C^\infty(\mathbb{R}))^{\mathbb{Z}}$. Consider now the differential operator that corresponds to a subspace $V \in \mathrm{Gr}_k(C^\infty(\mathbb{R}))^{\mathbb{Z}}$. It is easy to see that this operator is periodic, therefore we can consider the (periodic case of the) second Gelfand-Dickey Poisson structure on this set. The word-by-word repetition of the above arguments proves the following result.

THEOREM 2.3. *The Poisson structure on the $\mathbb{Z}$-invariant Grassmannian coincides with the second Gelfand-Dickey Poisson structure on the set of periodic differential operators. There is a (local) Poisson-Lie action of the group $\mathrm{GL}(C^\infty(\mathbb{R}))^{\mathbb{Z}}$ on the latter Poisson manifold.*

COROLLARY 2.3. *The differential operators with periodic coefficients form a Poisson-Lie subalgebra in the bialgebra $\mathfrak{gl}(C^\infty(\mathbb{R}))^{\mathbb{Z}}$. The action on the periodic Grassmannian is a Poisson-Lie action.*

We gained a lot because we considered not the operators in the space of functions on a circle, but the $\mathbb{Z}$-invariant operators on the line. Indeed, the kernel of the operator of the former type is a "function" on $S^1 \times S^1$, and there is no decomposition on the upper- and lower-triangular operators. However, in the case considered above the kernel is a function $K(x, y)\,dy$ such that

$$K(x, y) = K(x+1, y+1),$$

and the decomposition exists.

2.4. The matrix case. Here we want to make a little remark that what we have done so far can be easily generalized to the case of matrix operators. Instead of $\mathrm{GL}(C^\infty(\mathbb{R}))$ we must consider $\mathrm{GL}(C^\infty(\mathbb{R}, \mathbb{R}^n))$, where $C^\infty(\mathbb{R}, \mathbb{R}^n)$ is the space of vector-valued functions on $\mathbb{R}$. Instead of $\mathrm{Gr}_k(C^\infty(\mathbb{R}))$ we should consider the Grassmannian $\mathrm{Gr}_{nk}(C^\infty(\mathbb{R}, \mathbb{R}^n))$. Then instead of the set $\mathscr{D}_k$ of differential operators of order k we get the set of differential operators of order k with matrix coefficients of the size $n \times n$.

Similarly to the above, we identify the second Gelfand-Dickey Poisson structure on the set of matrix differential operators (it is given by the same formula (2.1)) of order k with the Poisson structure on the Grassmannian $\mathrm{Gr}_{kn}(C^{\infty}(\mathbb{R}, \mathbb{R}^n))$.

In contrast with the scalar case the case $k = 1$ is already very interesting. The set of differential operators in question is

$$\{\partial + A(x)\},$$

where A is a matrix function of x. Let us consider the periodic case. In this case this set is nothing else but a hyperplane $c = 1$ in the dual space to a central extension of the algebra of currents with values in $\mathfrak{gl}_n$. The Kirillov-Lie Poisson structure on this leaf coincides with the second Gelfand-Dickey Poisson structure, hence on the hyperplane $c = 2$ the Kirillov-Lie Poisson structure coincides with twice this structure. Therefore we get the following

COROLLARY 2.4. *There is a natural (local) Poisson-Lie action of the Poisson-Lie group* $\mathrm{GL}(C^{\infty}(\mathbb{R}, \mathbb{R}^n))^{\mathbb{Z}}$ *on a hyperplane* $c = 2$ *of the Kirillov-Lie Poisson structure on the dual space to the matrix current algebra.*

Let us consider this action a little bit more carefully. We can consider the subgroup of currents in $\mathrm{GL}(C^{\infty}(\mathbb{R}, \mathbb{R}^n))^{\mathbb{Z}}$, i.e., of operators of multiplication by a periodic matrix-valued function. It is easy to verify that the action of this subgroup coincides with the coadjoint action of this group on the dual space to the Lie algebra. (It is clear that the central extension acts in a trivial way, so we can consider the coadjoint action of the nonextended group.)

Actually, this subgroup is a Poisson-Lie subgroup with the trivial Poisson structure. This is compatible with the fact that the coadjoint action preserves the Kirillov-Lie Poisson structure. Moreover, we can consider the momentum mapping for the action of this subgroup on $\{\partial + A\}$ considered as an abstract Poisson manifold. It is easy to check that this mapping is identical on this manifold. Therefore the action of $\mathrm{GL}(C^{\infty}(\mathbb{R}, \mathbb{R}^n))^{\mathbb{Z}}$ on this (Poisson) manifold makes it possible to reconstruct the inclusion of this manifold into the dual space to the Lie algebra.

Moreover, we can also consider the Lie subalgebra of periodic vector fields on the line

$$\mathrm{Vect} \subset \mathfrak{gl}(C^{\infty}(\mathbb{R}, \mathbb{R}^n))^{\mathbb{Z}}.$$

This Lie subalgebra is a Poisson-Lie subalgebra of the Lie algebra of periodic differential operators, and, moreover, the corresponding Lie algebra bracket on the dual space is 0. (So the situation is similar to the above action of currents.) Therefore the action of this subalgebra is a trivial case of the Poisson-Lie action: it preserves the Poisson structure on the Grassmannian.

2.5. A conjecture on quantization: the Kac-Moody case. Let us consider a would-be quantization of Corollary 2.4. The quantization of the hyperplane in the dual space to the Lie algebra is something like the enveloping algebra with a fixed central charge. The quantization of the action of Vect on this hyperplane is a Sugavara inclusion, that gives us an (adjoint) action of the

Virasoro algebra on the enveloping algebra of the central extension of the currents algebra. In what follows we consider the Sugavara inclusion as an inclusion into the space of (inner) derivatives of this universal enveloping algebra.

Let us look on a possible quantization of the action of $\mathfrak{gl}(C^\infty(\mathbb{R}, \mathbb{R}^n))^{\mathbb{Z}}$. This should be an action of some big algebra on the enveloping algebra, but not by differentials, as in the case of Sugavara inclusion (where differentials were inner), but by transformations changing the multiplication law. (The Sugavara inclusion acts by differentials only because the action of Vect preserves the Poisson structure, which, in turn, is explained by the trivial Lie algebra structure on Vect^*.) Indeed, the quantization of

$$G \times X \to X$$

is (on the language of deformations $\mathscr{O}_q(X)$ of the algebra $\mathscr{O}(X)$ of functions on X)

$$\mathscr{O}_q(X) \to \mathscr{O}_q(X) \otimes \mathscr{O}_q(G),$$

or

$$\mathscr{O}_q(G)^* \otimes \mathscr{O}_q(X)^* \to \mathscr{O}_q(X)^*$$

(i.e., an action of $\mathscr{O}_q(G)^*$ in $\mathscr{O}_q(X)^*$), or, what is the same, an action of $\mathscr{O}_q(G)^*$ in $\mathscr{O}_q(X)$:

$$\mathscr{O}_q(G)^* \otimes \mathscr{O}_q(X) \to \mathscr{O}_q(X).$$

If the action of G is local (as above), we should change G to the infinitesimal neighborhood of $e \in G$, which results in the change of $\mathscr{O}_q(G)^*$ to $U_q(\mathfrak{g})$. In our situation we get the following

CONJECTURE. *There is an action*

$$U_q(\mathfrak{gl}(C^\infty(\mathbb{R}, \mathbb{R}^n))^{\mathbb{Z}}) \otimes U(\widehat{\mathfrak{gl}}_{n, c=2\varphi(q)}) \to U(\widehat{\mathfrak{gl}}_{n, c=2\varphi(q)})$$

such that the corresponding mapping

$$U_q(\mathfrak{gl}(C^\infty(\mathbb{R}, \mathbb{R}^n))^{\mathbb{Z}}) \otimes U(\widehat{\mathfrak{gl}}_{n, c=2\varphi(q)})^* \to U(\widehat{\mathfrak{gl}}_{n, c=2\varphi(q)})^*$$

is a morphism of coalgebras. Here φ is an unknown function, $\varphi'(1) = 1$.

REMARK 2.4. Due to usual anomalies, a possible modification is a change of some "obvious" value of $2\varphi(q)$ to $2\varphi(q) - h$, where h is the dual Coxeter number. It is interesting to compare this conjecture with the results of [4].

What we constructed in this paper is a prequantization of this action and a generalization of this prequantization to W_n-algebras.

2.6. The topological approach. In 2.0 and 2.1 we promised to give an explanation of how to modify the definitions of objects we consider to identify the space of linear functionals on smooth functions with the set of smooth functions itself. Let us recall that in the usual sense this set is the set of generalized functions. However, to define the Poisson structure on the sets we consider we need these functionals to be smooth functions.

Essentially we say that to avoid this difficulty we should consider a set of smooth functions on the set of pseudodifferential symbols that is much more fine than the set of continuous functions. Let us recall that by usual definition the set of *smooth linear* functions on a topological vector space V coincides with the set of *continuous linear* functions V^*. If we drop this restriction, then we get a much larger number of possible sheaves of smooth functions. A choice of such an object determines a subset of smooth linear functions

$$V^*_{\text{smooth}} \subset V^*.$$

The dual picture is the inclusion

$$V \subset V_{\text{antismooth}} \stackrel{\text{def}}{=} (V^*_{\text{smooth}})^*.$$

If $(V^*_{\text{smooth}})^{**} = V^*_{\text{smooth}}$, and $V^{**} = V$, then the latter inclusion determines the former. In this case we (by definition) call a function on V a smooth function if it is a restriction on V of a smooth "in a usual sense" function on $V_{\text{antismooth}}$. It is usually expressed by saying that we consider two different topologies on V (introduction of a different (and weaker) topology automatically adds additional elements to a vector space, therefore the vector space increases to $V_{\text{antismooth}}$).

Therefore we should work in the category of *rigged topological vector spaces*, i.e., spaces with several topologies. Here we define the inclusion

$$V \subset V_{\text{antismooth}}$$

corresponding to the Gelfand-Dickey structure. First of all, if we fix a topology on the space of allowed coefficients, then we can consider on the space of symbols the topology of a *direct limit of inverse limits*, i.e., if we denote by ΨDS_m the space of ΨDS of degree no more than m, then

$$\Psi DS = \operatorname*{ind\,lim}_{N\to\infty} \operatorname*{proj\,lim}_{M\to\infty} \Psi DS_N / \Psi DS_{-M}.$$

Since the space $\Psi DS_N/\Psi DS_{-M}$ is a finite product of spaces of coefficients, fixing a topology on smooth functions we fix a topology on ΨDS. Now we consider two topologies on the space of functions on the line:

$$C^\infty(\mathbb{R}) \subset C^{-\infty}(\mathbb{R})$$

(and the same for functions with finite support). Here $C^\infty(\mathbb{R})$ is a space of smooth functions on $\mathbb{R}$ with a coarse topology (i.e., with a topology of the projective limit corresponding to an increasing family of intervals), $C^{-\infty}(\mathbb{R})$ is the dual space to the space $C_0^\infty(\mathbb{R})$ of smooth functions with compact support, with the topology of an inductive limit with respect to this family of intervals (the space $C_0^\infty(\mathbb{R})$ is often denoted as $\mathscr{D}$, and the space $C^{-\infty}(\mathbb{R})$ as $\mathscr{D}'$).

This pair of topologies on functions defines a pair of topologies on the space of ΨDS. Taking the pair $V_+ \subset V_-$ as a dual space in this category to the pair $V_-^* \subset V_+^*$, we get a formal justification of what was said above: a differential of a "smooth function" on V_+, i.e., a smooth function on V_-, corresponds in a way described above to a pseudodifferential symbol with smooth coefficients, and not with generalized coefficients as in the naïve theory.

In the same way we can consider the ΨDS with compact support (or with rapidly decreasing coefficients) by introducing the pair $C_c^\infty(\mathbb{R}) \subset C_c^{-\infty}(\mathbb{R})$. ([9]) Such a play with topologies is standard in the theory of infinite-dimensional manifolds. Several other examples will be given in what follows. We should note that any smooth function on the smaller set of ΨDS (i.e., on the set of ΨDS with smooth coefficients) extends uniquely to the bigger one (i.e., of ΨDS with generalized coefficients) by definition, however, only the smaller set carries an algebra structure. ([10])

Now we discuss the (much simpler) theory for the group $\mathrm{GL}(C^\infty(\mathbb{R}))$. We should note however, that the only explanation we are going to give is the explanation of the choice of the dual space to the Lie algebra of this group. We cannot explain even such simple questions as what the relation between the Lie algebra we consider and the Lie group we consider is, or what the topology on this Lie algebra is. The only hope we have now is that the discussion below can justify the choice of the space we considered in 2.1 as a dual space to the Lie group. However, if we could find some good set of smooth functions on $\mathrm{GL}(C^\infty(\mathbb{R}))$ in the same sense as below, then the hypothetical Hopf algebra structure on this vector space could justify the algebraic discussion above. This is related to the fact that if we have a good description of functions on the group, and, in particular, of the dual space to the Lie algebra, then the language of Hopf algebras allows one to live quite comfortably without a reference to the Lie algebra.

We begin the discussion with a remark that formula (2.3) determines the r-matrix from the algebraic point of view. However, to be able to apply all the machinery described above we should verify that this is indeed an r-matrix, therefore we should show that this formula has something to do with $\mathfrak{g} \otimes \mathfrak{g}$. The problem is that this element is "outside" of any naturally defined notion of the tensor square for any naturally defined notion of the vector space $\mathfrak{gl}(C^\infty(\mathbb{R}))$.

([9]) This should be the choice if we considered the second Gelfand-Dickey structure in the context of integrable systems, since the usual *local Hamiltonians* are defined on this space.

([10]) Let us give here another example of a similar situation. In the theory of quantum groups one considers an algebra (or a Hopf algebra) of generalized functions on a group G with support in $e \in G$. This space carries a topology of an inductive limit. However, if one considers the dual Hopf algebra, it is not the dual space of ∞-jets of functions at $e \in G$, which carries a topology of a projective limit, but the subspace spanned by the matrix coefficients in finite-dimensional representation, that carries a topology of an inductive limit. If the group G is a linear algebraic group, the latter space coincides with polynomial functions on G.

In the finite-dimensional case, the dual space of a subspace is "smaller" than the dual space of the ambient space. The same is true for closed subspaces in the topological case. However, for a dense inclusion the dual space of the subspace is larger than the dual space of the ambient space. On the other hand, the largest possible tensor square is the space of bilinear functionals on the dual space. Therefore to put the given element into $\mathfrak{g} \otimes \mathfrak{g}$ we can take the smallest possible $\mathfrak{g}^*$, i.e., the largest possible $\mathfrak{g}$, and the largest possible notion of the tensor square. However, this is not sufficient.

To improve the situation we use the following trick: we take two different definitions of $\mathfrak{gl}(C^\infty(\mathbb{R}))$ that result in two topological vector spaces with an intersection that is dense in both spaces. We consider *the sum* of these two vector spaces as $\mathfrak{g}$. More precisely, we consider a linear functional on this sum to be an element of $\mathfrak{g}^*$. Although this sum does not carry a structure of a Lie algebra, however, below we propose an interpretation of this trick.

Denote these two Lie algebras by $\mathfrak{g}_1$ and $\mathfrak{g}_2$. Consider the corresponding Lie groups G_1 and G_2 and the group $G_{12} = G_1 \cap G_2$. Now call a function on G_{12} a *smooth function* if this function can be extended as a smooth function to both G_1 and G_2. ([11]) In this definition the tangent space to G_{12} at e is $\mathfrak{g}_1 \cap \mathfrak{g}_2$, but the cotangent space (i.e., the space of values of differentials of functions) is $\mathfrak{g}_1^* \cap \mathfrak{g}_2^* = (\mathfrak{g}_1 + \mathfrak{g}_2)^*$. Therefore if r is a bilinear functional on $\mathfrak{g}_1^* \cap \mathfrak{g}_2^*$, we can compute the value of r on any pair of differentials of functions. Therefore we can eventually define the Poisson bracket of (smooth!) functions on $G_1 \cap G_2$. Now we proceed with the description of the spaces $\mathfrak{g}_1$ and $\mathfrak{g}_2$.

If the pair $C^\infty(\mathbb{R}) \subset C^{-\infty}(\mathbb{R})$ represents the space of functions on a line, then instead of a group of continuous operators in the space of functions on $\mathbb{R}$ we should use the set of continuous operators in this pair, i.e., the set of pairs of maps

$$T_\infty, T_{-\infty}, \qquad T_\infty : C^\infty(\mathbb{R}) \to C^\infty(\mathbb{R}), \quad T_{-\infty} : C^{-\infty}(\mathbb{R}) \to C^{-\infty}(\mathbb{R}),$$
$$T_{-\infty}|_{C^\infty(\mathbb{R})} = T_\infty.$$

Hence the group we consider is $\mathrm{GL}(C^\infty(\mathbb{R})) \cap \mathrm{GL}(C^{-\infty}(\mathbb{R}))$, and the pair of group inclusions is

$$G_{12} = \mathrm{GL}(C^\infty(\mathbb{R})) \cap \mathrm{GL}(C^{-\infty}(\mathbb{R})) \subset \mathrm{GL}(C^\infty(\mathbb{R})) = G_1,$$
$$G_{12} = \mathrm{GL}(C^\infty(\mathbb{R})) \cap \mathrm{GL}(C^{-\infty}(\mathbb{R})) \subset \mathrm{GL}(C^{-\infty}(\mathbb{R})) = G_2.$$

In a similar way, instead of the dual space to a Lie algebra of continuous

([11]) Of course, we need to know first what a smooth function on, say, G_1 is. However, here we skip this discussion (until a better time).

operators in the space of functions we take the dual to a pair of inclusions

$$\mathfrak{g}_{12} = \mathfrak{gl}(C^{\infty}(\mathbb{R})) \cap \mathfrak{gl}(C^{-\infty}(\mathbb{R})) \subset \mathfrak{gl}(C^{\infty}(\mathbb{R})) = \mathfrak{g}_1,$$
$$\mathfrak{g}_{12} = \mathfrak{gl}(C^{\infty}(\mathbb{R})) \cap \mathfrak{gl}(C^{-\infty}(\mathbb{R})) \subset \mathfrak{gl}(C^{-\infty}(\mathbb{R})) = \mathfrak{g}_2,$$

i.e., the pair of inclusions

$$(\mathfrak{gl}(C^{\infty}(\mathbb{R})) \cap \mathfrak{gl}(C^{-\infty}(\mathbb{R})))^* \supset \mathfrak{gl}(C^{\infty}(\mathbb{R}))^*,$$
$$(\mathfrak{gl}(C^{\infty}(\mathbb{R})) \cap \mathfrak{gl}(C^{-\infty}(\mathbb{R})))^* \supset \mathfrak{gl}(C^{-\infty}(\mathbb{R}))^*.$$

We need all this machinery to find an appropriate definition of the space

$$\Lambda^2(\mathfrak{gl}(\text{functions on } \mathbb{R})).$$

Without it we could define the formulae for the r-matrix we want to construct, but we could not motivate a choice of (a large) space this matrix lies in. This is a usual difficulty in functional analysis. It is difficult to define a notion of a tensor product of a pair of spaces. It *is* possible in the case when one of these spaces is a *nuclear* space. However, in our case we want these spaces to be Lie algebras, and these two conditions (to be a Lie algebra and a nuclear space) seem to contradict each other in this particular case (at least we could not invent anything feasible).

So now we can *define* the space $\Lambda^2(\mathfrak{gl}(\text{functions on } \mathbb{R}))$ to be the space of skewsymmetric continuous bilinear forms on the space

$$\mathfrak{gl}(C^{\infty}(\mathbb{R}))^* \cap \mathfrak{gl}(C^{-\infty}(\mathbb{R}))^* \subset (\mathfrak{gl}(C^{\infty}(\mathbb{R})) \cap \mathfrak{gl}(C^{-\infty}(\mathbb{R})))^*.$$

This space is sufficiently small for the formula (2.3) to define a *continuous* bilinear form on it. Indeed, as we explained in 2.1, the kernels of covectors from this set are smooth with compact support in one variable. We hope that this sketch can help to justify our consideration of (2.3) as an r-matrix.

References

1. M. Adler, *On a trace functional for formal pseudodifferential operators and the Hamiltonian structure of Korteweg-deVries type equations*, Global Analysis, Proc. Biennial Sem. Canad. Math. Congr. (Univ. Calgary, Calgary, Alta. 1978), Lecture Notes in Math., Springer-Verlag, Berlin, Heidelberg, New York, 1979, pp. 1–16.
2. V. G. Drinfeld, *Hamiltonian structures on Lie groups, Lie bialgebras and the geometric meaning of classical Yang-Baxter equations*, Dokl. Akad. Nauk SSSR **268** (1983), no. 2, 285–287; English transl. in Soviet Math. Dokl. **27** (1983).
3. V. G. Drinfeld and V. V. Sokolov, *Lie algebras and equations of Korteweg-de Vries type*, Itogi Nauki i Tekhniki. Sovr. Probl. Mat. Noveĭshie Dostizheniya, vol. 24, 1984, pp. 81–180; English transl. in J. Soviet Math. **30** (1985), no. 2.
4. B. L. Feigin and Edward Frenkel, *Semi-infinite Weil complex and the Virasoro algebra*, Comm. Math. Phys. **137** (1991), no. 3, 617–639.
5. I. M. Gelfand and L. A. Dickey [L. A. Dikiĭ], *On the second Hamiltonian structure for equations of Korteweg-de Vries type*, Preprint, Institute for Applied Mathematics, Moscow, 1978.
6. Boris Khesin and Ilya Zakharevich, *The Gelfand-Dickey structure and an extension of the Lie algebra of pseudodifferential symbols*, in preparation.

7. Jiang-Hua Lu and Alan Weinstein, *Poisson Lie groups, dressing transformations, and Bruhat decompositions*, J. Differential Geom. **31** (1990), no. 2, 501–526.
8. A. O. Radul, *Nontrivial central extensions of Lie algebras of differential operators in two and higher dimensions*, Phys. Lett. B **265** (1991), no. 1-2, 86–91.
9. M. A. Semenov-Tyan-Shanskiĭ, *What is a classical r-matrix?*, Funktsional. Anal. i Prilozhen. **17** (1983), no. 4, 17–33; English transl., Functional Anal. Appl. **17** (1983), no. 4, 259–272.
10. ———, *Dressing transformations and Poisson group actions*, Publ. Res. Inst. Math. Sci. Kyoto Univ. **21** (1985), no. 6, 1237–1260.

Department of Mathematics, Massachusetts Institute of Technology, Cambridge, Massachusetts 02139

E-mail address: ilya@math.mit.edu